分子器件中的电子输运及应用

Electronic Transport in Molecular Device and Application

夏蔡娟　著

西北工业大学出版社

西　安

【内容简介】 本书介绍了构成微观或原子论物质理论基础的概念框架和凝聚态物理中常用的格林函数基本理论，强调了与电子特性的相关方面，尤其是电流。专注于当电压施加在分子器件上时流过器件的电流。该理论可用于预测新材料的结构、电气和机械性能以及化学反应速率等不同领域。本书主要介绍了该理论在分子器件输运方面的应用。

本书的目的是介绍电子输运理论的基本知识，可供凝聚态物理及相关领域的研究人员参考，并作为高等院校相关专业高年级本科生或研究生的教学用书。

图书在版编目(CIP)数据

分子器件中的电子输运及应用＝Electronic Transport in Molecular Device and Application：英文/夏蔡娟著．—西安：西北工业大学出版社，2018.11

ISBN 978-7-5612-6375-4

Ⅰ.①分… Ⅱ.①夏… Ⅲ.①分子电子学-研究-英文 Ⅳ.①TN01

中国版本图书馆CIP数据核字(2018)第261990号

FENZI QIJIAN ZHONG DE DIANZI SHUYUN JI YINGYONG

分 子 器 件 中 的 电 子 输 运 及 应 用

责任编辑： 朱辰浩　　**策划编辑：** 季　强

责任校对： 张　潼　　**装帧设计：** 李　飞

出版发行： 西北工业大学出版社

通信地址： 西安市友谊西路127号　　**邮编：** 710072

电　　话： (029)88491757，88493844

网　　址： www.nwpup.com

印 刷 者： 兴平市博闻印务有限公司

开　　本： 787 mm×960 mm　1/16

印　　张： 18.25

字　　数： 420千字

版　　次： 2018年11月第1版　2018年11月第1次印刷

定　　价： 78.00元

Preface

A hundred years ago the atomistic viewpoint was somewhat controversial and many renowned scientists of the day questioned the utility of postulating entities called atoms that no one could see. The microscopic theory of matter was largely developed in the course of the twentieth century following the advent of quantum mechanics and is gradually becoming an integral part of engineering disciplines, as we acquire the ability to engineer materials and devices on an atomic scale. This is a problem of great practical significance as electronic devices like transistors get downscaled to atomic dimensions. It is a rapidly evolving field of research and the specific examples I will use in this book may or may not be important. But the problem of current flow touches on some of the deepest issues of physics and the concepts we will discuss represent key fundamental concepts of quantum mechanics and non-equilibrium statistical mechanics that should be relevant to the analysis and design of nanoscale devices for many years into the future. This book is written very much in the spirit of a text-book that uses idealized examples to clarify general principles, rather than a research monograph that does justice to specific real-world issues.

This book consists of three parts. The first part is to introduce briefly the electronic transport theory. Although readers might be having good knowledge through the course of mathematical physics, a brief outline in the beginning of this book would be helpful for them to realize the mathematical properties of electronic transport theory. The second part is devoted to Green's functions. I found it a very good idea to incorporate the Green's functions in mathematics. This enables readers to have a comprehensive understanding of the Green's functions. The third part presents some of its applications.

My experience in learning this course, when I was a graduate student, revealed that is was not very easy to grasp the contents. Keeping that in view, this book was written mainly for beginners, where I put down all formulas and their derivation processes in detail and in such a manner that learners of the subject are benefited to the full extent.

I was having great respect for George Green, an outstanding mathematician, while writing this book. Thanks to the strong support of Xi'an Polytechnic University. Furthermore, I also acknowledge and express my deep sense of gratitude to individuals below for their valuable discussions and help：Professor Zhang Yingtang，Zhai Xuejun，Liu Desheng，Cheng Pengfei，Liu Hancheng，Zhang Dehua，Zhang Guoqing，Li Lianbi，Su Yaoheng and Wang Jun.

I would greatly appreciate any comments and suggestions for improvements. Although extreme care was taken to correct all the misprints，it is very likely that I have missed some of them. I shall be most grateful to those readers who are kind enough to bring to my notice any remaining mistakes，typographical or otherwise for remedial. Please feel free to contact me.

Author

2018. 08

Contents

Chapter 1 Electronic Transport

1.1 Schrödinger equation

Early in the twentieth century scientists were trying to build a model for atoms which were known to consist of negative particles called electrons surrounding a positive nucleus. A simple model pictures the electron (of mass m and charge $-q$) as orbiting the nucleus (with charge Zq) at a radius r (see Fig. 1.1.1) kept in place by electrostatic attraction, in much the same way that gravitational attraction keeps the planets in orbit around the sun.

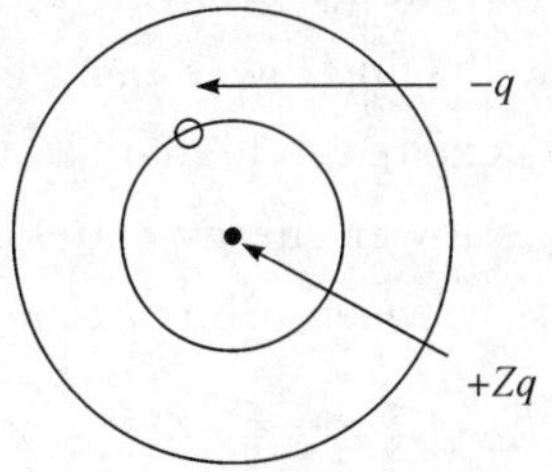

Fig. 1.1.1 Stationary orbits of an electron around a nucleus can be obtained by requiring their circumferences to be integer multiples of the de Broglie wavelength

$$\frac{Zq^2}{4\pi\varepsilon_0 r^2} = \frac{mv^2}{r} \rightarrow v = \sqrt{\frac{Zq^2}{4\pi\varepsilon_0 mr}} \quad (1.1.1)$$

Electrostatic force = Centripetal force

A faster electron describes an orbit with a smaller radius. The total energy of the electron is related to the radius of its orbit by the relatio.

$$E = -\frac{Zq^2}{4\pi\varepsilon_0 r^2} + \frac{mv^2}{2r} = -\frac{Zq^2}{8\pi\varepsilon_0 r^2} \quad (1.1.2)$$

Potential energy + Kinetic energy = Total energy

However, it was soon realized that this simple viewpoint was inadequate since, according to classical electrodynamics, an orbiting electron should radiate electromagnetic waves like an antenna, lose energy continuously and spiral into the nucleus. Classically it is impossible to come up with a stable structure for such a system except with the electron sitting right on top of the nucleus, in contradiction with experiment. It was apparent that a radical departure from classical physics was called for.

Bohr postulated that electrons could be described by stable orbits around the nucleus at specific distances from the nucleus corresponding to specific values of angular momenta. It was later realized that these distances could be determined by endowing the electrons with a wavelike character having a de Broglie wavelength equal to (h/mv), h being the Planck constant. One could then argue that the circumference of an orbit had to be an integer multiple of wavelengths in order to be stable:

$$2\pi r = n(h/mv) \tag{1.1.3}$$

Combining Eq. (1.1.3) with Eq. (1.1.1) and Eq. (1.1.2) we obtain the radius and energy of stable orbits respectively:

(Bohr radius) $$r_n = (n^2/Z)a_0 \tag{1.1.4}$$

where $a_0 = 4\pi\varepsilon_0\hbar^2/mq^2 = 0.052,9$ nm (1.1.5)

$$E_n = -(Z^2/n^2)E_0 \tag{1.1.6a}$$

where $E_0 = q^2/8\pi\varepsilon_0 a_0 = 13.6$ eV(1 Rydberg) (1.1.6b)

Once the electron is in its lowest energy orbit $(n = 1)$ it cannot lose any more energy because there are no stationary orbits having lower energies available [see Fig. 1.1.2(a)]. If we beat up the atom, the electron is excited to higher stationary orbits [see Fig. 1.1.2(b)]. When it subsequently jumps down to lower energy states, it emits photons whose energy $h\nu$ corresponds to the energy difference between orbits m and n:

$$h\nu = E_m - E_n = E_0 Z^2\left(\frac{1}{n^2} - \frac{1}{m^2}\right) \tag{1.1.7}$$

Experimentally it had been observed that the light emitted by a hydrogen atom indeed consisted of discrete frequencies that were described by this relation with integer values of m and n. This striking agreement with experiment suggested that there was some truth to this simple picture, generally known as the Bohr model.

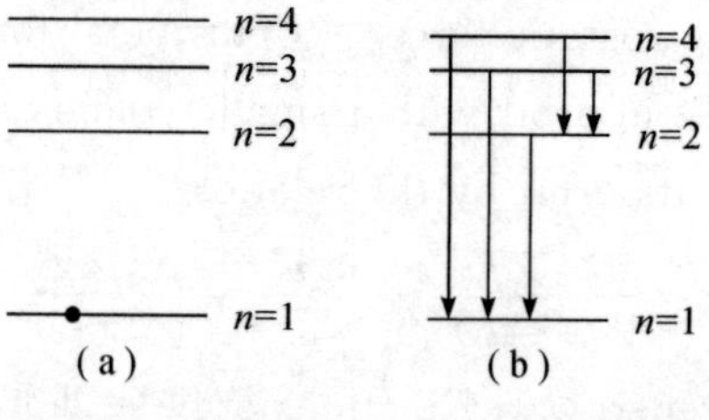

Fig. 1.1.2 Electronic transition states

(a) left to itself, the electron relaxes to its lowest energy orbit($n=1$);

(b) if we heat up the atom. the electron is excited to higher stationary orbits. When it subsequently jumps down to lower energy states, it emits photons whose energy $h\nu$ corresponds to the energy difference between the initial and final orbits

The Schrödinger equation put this heuristic insight on a formal quantitative basis allowing one to calculate the energy levels for any confining potential $U(r)$.

$$\mathrm{i}\hbar = \left(-\frac{\hbar^2}{2m}\nabla^2 + U(r)\right)\psi \tag{1.1.8}$$

How does this equation lead to discrete energy levels? Mathematically, one can show that if we assume a potential $U(r) = -Zq^2/4\pi\varepsilon_0 r$ appropriate for a nucleus of charge $+Zq$, then the solutions to this equation can be labeled with three indices n, l and m

$$\psi(r,t) = \phi_{nlm}(r)\exp(-\mathrm{i}E_n t/\hbar) \tag{1.1.9}$$

where the energy E_n depends only on the index n and is given by $E_n = -(Z^2/n^2)E_0$ in agreement with the heuristic result obtained earlier [see Eq. (1.1.6a)]. The Schrödinger equation provides a formal wave equation for the electron not unlike the equation that describes, for example, an acoustic wave in a sound box. The energy E of the electron plays a role similar to that played by the frequency of the acoustic wave. It is well-known that a sound box resonates at specific frequencies determined by the size and shape of the box. Similarly an electron wave in an atomic box "resonates" at specific energies determined by the size and shape of the box as defined by the potential energy $U(r)$. Let us elaborate on this point a little further.

To keep things simple let us consider the vibrations $u(x,t)$ of a one-dimensional (1D) string described by the 1D wave equation:

$$\frac{\partial^2 u}{\partial t^2} = v^2 \frac{\partial^2 u}{\partial x^2} \tag{1.1.10}$$

The solutions to this equation can be written in the form of plane waves with a linear dispersion $\omega = \pm vk$:

$$u = u_0 \exp(\mathrm{i}kx)\exp(-\mathrm{i}\omega t) \rightarrow \omega^2 = v^2 k^2 \tag{1.1.11}$$

What happens if we clamp the two ends so that the displacement there is forced to be zero (see Fig. 1.1.3)? We have to superpose solutions with $+k$ and $-k$ to obtain standing wave solutions. The allowed values of k are quantized leading to discrete resonant frequencies:

$$u = u_0 \sin(kx)\exp(-\mathrm{i}\omega t) \rightarrow k = \frac{n\pi}{L} \rightarrow \omega = n\pi v/L \tag{1.1.12}$$

Fig. 1.1.3 Standing waves

(a) acoustic waves in a "guitar" string with the displacement clamped to zero at either end;

(b) electron waves in a one-dimensional box with the wavefunction clamped to zero at both ends by an infinite potential

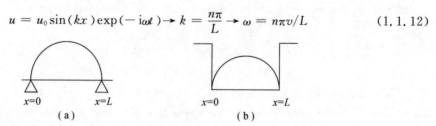

Well, it's the same way with the Schrödinger equation. If there is no confining potential ($U = 0$), we can write the solutions to the 1D Schrödinger equation:

$$\mathrm{i}\hbar \frac{\partial \psi}{\partial t} = -\frac{\hbar^2}{2m}\frac{\partial^2 \psi}{\partial x^2} \tag{1.1.13}$$

in the form of plane waves with a parabolic dispersion law $E = \hbar^2 k^2/2m$:

$$\psi = \psi_0 \exp(\mathrm{i}kx)\exp(-\mathrm{i}Et/\hbar) \rightarrow E = \hbar^2 k^2/2m \tag{1.1.14}$$

If we fix the two ends we get standing waves with quantized k and resonant frequency.

$$\psi = \psi_0 \sin(kx)\exp(-\mathrm{i}Et/\hbar) \rightarrow k = n\pi/L \rightarrow E = \hbar^2 \pi^2 n^2/2mL^2 \tag{1.1.15}$$

Atomic "boxes" are of course defined by potentials $U(\boldsymbol{r})$ that are more complicated than the simple rectangular 1D potential shown in Fig. 1.1.2(b), but the essential point is the same: anytime we confine a wave to a box, the frequency or energy is discretized because of the need for the wave to "fit" inside the box.

Another kind of box that we will often use is a ring (see Fig. 1.1.4) where the end point at $x = L$ is connected back to the first point at $x = 0$ and there are no ends. Real boxes are seldom in this form but this idealization is often used since it simplifies the mathematics. The justification for this assumption is that if we are interested in the properties in the interior of the box, then what we assume at the ends (or surfaces) should make no real difference and we could assume anything that makes our calculations simpler. However, this may not be a valid argument for "nanostructures" where the actual surface conditions can and do affect what an experimentalist measures.

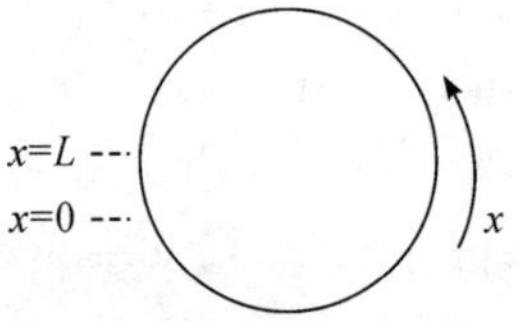

Fig. 1.1.4 Standing waves in a ring

Anyway, for a periodic box the eigenfunctions are given by (see Eq. (1.1.15))

$$\psi = \psi_0 \sin(kx)\exp(-\mathrm{i}Et/\hbar)$$

$$\text{and } \psi = \psi_0 \cos(kx)\exp(-\mathrm{i}Et/\hbar)$$

$$\text{with } k = 2n\pi/L \rightarrow E = 2\hbar^2 \pi^2 n^2/mL^2 \tag{1.1.16}$$

The values of k are spaced by $2\pi/L$ instead of π/L, so that there are half as many allowed values. But for each value of k there is a sine and a cosine function which have the same eigenvalue, so that the eigenvalues now come in pairs.

An important point to note is that whenever we have degenerate eigenstates, that is, two or more eigenfunctions with the same eigenvalue, any linear combination of these eigenfunctions is also an eigenfunction with the same eigenvalue. So we could just as well write the eigenstates as

$$\psi = \psi_0 \exp(+ikx)\exp(-\mathrm{i}Et/\hbar)$$

and
$$\psi = \psi_0 \exp(-ikx)\exp(-\mathrm{i}Et/\hbar)$$

with
$$k = 2n\pi/L \rightarrow E = 2\hbar^2 \pi^2 n^2/mL^2 \qquad (1.1.17)$$

This is done quite commonly in analytical calculations and the first of these is viewed as the $+k$ state traveling in the positive x-direction while the second is viewed as the $-k$ state traveling in the negative x-direction.

An electron with a wavefunction $\psi(x,t)$ has a probability of $\psi^* \psi \mathrm{d}V$ of being found in a volume $\mathrm{d}V$. When a number of electrons are present we could add up $\psi^* \psi$ for all the electrons to obtain the average electron density $n(x,t)$. What is the corresponding quantity we should sum to obtain the probability current density $J(x,t)$?

The appropriate expression for the probability current density

$$J = \frac{\mathrm{i}\hbar}{2m}\left(\psi \frac{\partial \psi^*}{\partial x} - \psi^* \frac{\partial \psi}{\partial x}\right) \qquad (1.1.18)$$

is motivated by the observation that as long as the wavefunction $\psi(x,t)$ obeys the Schrödinger equation, it can be shown that

$$\frac{\partial J}{\partial x} + \frac{\partial n}{\partial t} = 0 \qquad (1.1.19)$$

if J is given by Eq. (1.1.18) and $n = \psi^* \psi$. The ensures that the continuity equation is satisfied regardless of the detailed dynamics of the wavefunction. The electrical current density is obtained by multiplying J by the charge $(-q)$ of an electron.

It is straightforward to check that the "$+k$" and "$-k$" states in Eq. (1.1.17) carry equal and opposite non-zero currents proportional to the electron density

$$J = (\hbar k/m)\, \psi\psi^* \qquad (1.1.20)$$

suggesting that we associate $(\hbar k/m)$ with the velocity v of the electron (since we expect J to equal nv). However, this is true only for the plane wave functions in Eq. (1.1.17). The cosine and sine states in Eq. (1.1.16), for example, carry zero current. Indeed Eq. (1.1.18) will predict zero current for any real wavefunction.

The Schrödinger equation for a hydrogen atom can be solved analytically, but most other practical problems require a numerical solution. In this section I will describe one way of obtaining a numerical solution to the Schrödinger equation. Most numerical methods have one thing in common—they use some trick to convert the wavefunction $\psi(r,t)$ into a column vector $\boldsymbol{\psi}(t)$ and the differential operator H_{op} into a matrix H so that the Schrödinger equation is converted from a partial differential equation into a matrix equation

$$\mathrm{i}\hbar \frac{\partial}{\partial t}\psi(r,t) = H_{\mathrm{op}}\psi(r,t) \qquad \mathrm{i}\hbar \frac{\mathrm{d}}{\mathrm{d}t}\boldsymbol{\psi}(t) = \boldsymbol{H}\boldsymbol{\psi}(t)$$

This conversion can be done in many ways, but the simplest one is to choose a discrete lattice. To see how this is done let us for simplicity consider just one dimension and discretize the position variable x into a lattice as shown in Fig. 1.1.5: $x_a = na$.

We can represent the wavefunction $\psi(x,t)$ by a column vector $[\psi_1(t)\ \psi_2(t)\ \cdots]^{\mathrm{T}}$ ("T" denotes transpose) containing its values around each of the lattice points at time t. Suppressing the time variable t for clarity, we can write

$$[\psi_1 \quad \psi_2 \quad \cdots] = [\psi(x_1) \quad \psi(x_2) \quad \cdots]$$

This representation becomes exact only in the limit $a \to 0$, but as long as a is smaller than the spatial scale on which ψ varies, we can expect it to be reasonably accurate.

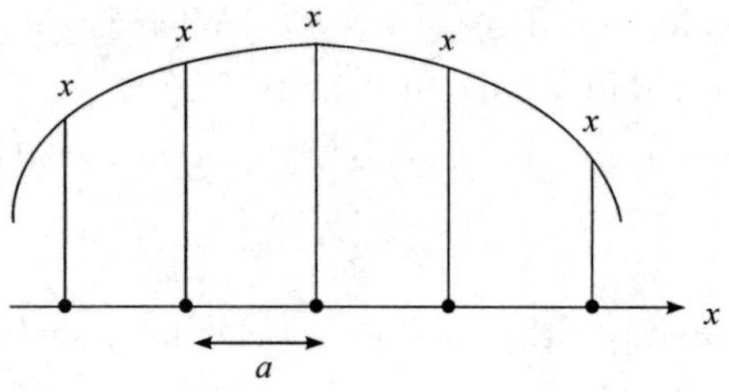

Fig. 1.1.5 A continuous function can be represented by its values at a set of points on a discrete lattice

The next step is to obtain the matrix representing the Hamiltonian operator

$$H_{\mathrm{op}} \equiv -\frac{\hbar^2}{2m}\frac{\mathrm{d}^2\psi}{\mathrm{d}x^2} + U(x) \tag{1.1.21}$$

Basically what we are doing is to turn a differential equation into a difference equation. There is a standard procedure for doing this—the finite difference technique:

$$\left(\frac{\partial^2\psi}{\partial x^2}\right)_{x=x_n} \to \frac{1}{a^2}[\psi(x_{n+1}) - 2\psi(x_n) + \psi(x_{n-1})] \tag{1.1.22}$$

and

$$U(x)\psi(x) \to U(x_n)\psi(x_n)$$

This allows us to write [note: $t_0 \equiv \hbar^2/2ma^2$ and $U_n = U(x_n)$]

$$\mathrm{i}\hbar\frac{\mathrm{d}\psi_n}{\mathrm{d}t} = (H_{\mathrm{op}}\psi)_{x=x_n} = (U_n + 2t_0)\psi_n - t_0\psi_{n-1} - t_0\psi_{n+1} =$$

$$\sum_m [(U_n + 2t_0)\delta_{n,m} - t_0\delta_{n,m+1} - t_0\delta_{n,m-1}]\psi_m \tag{1.1.23}$$

where $\delta_{n,m}$ is the Kronecker delta, which is one if $n = m$ and zero if $n \neq m$. We can write Eq. (1.1.23) as a matrix equation:

$$\mathrm{i}\hbar\frac{\mathrm{d}}{\mathrm{d}t}\boldsymbol{\psi}(t) = \boldsymbol{H}\boldsymbol{\psi}(t) \tag{1.1.24}$$

The elements of the Hamiltonian matrix are given by

$$H_{n,m} = [U_n + 2t_0]\delta_{n,m} - t_0\,\delta_{n,m+1} - t_0\,\delta_{n,m-1} \tag{1.1.25}$$

where $t_0 = \hbar^2/2ma^2$ and $U_n = U(x_n)$. This means that the matrix representing $\boldsymbol{H}$ looks like this

$$\boldsymbol{H} = \begin{bmatrix} 2t_0 + U_1 & -t_0 & 0 & 0 \\ -t_0 & 2t_0 + U_2 & 0 & 0 \\ \vdots & \vdots & \vdots & \vdots \\ 0 & 0 & 2t_0 + U_{N-1} & -t_0 \\ 0 & 0 & -t_0 & 2t_0 + U_N \end{bmatrix} \tag{1.1.26}$$

For a given potential function $U(x)$, it is straightforward to set up this matrix, once we have chosen an appropriate lattice spacing a.

Now that we have converted the Schrödinger equation into a matrix equation [see Eq. (1.1.24)]

$$\mathrm{i}\hbar \frac{\mathrm{d}}{\mathrm{d}t}\boldsymbol{\psi}(t) = \boldsymbol{H}\boldsymbol{\psi}(t)$$

How do we calculate $\boldsymbol{\psi}(t)$ given some initial state $\boldsymbol{\psi}(0)$? The standard procedure is to find the eigenvalues E_α and eigenvectors $\boldsymbol{\alpha}$ of the matrix $\boldsymbol{H}$:

$$\boldsymbol{H}\boldsymbol{\alpha} = E_\alpha \boldsymbol{\alpha} \tag{1.1.27}$$

Making use of Eq. (1.1.24) it is easy to show that the wavefunction $\boldsymbol{\psi}(t) = \mathrm{e}^{-\mathrm{i}E_\alpha t/\hbar}\boldsymbol{\alpha}$ satisfies Eq. (1.1.24). Since Eq. (1.1.24) is linear, any superposition of such solutions

$$\boldsymbol{\psi}(t) = \sum_\alpha C_\alpha\, \mathrm{e}^{-\mathrm{i}E_\alpha t/\hbar}\boldsymbol{\alpha} \tag{1.1.28}$$

is also a solution. It can be shown that this form, Eq. (1.1.28), is "complete", that is, any solution to Eq. (1.1.24) can be written in this form. Given an initial state we can figure out the coefficients C_α. The wavefunction at subsequent times t is then given by Eq. (1.1.28). Later we will discuss how we can figure out the coefficients. For the moment we are just trying to make the point that the dynamics of the system are easy to visualize or describe in terms of the eigenvalues (which are the energy levels that we talked about earlier) and the corresponding eigenvectors (which are the wavefunctions associated with those levels) of $\boldsymbol{H}$. That is why the first step in discussing any system is to write down the matrix $\boldsymbol{H}$ and to find its eigenvalues and eigenvectors.

1.2 Basis functions

We have seen that it is straightforward to calculate the energy levels for atoms using the SCF method, because the spherical symmetry effectively reduces it to a one-dimensional problem. Molecules, on the other hand, do not have this spherical symmetry and a more efficient approach is needed to make the problem numerically tractable. The concept of basis functions provides a convenient computational tool for solving the Schrödinger equation (or any differential equation for that matter). At the same time it is also a very important conceptual tool that is fundamental to the quantum mechanical viewpoint. In this chapter we attempt to convey both these aspects.

The basic idea is that the wavefunction can, in general, be expressed in terms of a set of basis functions, $u_m(r)\,\Phi(r) = \sum_{m=1}^{M} \phi_m u_m(r)$. We can then represent the wavefunction by a column vector consisting of the expansion coefficients $\boldsymbol{\Phi}(r) \rightarrow [\phi_1\ \phi_2 \cdots \phi_M]^{\mathrm{T}}$, "T" denotes transpose. We represented the wavefunction by its values at different points on a discrete lattice: $\boldsymbol{\Phi}(r) \rightarrow [\Phi(r_1)\ \Phi(r_2)\ \cdots\ \Phi(r_M)]^{\mathrm{T}}$. However, the difference is that now we have the freedom to choose the basis functions $u_m(r)$: if we choose them to look much like our expected wavefunction, we can represent the wavefunction accurately with just a few terms, thereby reducing the size of the resulting matrix $\boldsymbol{H}$ greatly. This makes the approach useful as a computational tool (similar in spirit to the concept of "shape functions" in the finite element method [Ramdas Ram-Mohan, 2002; White et al., 1989)].

The basic formulation can be stated fairly simply. We write the wavefunction in terms of any set of basis functions $u_m(r)$

$$\Phi_m(r) = \sum_m \phi_m u_m(r) \tag{1.2.1}$$

and substitute it into the Schrödinger equation $E\boldsymbol{\Phi}(r) = H_{\mathrm{op}}\boldsymbol{\Phi}(r)$ to obtain

$$E\sum_m \phi_m u_m(r) = \sum_m \phi_m H_{\mathrm{op}} u_m(r)$$

Multiply both sides by $u_n^*(\boldsymbol{r})$ and integrate over all $\boldsymbol{r}$ to yield

$$E\sum_m S_{nm}\phi_m = \sum_m H_{nm}\phi_m$$

which can be written as a matrix equation

$$\boldsymbol{ES\phi} = \boldsymbol{H\phi} \tag{1.2.2}$$

where

$$S_{nm} = \int \mathrm{d}r\, u_n^*(r) u_m(r) \tag{1.2.3a}$$

$$H_{nm} = \int \mathrm{d}r\, u_n^*(r) H_{\mathrm{op}} u_m(r) \tag{1.2.3b}$$

To proceed further we have to evaluate the integrals and that is the most time-consuming step in the process. But once the matrix elements have been calculated, it is straight forward to obtain the eigenvalues E_α and Φ_α of the matrix. The eigen-functions can then be written down in "real space" by substituting the coefficients back into the original expansion in Eq. (1.2.1):

$$\Phi_\alpha(r) = \frac{1}{\sqrt{Z_\alpha}} \sum_m \phi_{m\alpha} u_m(r), \quad \Phi_\alpha^*(r) = \frac{1}{\sqrt{Z_\alpha}} \sum_m \phi_{n\alpha}^* u_n^*(r) \tag{1.2.4}$$

where Z_α is a constant chosen to ensure proper normalization:

$$I = \int \mathrm{d}r \Phi_\alpha^*(r) \Phi_\alpha(r) \rightarrow Z_\alpha = \sum_n \sum_m \phi_{n\alpha}^* \phi_{m\alpha} S_{nm} \tag{1.2.5}$$

Eq. (1.2.1)- Ep. (1.2.5) summarize the basic mathematical relations involved in the use of basis functions.

To understand the underlying physics and how this works in practice let us look at a specific example. We stated that the lowest energy eigenvalue of the Schrödinger equation including the two nuclear potentials (see Fig. 1.2.1) but excluding the self-consistent potential

$$E_{\alpha 0} \Phi_{\alpha 0}(r) = \left(-\frac{\hbar^2}{2m} \nabla^2 + U_N(r) + U_{N'}(r) \right) \Phi_{\alpha 0}(r) \tag{1.2.6}$$

is approximately given by ($E_1 \equiv - E_0 = - 13.6$ eV)

$$E_{B0} = E_1 + \frac{a+b}{1+s} \tag{1.2.7}$$

where

$$a = -2E_0 \frac{1-(1+R_0)e^{-2R_0}}{R_0}, \qquad b = -2E_0(1+R_0)e^{-R_0}$$

$$s = e^{-R_0}[1 + R_0 + (R_0^2/3)], \qquad R_0 \equiv R/a_0$$

R being the center-to-center distance between the hydrogen atoms.

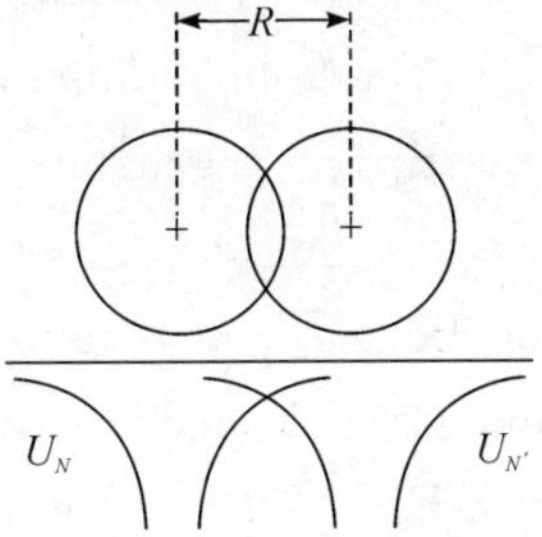

Fig. 1.2.1 U_N and $U_{N'}$ are the Coulombic potentials due to the left and right nuclei of a H_2 molecule respectively

We will now use the concept of basis functions to show how this result is obtained from Eq. (1.2.6).

Note that the potential $U(r) = U_N(r) + U_{N'}(r)$ in Eq. (1.2.6) is not spherically symmetric. This means that we cannot simply solve the radial Schrödinger equation. In general, we have to solve the full three-dimensional Schrödinger equation, which is numerically quite challenging; the problem is made tractable by using basis functions to expand the wavefunction. In the present case we can use just two basis functions

$$\Phi_{\alpha 0}(r) = \phi_L u_L + \phi_R u_R(r) \tag{1.2.8}$$

where $u_L(r)$ and $u_R(r)$ represent a hydrogenic 1s orbital centered around the left and right nuclei respectively (see Fig. 1.2.2).

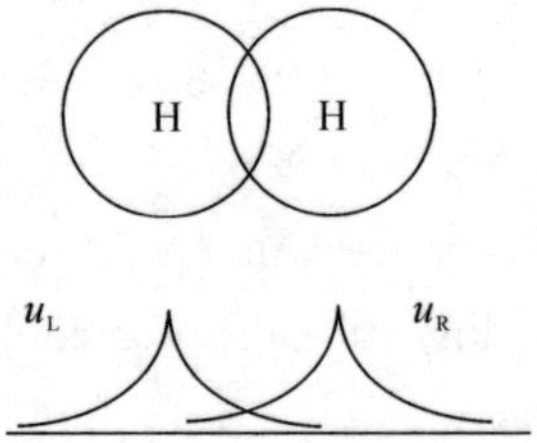

Fig. 1.2.2 A convenient basis set for the H_2 molecule consists of two 1s orbitals centered around the left and right nuclei respectively

This means that

$$E_1 u_L(r) = \left[-\frac{\hbar^2}{2m}\nabla^2 + U_N(r)\right]u_L(r) \tag{1.2.9a}$$

and

$$E_1 u_R(r) = \left[-\frac{\hbar^2}{2m}\nabla^2 + U_N(r)\right]u_R(r) \tag{1.2.9b}$$

The ansatz in Eq. (1.2.8) is motivated by the observation that it clearly describes the eigenstates correctly if we move the two atoms far apart: the eigenstates are then given by $(\phi_L\ \phi_R) = (1\ \ 0)$ and $(\phi_L\ \phi_R) = (0\ \ 1)$. It seems reasonable to expect that if the bond length R is not too short (compared to the Bohr radius a_0), Eq. (1.2.8) will still provide a reasonably accurate description of the correct eigenstates with an appropriate choice of the coefficients ϕ_L and ϕ_R.

Since we have used only two functions u_L and u_R to express our wavefunction, the matrices $\boldsymbol{S}$ and $\boldsymbol{H}$ in Eq. (1.2.2) are simple (2×2) matrices whose elements can be written down from Eqs. (1.2.3a, b) making use of Eqs. (1.2.9a, b):

$$\boldsymbol{S} = \begin{bmatrix} 1 & s \\ s & 1 \end{bmatrix} \text{ and } \boldsymbol{H} = \begin{bmatrix} E_1 + a & E_1 s + b \\ E_1 s + b & E_1 + a \end{bmatrix} \tag{1.2.10}$$

where

$$s = \int \mathrm{d}r\, u_L^*(r) u_R(r) = \int \mathrm{d}r\, u_R^*(r) u_L(r) \tag{1.2.11a}$$

$$a = \int \mathrm{d}r\, u_L^*(r) U_{N'}(r) u_L(r) = \int \mathrm{d}r\, u_L^*(r) U_N(r) u_L(r) \tag{1.2.11b}$$

$$b = \int \mathrm{d}r\, u_L^*(r) U_N(r) u_R(r) = \int \mathrm{d}r\, u_L^*(r) U_{N'}(r) u_R(r) =$$

$$\int \mathrm{d}\,r\,u_{\mathrm{R}}^{*}(r)U_{N}(r)u_{\mathrm{L}}(r)=\int \mathrm{d}\,r\,u_{\mathrm{R}}^{*}(r)U_{N'}(r)u_{\mathrm{L}}(r) \tag{1.2.11c}$$

Hence Eq. (1.2.2) becomes

$$E\begin{bmatrix}\phi_{\mathrm{L}}\\ \phi_{\mathrm{R}}\end{bmatrix}=\begin{bmatrix}1 & s\\ s & 1\end{bmatrix}^{-1}\begin{bmatrix}E_1+a & E_1 s+b\\ E_1 s+b & E_1+a\end{bmatrix}\begin{bmatrix}\phi_{\mathrm{L}}\\ \phi_{\mathrm{R}}\end{bmatrix} \tag{1.2.12}$$

from which it is straightforward to write down the two eigenvalues—the lower one is called the bonding level (B) and the higher one is called the anti-bonding level (A):

$$E_B=E_1+\frac{a+b}{1+s}\quad\text{and}\quad E_A=E_1+\frac{a-b}{1-s} \tag{1.2.13}$$

The quantities a, b and s can be evaluated by plugging in the known basis functions $u_{\mathrm{L}}(r)$, $u_{\mathrm{R}}(r)$ and the nuclear potentials $U_N(r)$ and $U_{N'}(r)$ into Eqs. (1.2.11a, b, c). The integrals can be performed analytically to yield the results stated earlier in Eq. (1.2.7).

The wavefunctions corresponding to the bonding and anti-bonding levels are given by

$$[\phi_{\mathrm{L}}\ \ \phi_{\mathrm{R}}]_B=[1\ 1],\ [\phi_{\mathrm{L}}\ \ \phi_{\mathrm{R}}]_A=[1\ -1]$$

which represent a symmetric (B) [see Fig 1.2.3(a)] and an antisymmetric (A) [see Fig. 1.2.3(b)] combination of two 1s orbitals centered around the two nuclei respectively. Both electrons in a H_2 molecule occupy the symmetric or bonding state whose wavefunction can be written as

$$\Phi_{B0}(r)=\frac{1}{\sqrt{Z}}[u_{\mathrm{L}}(r)+u_{\mathrm{R}}(r)] \tag{1.2.14}$$

where

$$u_{\mathrm{L}}(r)=\frac{1}{\sqrt{\pi a_0^3}}\exp\left(\frac{-|r-r_{\mathrm{L}}|}{a_0}\right),\quad r_{\mathrm{L}}=-(R_0/2)\hat{z}$$

$$u_{\mathrm{R}}(r)=\frac{1}{\sqrt{\pi a_0^3}}\exp\left(\frac{-|r-r_{\mathrm{R}}|}{a_0}\right),\quad r_{\mathrm{R}}=+(R_0/2)\hat{z}$$

The constant Z has to be chosen to ensure correct normalization of the wavefunction:

$$I=\int \mathrm{d}\,r\,\Phi_{B0}^{*}(r)\Phi_{B0}(r)=\frac{2(1+s)}{Z}\rightarrow Z=2(1+s)$$

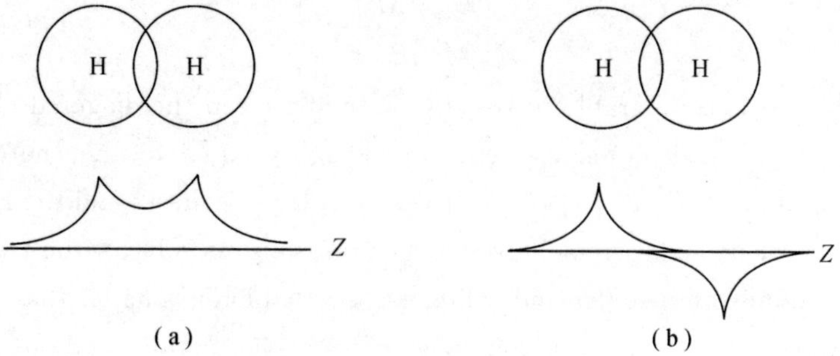

Fig. 1.2.3 A convenient basis set for H_2

The electron density $n(r)$ in a H_2 molecule is given by $|\Phi_{B0}(r)|^2$, multiplied by two since we have two electrons (one up-spin and one down-spin) with this wavefunction. Fig. 1.2.4 shows a plot of the electron density along a line joining the two nuclei.

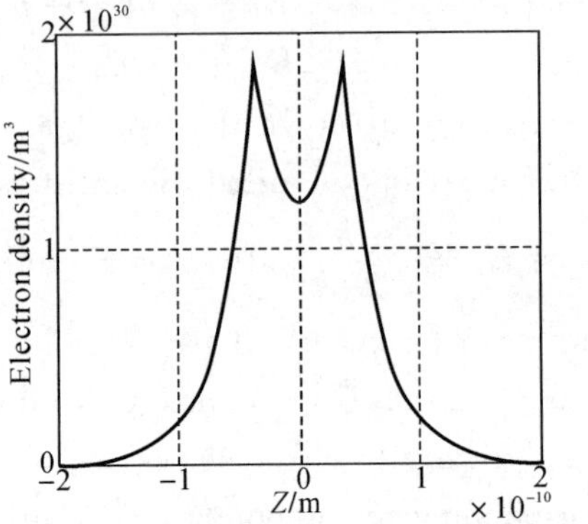

Fig. 1.2.4 Plot of electron density along the axis joining two hydrogen atoms assuming they are separated by the equilibrium bond distance of $R = 0.074$ nm

If we were to start from the Schrödinger equation and use a discrete lattice representation, we would need a fairly large number of basis functions per atom. For example if the lattice points are spaced by 0.5 Å(1 Å$=10^{-10}$ m) and the size of an atom is 2.5 Å, then we need $5^3 = 125$ lattice points (each of which represents a basis function), since the problem is a three-dimensional one. What do we lose by using only one basis function instead of 125? The answer is that our results are accurate only over a limited range of energies.

To see this, suppose we were to use not just the 1s orbital as we did previously, but also the 2s, $2p_x$, $2p_y$, $2p_z$, 3s, $3p_x$, $3p_y$ and $3p_z$ orbitals (see Fig. 1.2.5). We argue that the lowest eigenstates will still be essentially made up of 1s wavefunctions and will involve negligible amounts of the other wavefunctions, so that fairly accurate results can be obtained with just one basis function per atom. The reason is that an off-diagonal matrix element M modifies the eigenstates of a matrix

$$\begin{bmatrix} E_1 & M \\ M & E_2 \end{bmatrix}$$

significantly only if it is comparable to the difference between the diagonal elements that is, if $M \geqslant |E_1 - E_2|$. The diagonal elements are roughly equal to the energy levels of the isolated atoms, so that $|E_1 - E_2|$ is 10 eV if we consider say the 1s and the 2s levels of a hydrogen atom. The off-diagonal element M depends on the proximity of the two atoms and for typical covalently bonded molecules and solids is 2 eV, which is smaller than $|E_1 - E_2|$. As a result the bonding level is primarily composed of 1s wavefunctions and our treatment based on a 2×2 matrix is fairly accurate. But a proper treatment of the higher energy levels would require more basis functions to be included.

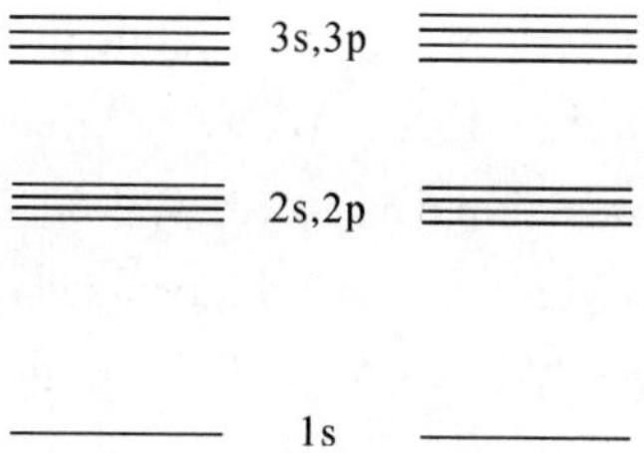

Fig. 1. 2. 5 Eigenstates of a H_2 molecule when the atoms are not too close. All these states could be used as basis functions for a more accurate treatment of the hydrogen molecule

The concept of basis functions is widely used for ab initio calculations where the Schrödinger equation is solved directly including a self-consistent field. For large molecules or solids such calculations can be computationally quite intensive due to the large number of basis functions involved and the integrals that have to be evaluated to obtain the matrix elements. The integrals arising from the self-consistent field are particularly time consuming. For this reason, semi-empirical approaches are widely used where the matrix elements are adjusted through a combination of theory and experiment. Such semi-empirical approaches can be very useful if the parameters turn out to be "transferable", that is, if we can obtain them by fitting one set of observations and then use them to make predictions in other situations. For example, we could calculate suitable parameters to fit the known energy levels of an infinite solid and then use these parameters to calculate the energy levels in a finite nanostructure carved out of that solid.

We have mentioned that all practical methods for solving the energy levels of molecules and solids usually involve some sort of expansion in basis functions. However, the concept of basis functions is more than a computational tool. It represents an important conceptual tool for visualizing the physics and developing an intuition for what to expect. Indeed the concept of a wavefunction as a superposition of basis functions is central to the entire structure of quantum mechanics as we will try to explain next.

It is useful to compare Eq. (1. 2. 1) with the expression for an ordinary three-dimensional vector **V** in terms of the three unit vectors x, y and z.

$$\mathbf{V} = \mathbf{V}_x \hat{x} + \mathbf{V}_y \hat{y} + \mathbf{V}_z \hat{z} \leftrightarrow \Phi = \phi_1 u_1 + \phi_2 u_2 + \phi_3 u_3 + \cdots$$

We can view the wavefunction as a state vector in an N-dimensional space called the Hilbert space, N being the total number of basis functions $u_m(r)$. The $\varphi_m s$ in Eq. (1. 2. 1) are like the components of Φ, while the $u_m(r)s$ are the associated unit vectors along the N coordinate axes. Choosing a different set of basis functions $u_m(r)$ is like choosing a different coordinate system: the components ϕ_m along the different axes all change, though the state vector remains the same. In principle, N is infinite, but in practice we can often get accurate results with a manageably finite value of N. We have tried to depict this analogy, but it is difficult to do justice to an N-dimensional vector ($N > 3$) on two-dimensional paper. In the Dirac notation, which is very convenient and widely used, the state vector associated with wavefunction $\Phi(r)$ is denoted by a "ket" $|\Phi\rangle$ and the unit vectors associated with the basis functions $u_m(r)$ are also written as kets $|m\rangle$. In this notation the expansion in terms of basis functions [see Eq. (1.2.1)] is written as

$$\Phi(r) = \sum_m \phi_m u_m(r) \xrightarrow{\text{(Dirac notation)}} |\Phi\rangle = \sum_m \phi_m |m\rangle \qquad (1.2.15)$$

Scalar product: A central concept in vector algebra is that of the scalar product:

$$\mathbf{A} \cdot \mathbf{B} = \mathbf{A}_x \mathbf{B}_x + \mathbf{A}_y \mathbf{B}_y + \mathbf{A}_z \mathbf{B}_z = \sum_{m=x,y,z} \mathbf{A}_m \mathbf{B}_m$$

The corresponding concept in Hilbert space is that of the overlap of any two functions $f(r)$ and $g(r)$:

$$\int \mathrm{d}r f^*(r) g(r) \xrightarrow{\text{(Dirac notation)}} \langle f \mid g\rangle$$

The similarity of the overlap integral to a scalar product can be seen by discretizing the integral:

$$\int \mathrm{d}r f^*(r) g(r) \approx a^3 \sum_m f^*(r_m) g(r_m)$$

In the discrete lattice representation (see Fig. 1.2.6) the "component" f_m of $f(r)$ along $u_m(r)$ is given by $f(r_m)$ just as $\mathbf{A}_\mathrm{m}$ represents the component of the vector $\mathbf{A}$ along $\hat{m}$:

$$\int \mathrm{d}r f^*(r) g(r) \approx a^3 \sum_m f_m^* g_m \quad cf.\ \mathbf{A} \cdot \mathbf{B} = \sum_{m=x,y,z} \mathbf{A}_m \mathbf{B}_m$$

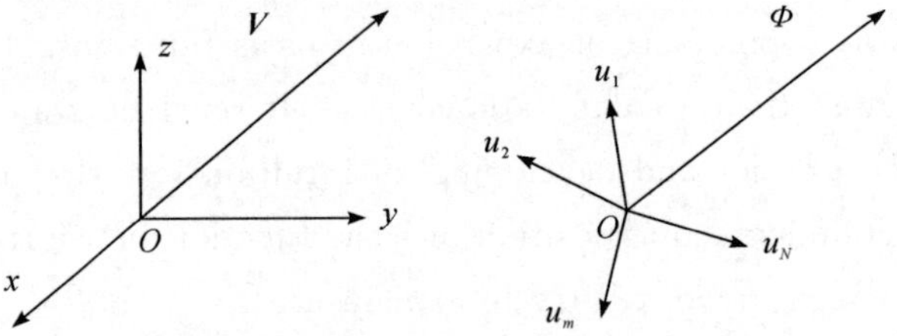

(a) (b)

Fig. 1.2.6 The vector and state vector

(a) an ordinary vector **V** in three-dimensional space can be expressed in terms of its components along x, y and z; (b) $\Phi(r)$ can be expressed in terms of its components along the basis functions $u_m(r)$

One difference here is that we take the complex conjugate of one of the functions (this is not important if we are dealing with real functions) which is represented by the "bra" $\langle f|$ as opposed to the "ket" $|g\rangle$. The scalar product is represented by juxtaposing a "bra" and a "ket" as in $\langle f \mid g\rangle$.

Coordinate systems are said to be orthogonal if $\hat{n} \cdot \hat{m} = \delta_{nm}$, where the indices m and n stand for x, y or z and δ_{nm} is the Kronecker delta which is defined as

$$\left.\begin{aligned} \delta_{nm} &= 1, \text{ if } \quad n = m \\ \delta_{nm} &= 0, \text{ if } \quad n \neq m \end{aligned}\right\} \tag{1.2.16}$$

This is usually true (for example, $\hat{x} \cdot \hat{y} = \hat{y} \cdot \hat{z} = \hat{z} \cdot \hat{x} = 0$) but it is possible to work with non-orthogonal coordinate systems too. Similarly the basis functions $u_m(r)$ are said to be orthogonal if the following relation is satisfied:

$$\int \mathrm{d}r\, u_n^*(r) u_m(r) = \delta_{nm} \xrightarrow{\text{(Dirac notation)}} \langle n \mid m\rangle = \delta_{nm} \tag{1.2.17}$$

Note that the basis functions we used for the hydrogen molecule are non-orthogonal since

$$\int \mathrm{d}r\, u_{\mathrm{L}}^*(r) u_{\mathrm{R}}(r) \equiv s = \mathrm{e}^{-R_0}[1 + R_0 + (R_0^2/3)] \neq 0$$

In general

$$\int \mathrm{d}r\, u_n^*(r) u_m(r) = S_{nm} \xrightarrow{\text{(Dirac notation)}} \langle n \mid m\rangle = S_{nm} \tag{1.2.18}$$

Orthogonalization: Given a non-orthogonal set of basis functions $\{u_n(r)\}$, we can define another set

$$\tilde{u}_i(r) = \sum_n [\boldsymbol{S}^{-1/2}]_{ni}\, u_n(r) \tag{1.2.19}$$

which will be orthogonal. This is shown as follows

$$\int \mathrm{d}r\, \tilde{u}_i^*(r) \tilde{u}_j(r) = \sum_n \sum_m [\boldsymbol{S}^{-1/2}]_{in} S_{nm} [\boldsymbol{S}^{-1/2}]_{mj} = [\boldsymbol{S}^{-1/2} \boldsymbol{S} \boldsymbol{S}^{-1/2}]_{ij} = \delta_{ij}$$

where we have made use of Eq. (1.2.17). This means that if we use the new set $\{u_i(r)\}$ as our basis, then the overlap matrix $\boldsymbol{S} = \boldsymbol{I}$, where $\boldsymbol{I}$ is the identity matrix which is a diagonal matrix with ones on the diagonal. This is a property of orthogonal basis functions which makes them conceptually easier to deal with.

Even if we start with a non-orthogonal basis, it is often convenient to orthogonalize it. What we might lose in the process is the local nature of the original basis which makes it convenient to visualize the physics. For example, the $\{u_n(r)\}$ we used for the hydrogen molecule were localized on the left and right hydrogen atoms respectively But the orthogonalized basis $\{u_i(r)\}$ will be linear combinations of the two and thus less local than $\{u_n(r)\}$. As a rule, it is difficult to find basis functions that are both local and orthogonal. From here on we will generally assume that the basis functions we use are orthogonal.

An operator like H_{op} acting on a state vector changes it into a different state vector—we could say that it "rotates" the vector. With ordinary vectors we can represent a rotation by a matrix

$$\begin{bmatrix} A'_x \\ A'_y \end{bmatrix} = \begin{bmatrix} R_{xx} & R_{xy} \\ R_{yx} & R_{yy} \end{bmatrix} \begin{bmatrix} A_x \\ A_y \end{bmatrix}$$

where for simplicity we have assumed a two-dimensional vector. How do we write down the matrix $\boldsymbol{R}$ corresponding to an operator R_{op}? The general principle is the following $R_{nm} = \hat{n} \cdot (R_{op}\, \hat{m})$. For example, suppose we consider an operator that rotates a vector by an angle θ. We then obtain

$$R_{xx} = \hat{x} \cdot (R_{op}\, \hat{x}) = \hat{x} \cdot (\hat{x}\cos\theta + \hat{y}\sin\theta) = \cos\theta$$

$$R_{yx} = \hat{y} \cdot (R_{op}\, \hat{x}) = \hat{y} \cdot (\hat{x}\cos\theta + \hat{y}\sin\theta) = \sin\theta$$

$$R_{xy} = \hat{x} \cdot (R_{op}\, \hat{y}) = \hat{x} \cdot (-\hat{x}\sin\theta + \hat{y}\cos\theta) = -\sin\theta$$

$$R_{yy} = \hat{y} \cdot (R_{op}\, \hat{y}) = \hat{y} \cdot (-\hat{x}\sin\theta + \hat{y}\cos\theta) = \cos\theta$$

The matrix representation for any operator H_{op} in Hilbert space is written using a similar prescription:

$$A_{nm} = \int \mathrm{d}\boldsymbol{r}\, u_n^*(\boldsymbol{r})[A_{op}\, u_m(\boldsymbol{r})] \xrightarrow{\text{(Dirac notation)}} A_{nm} = \langle n \mid A_{op} m \rangle \tag{1.2.20}$$

Constant operator: What is the matrix representing a constant operator, one that simply multiplies a state vector by a constant $\boldsymbol{C}$? In general, the answer is

$$\boldsymbol{C}_{nm} = \boldsymbol{C}\int \mathrm{d}\boldsymbol{r}\, u_n^*(\boldsymbol{r}) u_m(\boldsymbol{r}) = \boldsymbol{C}S_{nm} \tag{1.2.21}$$

which reduces to $\boldsymbol{C}$ for orthogonal bases.

The matrix representation of the Schrödinger equation obtained in the last section

$$E\Phi(\boldsymbol{r}) = H_{op}\Phi(\boldsymbol{r}) \rightarrow \boldsymbol{ES\Phi} = \boldsymbol{H\Phi} \tag{1.2.22}$$

can now be understood in terms of the concepts described in this section. Like the rotation operator in vector space, any differential operator in Hilbert space has a matrix representation. Once we have chosen a set of basis functions, H_{op} becomes the matrix $\boldsymbol{H}$ while the constant E becomes the matrix $\boldsymbol{ES}$:

$$S_{nm} = \langle n \mid m \rangle \equiv \int \mathrm{d}\boldsymbol{r}\, u_n^*(\boldsymbol{r}) u_m(\boldsymbol{r}) \tag{1.2.23a}$$

$$H_{nm} = \langle n \mid H_{op} m \rangle \equiv \int \mathrm{d}\boldsymbol{r}\, u_n^*(\boldsymbol{r})[H_{op}\, u_m(\boldsymbol{r})] \tag{1.2.23b}$$

We could orthogonalize the basis set following Eq. (1.2.20), so that in terms of the orthogonal basis $\{u_i(\boldsymbol{r})\}$, the Schr ö dinger equation has the form of a standard matrix eigenvalue equation: $E\tilde{\boldsymbol{\phi}} = \tilde{\boldsymbol{H}}\tilde{\boldsymbol{\phi}}$, where the matrix elements of $\boldsymbol{H}$ are given by

$$\widetilde{H}_{ij} = \int \mathrm{d}r\, \tilde{u}_i^*(r)[H_{\mathrm{op}}\, \tilde{u}_j(r)]$$

Suppose we have expanded our wavefunction in one basis and would like to change to a different basis:

$$\Phi(r) = \sum_m \phi_m u_m(r) \rightarrow \Phi(r) = \sum_i \phi'_i u'_i(r) \tag{1.2.24}$$

Such a transformation can be described by a transformation matrix $\boldsymbol{C}$ obtained by writing the new basis in terms of the old basis:

$$u'_i(r) = \sum_m C_{mi} u_m(r) \tag{1.2.25}$$

From Eq. (1.2.24) and Eq. (1.2.25) we can show that

$$\phi_m = \sum_m C_{mi} \phi'_i \rightarrow \boldsymbol{\phi} = \boldsymbol{C\phi}' \tag{1.2.26a}$$

Similarly we can show that any matrix $\boldsymbol{A}'$ in the new representation is related to the matrix $\boldsymbol{A}$ in the old representation by

$$\boldsymbol{A}'_{ij} \sum_j \sum_i \boldsymbol{C}'_{nj} \boldsymbol{A}_{nm} \boldsymbol{C}_{mi} \rightarrow \boldsymbol{A}' = \boldsymbol{C}^+ \boldsymbol{AC} \tag{1.2.26b}$$

There is a special class of transformations which conserves the "norm" of a state vector, that is

$$\sum_m \phi_m^* \phi_m = \sum_i \phi'^*_i \phi'_i \rightarrow \boldsymbol{\phi}^+ \boldsymbol{\phi} = \boldsymbol{\phi}'^+ + \boldsymbol{\phi}' \tag{1.2.27}$$

Substituting for $\boldsymbol{\phi}$ from Eq. (1.2.26a) into Eq. (1.2.27)

$$\boldsymbol{\phi}'^+ \boldsymbol{C}^+ \boldsymbol{C} \boldsymbol{\phi}' = \boldsymbol{\phi}'^+ + \boldsymbol{\phi}' \rightarrow \boldsymbol{C}^+ \boldsymbol{C} = \boldsymbol{I} \tag{1.2.28}$$

A matrix $\boldsymbol{C}$ that satisfies this condition [see Eq. (1.2.28)] is said to be unitary and the corresponding transformation is called a unitary transformation.

Note that for a unitary transformation, $\boldsymbol{C}^+ = \boldsymbol{C}^{-1}$, allowing us to write an inverse transformation from Eq. (1.2.26b) simply as $\boldsymbol{A} = \boldsymbol{CA}'\boldsymbol{C}^+$.

The matrix $\boldsymbol{A}$ representing a Hermitian operator A_{op} is Hermitian (in any representation) which means that it is equal to its conjugate transpose $\boldsymbol{A}^+$:

$$\boldsymbol{A} = \boldsymbol{A}^+, \quad \text{i.e.} \quad \boldsymbol{A}_{mn} = \boldsymbol{A}_{nm}^* \tag{1.2.29}$$

If A_{op} is a function like $U(r)$ then it is easy to show that it will be Hermitian as long as it is real: $U_{mn}^* = \left[\left[\int \mathrm{d}r\, u_m^*(r) U(r)\, u_n(r)\right]^*\right] = U_{nm}$.

If A_{op} is a differential operator like $\mathrm{d}/\mathrm{d}x$ or $\mathrm{d}^2/\mathrm{d}x^2$ then it takes a little more work to check if it is Hermitian or not. An easier approach shows the matrix representing $\mathrm{d}^2/\mathrm{d}x^2$ and it is clearly Hermitian in this representation. Also, it can be shown that a matrix that is Hermitian in one representation will remain Hermitian in any other representation. The Hamiltonian operator is Hermitian since it is a sum of Hermitian operators like $\partial^2/\partial x^2$, $\partial^2/\partial y^2$, and

$U(r)$. An important requirement of quantum mechanics is that the eigenvalues corresponding to any operator A_{op} presenting any observable must be real. This is ensured by requiring all such operators A_{op} to be Hermitian (not just the Hamiltonian operator H_{op} which represents the energy) since the eigenvalues of a Hermitian matrix are real.

Another useful property of a Hermitian matrix is that if we form a matrix $\mathbf{V}$ out of all the normalized eigenvectors

$$\mathbf{V} = [\mathbf{V}_1\ \mathbf{V}_2 \cdots]$$

then this matrix will be unitary, that is, $\mathbf{V}^+\mathbf{V} = \boldsymbol{I}$. Such a unitary matrix can be used to transform all column vectors $\boldsymbol{\phi}$ and matrices $\boldsymbol{M}$ to a new basis that uses the eigenvectors as the basis:

$$\boldsymbol{\phi}_{\text{new}} = \mathbf{V}^+ \boldsymbol{\phi}_{\text{old}} \leftrightarrow \boldsymbol{\phi}_{\text{old}} = \mathbf{V}^+ \boldsymbol{\phi}_{\text{new}} \tag{1.2.30}$$

$$\boldsymbol{M}_{\text{new}} = \mathbf{V}^+ \boldsymbol{M}_{\text{old}}\mathbf{V} \leftrightarrow \boldsymbol{M}_{\text{old}} = \mathbf{V}\boldsymbol{M}_{\text{new}}\mathbf{V}^+$$

If $\mathbf{V}$ is the eigenvector matrix corresponding to a Hermitian matrix like $\boldsymbol{H}$, then the new representation of $\boldsymbol{H}$ will be diagonal with the eigenvalues E_m along the diagonal:

$$\boldsymbol{H}' = \mathbf{V}^+\boldsymbol{H}\mathbf{V} = \begin{bmatrix} E_1 & 0 & 0 & 0 & \cdots \\ 0 & E_2 & 0 & 0 & \cdots \\ 0 & 0 & E_3 & 0 & \cdots \\ \cdots & \cdots & \cdots & \cdots & E_m \end{bmatrix} \tag{1.2.31}$$

For this reason the process of finding eigenfunctions and eigenvalues is often referred to as diagonalization.

1.3 Density matrix

In 1.1, I described a very simple model for current flow, namely a single level ε which communicates with two contacts, labeled the source and the drain. The strength of the coupling to the source (or the drain) was characterized by the rate $\gamma_1/\hbar$ (or $\gamma_2/\hbar$) at which an electron initially occupying the level would escape into the source(or the drain).

I pointed out that the flow of current is due to the difference in "agenda" between the source and the drain, each of which is in a state of local equilibrium, but maintained at two different electrochemical potentials and hence with two distinct Fermi functions:

$$f_1(E) = f_0(E-\mu_1) = \frac{1}{\exp[(E-\mu_1)/k_BT]+1} \tag{1.3.1a}$$

$$f_2(E) = f_0(E-\mu_2) = \frac{1}{\exp[(E-\mu_2)/k_BT]+1} \tag{1.3.1b}$$

by the applied bias V : $\mu_2 - \mu_1 = -qV$. The source would like the number of electrons occupying the level to be equal to $f_1(\varepsilon)$ while the drain would like to see this number be $f_2(\varepsilon)$.

The actual steady-state number of electrons N lies somewhere between the two and the source keeps pumping in electrons while the drain keeps pulling them out, each hoping to establish equilibrium with itself. In the process, a current flows in the external circuit (see Fig. 1.3.1). My purpose in this chapter is essentially to carry out a generalized version of this treatment applicable to an arbitrary multi-level device (see Fig 1.3.2) whose energy levels and coupling are described by matrices rather than ordinary numbers.

$\varepsilon \to \boldsymbol{H}$ Hamiltonian matrix

$\gamma_{1,2} \to \boldsymbol{\Gamma}_{1,2}$ Broadening matrices $\boldsymbol{\Gamma}_{1,2} = \boldsymbol{I}(\boldsymbol{\Sigma}_{1,2} - \boldsymbol{\Sigma}_{1,2}^{+})$

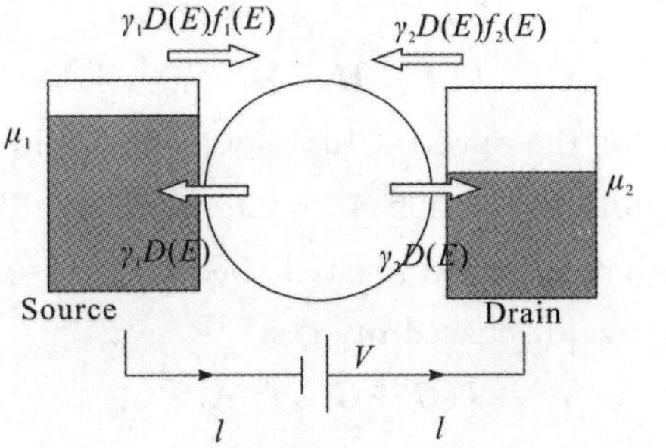

Fig. 1.3.1 Flux of electrons into and a channel: independent level model

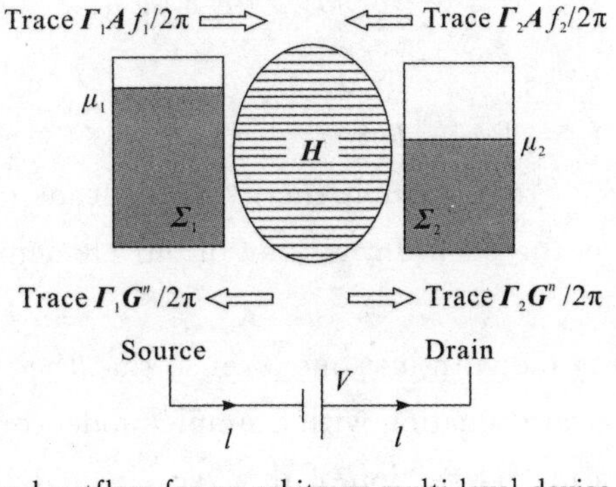

Fig. 1.3.2 Inflow and outflow for an arbitrary multi-level device whose energy levels are described by a Hamiltonian matrix $\boldsymbol{H}$ and whose coupling to the source and drain contacts is described by self-energy matrices

We saw that connecting a device to a reservoir broadens its energy levels and it is convenient to talk in terms of a continuous independent energy variable, rather than a discrete set of eigenstates.

$$\boldsymbol{\rho} = \int_{-\infty}^{+\infty} (\mathrm{d}E/2\pi)[\boldsymbol{G}^n(E)] \tag{1.3.2a}$$

where, in equilibrium,

$$\boldsymbol{G}^n(E)_{\text{eq}} = \boldsymbol{A}(E) f_0(E-\mu) \tag{1.3.2b}$$

Just as the spectral function $\boldsymbol{A}$ represents the matrix version of the density of states per unit energy, the correlation function $\boldsymbol{G}^n$ is the matrix version of the electron density per unit energy.

The first result we will prove is that when the device is connected to two contacts with two distinct Fermi functions $f_1(E)$ and $f_2(E)$, the density matrix is given by Eq. (1.3.2a) with (dropping the argument E for clarity)

$$\boldsymbol{G}^n = \boldsymbol{A}_1 f_1 + \boldsymbol{A}_2 f_2 \tag{1.3.3}$$

where

$$\boldsymbol{A}_1 = \boldsymbol{G}\boldsymbol{\Gamma}_1 \boldsymbol{G}^+ ,\ \boldsymbol{A}_2 = \boldsymbol{G}\boldsymbol{\Gamma}_2 \boldsymbol{G}^+ \tag{1.3.4}$$

$$\boldsymbol{G} = (E\boldsymbol{I} - \boldsymbol{H} - \boldsymbol{\Sigma}_1 - \boldsymbol{\Sigma}_2)^{-1} \tag{1.3.5}$$

suggesting that a fraction $\boldsymbol{A}_1$ of the spectral function remains in equilibrium with the source Fermi function f_1, while another fraction $\boldsymbol{A}_2$ remains in equilibrium with the drain Fermi function f_2. We will show that these two partial spectral functions indeed add up to give the total spectral function $\boldsymbol{A}$ that we discussed in

$$\boldsymbol{A} \equiv \boldsymbol{I}(\boldsymbol{G} - \boldsymbol{G}^+) = \boldsymbol{A}_1 + \boldsymbol{A}_2 \tag{1.3.6}$$

We will show that the current I_i at terminal i can be written in the form

$$I_i = (-q/h)\int_{-\infty}^{+\infty} \mathrm{d}E\, \widetilde{I}_i(E) \tag{1.3.7a}$$

with

$$\widetilde{I}_i = \mathrm{Trace}(\boldsymbol{\Gamma}_i \boldsymbol{A}) f_1 - \mathrm{Trace}(\boldsymbol{\Gamma}_i \boldsymbol{G}^n) \tag{1.3.7b}$$

representing a dimensionless current per unit energy. This leads to the picture shown in Fig. 1.3.2 which can be viewed as the quantum version of our elementary picture from (see Fig. 1.3.1).

We went through an example with just one level so that the electron density and current could all be calculated from a rate equation with a simple model for broadening. I then indicated that in general we need a matrix version of this "scalar model".

It is instructive to check that the full "matrix model" we have stated above(and will derive in this chapter) reduces to our old results when we specialize to a one-level system so that all the matrices reduce to pure numbers.

From Eq. (1.3.5), $\boldsymbol{G}(E) = (E - \varepsilon + (\mathrm{i}\boldsymbol{\Gamma}/2))^{-1}$

From Eq. (1.3.4), $\boldsymbol{A}_1(E) = \dfrac{\boldsymbol{\Gamma}_1}{(E-\varepsilon)^2 + (\boldsymbol{\Gamma}/2)^2}$, $\quad \boldsymbol{A}_2(E) = \dfrac{\boldsymbol{\Gamma}_2}{(E-\varepsilon)^2 + (\boldsymbol{\Gamma}/2)^2}$

From Eq. (1.3.6), $\boldsymbol{A}(E) = \dfrac{\boldsymbol{\Gamma}}{(E-\varepsilon)^2 + (\boldsymbol{\Gamma}/2)^2}$

From Eq. (1.3.3), $\boldsymbol{G}^n(E) = \boldsymbol{A}(E)\left[\dfrac{\boldsymbol{\Gamma}_1}{\boldsymbol{\Gamma}} f_1(E) + \dfrac{\boldsymbol{\Gamma}_2}{\boldsymbol{\Gamma}} f_2(E)\right]$

Similarly, from Eq. (1.3.7) the current at the two terminals is given by:

$$I_1 = \frac{q}{h}\int_{-\infty}^{+\infty} \mathrm{d}E\,\boldsymbol{\Gamma}_1[\mathbf{A}(E)f_1(E) - \mathbf{G}^n(E)]$$

$$I_2 = \frac{q}{h}\int_{-\infty}^{+\infty} \mathrm{d}E\,\boldsymbol{\Gamma}_2[\mathbf{A}(E)f_2(E) - \mathbf{G}^n(E)]$$

Eq. (1.3.7) can be combined with Eq. (1.3.3) and Eq. (1.3.6) to write

$$\bar{I}_1 = -\bar{I}_2 = \bar{T}(E)(f_1(E) - f_2(E))$$

where

$$\bar{T}(E) \equiv \mathrm{Trace}(\boldsymbol{\Gamma}_1\,\mathbf{A}_2) = \mathrm{Trace}(\boldsymbol{\Gamma}_2\,\mathbf{A}_1) \tag{1.3.8}$$

The current I in the external circuit is given by

$$I = (q/h)\int_{-\infty}^{+\infty} \mathrm{d}E\,\bar{T}(E)(f_1(E) - f_2(E)) \tag{1.3.9}$$

The quantity $\bar{T}(E)$ appearing in the current equation [see Eq. (1.3.9)] is called the transmission function, which tells us the rate at which electrons transmit from the source to the drain contacts by propagating through the device. Knowing the device Hamiltonian $\boldsymbol{H}$ and its coupling to the contacts described by the self-energy matrices $\boldsymbol{\Sigma}_{1,2}$, we can calculate the current either from Eq. (1.3.7) or from Eq. (1.3.9). This procedure can be used to analyze any device as long as the evolution of electrons through the device is coherent. Let me explain what that means.

The propagation of electrons is said to be coherent if it does not suffer phase-breaking scattering processes that cause a change in the state of an external object. For example, if an electron were to be deflected from a rigid (that is unchangeable) defect in the lattice, the propagation would still be considered coherent. The effect could be incorporated through an appropriate defect potential in the Hamiltonian $\boldsymbol{H}$ and we could still calculate the current from Fig. 1.3.2. But, if the electron transferred some energy to the atomic lattice causing it to start vibrating that would constitute a phase-breaking process and the effect cannot be included in $\boldsymbol{H}$. How it can be included is the subject of 1.1.

I should mention here that coherent transport is commonly treated using the transmission formalism which starts with the assumption that the device is connected to the contacts by two ideal leads which can be viewed as multi-moded quantum wires so that one can calculate an $\boldsymbol{s}$-matrix for the device (see Fig. 1.3.3), somewhat like a microwave waveguide, The transmission matrix $\boldsymbol{s}_{21}$ (or $\boldsymbol{s}_{12}$) is of size $M \times N$ (or $N \times M$) if lead 1 has N modes and lead 2 has M modes and the transmission function is obtained from its trace: $\bar{T}(E) = \mathrm{Trace}(\boldsymbol{s}_{12}\,\boldsymbol{s}_{21}^{+}) = \mathrm{Trace}(\boldsymbol{s}_{21}\,\boldsymbol{s}_{21}^{+})$. This approach is widely used and seems quite appealing especially to those familiar with the concept of s-matrices in microwave waveguides.

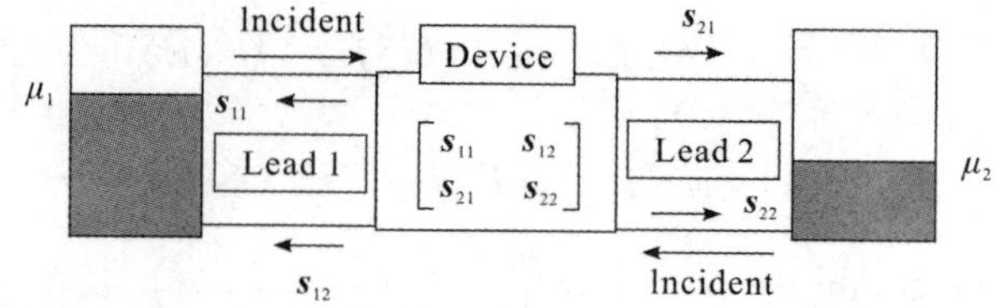

Fig. 1.3.3 The transmission formalism assumes the device to be connected via ideal multi-moded quantum wires to the contacts and transmission function is related to the s-matrix between these leads

For coherent transport, one can calculate the transmission from the Green's function method, using the relation

$$\bar{T}(E) \equiv \text{Trace}(\boldsymbol{\Gamma}_1 \boldsymbol{G} \boldsymbol{\Gamma}_2 \boldsymbol{G}^{+}) = \text{Trace}(\boldsymbol{\Gamma}_2 \boldsymbol{G} \boldsymbol{\Gamma}_1 \boldsymbol{G}^{+}) \qquad (1.3.10)$$

obtained by combining Eq. (1.3.8) with Eq. (1.3.4). We will derive all the equations [Eq. (1.3.2a)– Eq. (1.3.7a)] given in this section. But for the moment let me just try to justify the expression for the transmission [see Eq. (1.3.10)] using a simple example. Consider now a simple 1D wire modeled with a discrete lattice (see Fig. 1.3.4) We wish to calculate the transmission coefficient

$$\bar{T}(E) = (\nu_2/\nu_1)\,|t|^2 \qquad (1.3.11)$$

where the ratio of velocities (ν_2/ν_1) is included because the transmission is equal to the ratio of the transmitted to the incident current, and the current is proportional to the velocity times the probability $|\boldsymbol{\psi}|^2$.

To calculate the transmission from the Green's function approach, we start from the Schrödinger equation $(\boldsymbol{EI} - \boldsymbol{H})\boldsymbol{\psi} = \{0\}$, describing the entire infinite system and use the same approach to eliminate the semi-infinite leads

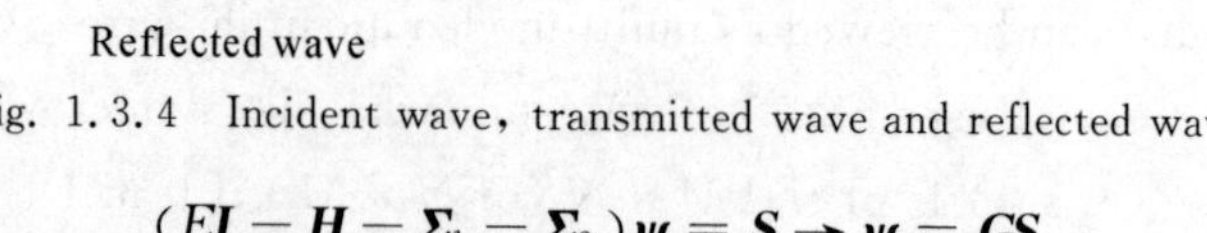

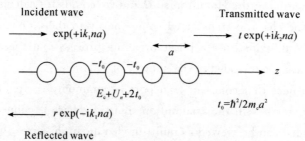

Fig. 1.3.4 Incident wave, transmitted wave and reflected wave

$$(\boldsymbol{EI} - \boldsymbol{H} - \boldsymbol{\Sigma}_1 - \boldsymbol{\Sigma}_2)\boldsymbol{\psi} = \boldsymbol{S} \rightarrow \boldsymbol{\psi} = \boldsymbol{GS} \qquad (1.3.12)$$

where $\boldsymbol{G}$ is given by Eq. (1.3.5). $\boldsymbol{\Sigma}_1$ and $\boldsymbol{\Sigma}_2$ are matrices that represent the effects of the two leads: each has only one non-zero element:

$$\Sigma_1(1,1) = -t_0 \exp(\mathrm{i} k_1 a)\ ,\ \Sigma_2(N,N) = -t_0 \exp(\mathrm{i} k_2 a)$$

corresponding to the end point of the channel (1 or N) where the lead is connected. The source term $\boldsymbol{S}$ is a column vector with just one non-zero element corresponding to the end point (1) on which the electron wave is incident:

$$S(1) = \mathrm{i}2\, t_0 \sin k_1 a = \mathrm{i}(\hbar \nu_1 / a)$$

Note that in general for the same energy E, the k-values (and hence the velocities) can be different at the two ends of the lattice since the potential energy U is different:

$$E = E_c + U_1 + 2\, t_0 \cos k_1 a = E_c + U_N + 2\, t_0 \cos k_2 a$$

From Eq. (1.3.12) we can write $t = \psi(N) = G(N,1)S(1)$ so that from Eq. (1.3.11)

$$\bar{T}(E) = (\hbar \nu_1 / a)(\hbar \nu_2 / a)\ |G(1,N)|^2$$

which is exactly what we get from the general expression in Eq. (1.3.10).

This simple example is designed to illustrate the relation between the Green's function and transmission points of view. I believe the advantages of the Green's function formulation are threefold.

(1) The generality of the derivation shows that the basic results apply to arbitrarily shaped channels described by $\boldsymbol{H}$ with arbitrarily shaped contacts described by $\boldsymbol{\Sigma}_1$, $\boldsymbol{\Sigma}_2$. This partitioning of the channels from the contacts is very useful when dealing with more complicated structures.

(2) The Green's function approach allows us to calculate the density matrix (hence the electron density) as well. This can be done within the transmission formalism, but less straightforwardly.

(3) The Green's function approach can handle incoherent transport with phase-breaking scattering. Phase-breaking processes car only be included phenomenologically within the transmission formalism (Buttiker, 1988). We will derive the expressions for the density matrix, Eq. (1.3.3)- Eq. (1.3.6), the expression for the current, Eq. (1.3.7).

Next we will derive the results for the non-equilibrium density matrix [Eq. (1.3.3)- Eq. (1.3.6)] for a channel connected to two contacts. I would like to start by revisiting the problem of a channel connected to one contact and clearing up a conceptual issue, before we take on the real problem with two contacts. We started from a Schrödinger equation for the composite contact-channel system

$$\begin{bmatrix} E\boldsymbol{I}_R - \boldsymbol{H}_R + \mathrm{i}\boldsymbol{\eta} & -\boldsymbol{\tau}^+ \\ -\boldsymbol{\tau} & E\boldsymbol{I} - \boldsymbol{H} \end{bmatrix} \begin{bmatrix} \boldsymbol{\Phi}_R + \boldsymbol{\chi} \\ \boldsymbol{\psi} \end{bmatrix} = \begin{bmatrix} \boldsymbol{S}_R \\ 0 \end{bmatrix} \tag{1.3.13}$$

and showed that the scattered waves $\boldsymbol{\psi}$ and $\boldsymbol{\chi}$ can be viewed as arising from the "spilling over" of the wavefunction $\boldsymbol{\Phi}_R$ in the isolated contact (see Fig. 1.3.5). Using straightforward matrix algebra we obtained

$$\boldsymbol{\chi} = \boldsymbol{G}_R \boldsymbol{\tau}^+ \boldsymbol{\psi} \tag{1.3.14}$$

where

$$\boldsymbol{G}_R \equiv (E\boldsymbol{I}_R - \boldsymbol{H}_R + \mathrm{i}\boldsymbol{\eta})^{-1} \tag{1.3.15}$$

$$\boldsymbol{\psi} = \boldsymbol{G}\boldsymbol{S} \tag{1.3.16}$$

$$\boldsymbol{G} \equiv (E\boldsymbol{I} - \boldsymbol{H} - \boldsymbol{\Sigma})^{-1} \tag{1.3.17}$$

$$\boldsymbol{\Sigma} \equiv \boldsymbol{\tau}\boldsymbol{G}_R \boldsymbol{\tau}^+ \tag{1.3.18}$$

$$\boldsymbol{S} = \boldsymbol{\tau}\boldsymbol{\Phi}_R \tag{1.3.19}$$

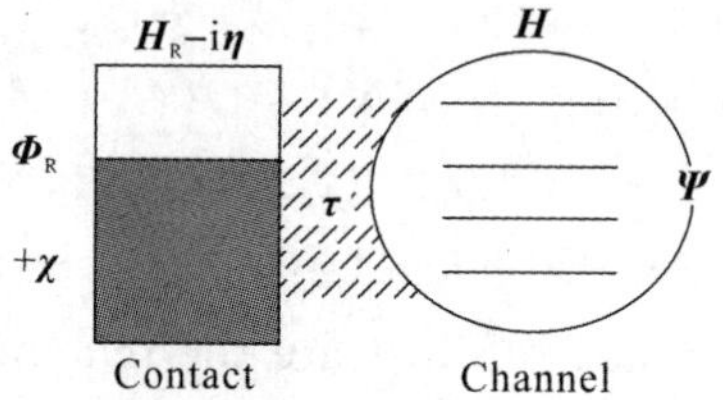

Fig. 1.3.5 The isolated contact of device

Since there is only one contact this is really an equilibrium problem and the density matrix is obtained simply by filling up the spectral function

$$\boldsymbol{A}(E) = \mathrm{i}(\boldsymbol{G} - \boldsymbol{G}^+) \tag{1.3.20}$$

according to the Fermi function as stated in Eqs. (1.3.2a, b). What I would like do now is to obtain this result in a completely different way. I will assume that the source waves $\boldsymbol{\Phi}_R$ from the contact are filled according to the Fermi function and the channel itself is filled simply by the spilling over of these wavefunctions. I will show that the resulting density matrix in the channel is identical to what we obtained earlier. Once we are clear about the approach we will extend it to the real problem with two contacts.

Before we connect the contact to the device, the electrons will occupy the contact eigenstates α according to its Fermi function, so that we can write down the density matrix for the contact as

$$\boldsymbol{\rho}_R(r, r') = \sum_{\alpha} \boldsymbol{\phi}_\alpha(r) f_0(\varepsilon_\alpha - \mu) \boldsymbol{\phi}_\alpha^*(r')$$

or in matrix notation as

$$\boldsymbol{\rho}_R = \sum_{\alpha} f_0(\varepsilon_\alpha - \mu) \boldsymbol{\phi}_\alpha \boldsymbol{\phi}_\alpha^+ \tag{1.3.21}$$

Now we wish to calculate the device density matrix by calculating the response of the device to the excitation $\boldsymbol{\tau\Phi}$ from the contact. We can write the source term due to each contact eigenstate α as $\boldsymbol{S}_\alpha = \boldsymbol{\tau}\boldsymbol{\phi}_\alpha$, find the resulting device wavefunction and then obtain the device density matrix by adding up the individual components weighted by the appropriate Fermi factors for the original contact eigenstate α:

$$\boldsymbol{\rho} = \sum_{\alpha} f_0(\varepsilon_\alpha - \mu)\boldsymbol{\psi}_\alpha \boldsymbol{\psi}_\alpha^+ = \int \mathrm{d}E f_0(E-\mu)\sum_{\alpha}\delta(E-\varepsilon_\alpha)\boldsymbol{\psi}_\alpha \{\boldsymbol{\psi}_\alpha\}^+ =$$

$$\int \mathrm{d}E f_0(E-\mu)\boldsymbol{G\tau}[\sum_{\alpha}\delta(E-\varepsilon_\alpha)\boldsymbol{\phi}_\alpha \boldsymbol{\phi}_\alpha^+]\boldsymbol{\tau}^+ \boldsymbol{G}^+ = \int \frac{\mathrm{d}E}{2\pi} f_0(E-\mu)\boldsymbol{G\tau} \boldsymbol{A}_\mathrm{R} \boldsymbol{\tau}^+ \boldsymbol{G}^+ \tag{1.3.22}$$

making use of the expression for the spectral function in the contact:

$$\boldsymbol{A}_\mathrm{R}(E) = \sum_{\alpha}\delta(E-\varepsilon_\alpha)\boldsymbol{\phi}_\alpha \boldsymbol{\phi}_\alpha^+ \tag{1.3.23}$$

$$\boldsymbol{\Gamma} = \mathrm{i}(\boldsymbol{\Sigma} - \boldsymbol{\Sigma}^+) = \boldsymbol{\tau} \boldsymbol{A}_\mathrm{R} \boldsymbol{\tau}^+ \tag{1.3.24}$$

$$\boldsymbol{G}^n = \boldsymbol{G\Gamma G}^+ f_0(E-\mu) \tag{1.3.25}$$

where we have made use of Eq. (1.3.2a). To show that this is the same as our earlier result (see Eqs. (1.3.2a,b)), we need the following important identity. If

$$\boldsymbol{G} = (E\boldsymbol{I} - \boldsymbol{H} - \boldsymbol{\Sigma})^{-1} \quad , \quad \boldsymbol{\Gamma} = \mathrm{i}(\boldsymbol{\Sigma} - \boldsymbol{\Sigma}^+)$$

then

$$\boldsymbol{A} \equiv i(\boldsymbol{G} - \boldsymbol{G}^+) = \boldsymbol{G\Gamma G}^+ = \boldsymbol{G}^+\boldsymbol{\Gamma G} \tag{1.3.26}$$

This is shown by writing $(\boldsymbol{G}^+)^{-1} - \boldsymbol{G}^{-1} = \boldsymbol{\Sigma} - \boldsymbol{\Sigma}^+ = -\mathrm{i}\boldsymbol{\Gamma}$, then pre-multiplying with $\boldsymbol{G}$ and post-multiplying with $\boldsymbol{G}^+$ to obtain

$$\boldsymbol{G} - \boldsymbol{G}^+ = -\mathrm{i}\boldsymbol{G\Gamma G}^+ \rightarrow \boldsymbol{A} = \boldsymbol{G\Gamma G}^+$$

Alternatively if we pre-multiply with $\boldsymbol{G}^+$ and post-multiply with $\boldsymbol{G}$ we obtain

$$\boldsymbol{G} - \boldsymbol{G}^+ = -\mathrm{i}\,\boldsymbol{G}^+\boldsymbol{\Gamma G} \rightarrow \boldsymbol{A} = \boldsymbol{G}^+\boldsymbol{\Gamma G}$$

I have used this simple one-contact problem to illustrate the important physical principle that the different eigenstates are uncorrelated and so we should calculate their contributions to the density matrix independently and then add them up.

This is a little bit like Young's two-slit experiment shown below. If the two slits are illuminated coherently then the intensity on the screen will show an interference pattern.

But if the slits are illuminated incoherently then the intensity on the screen is simply the sum of the intensities we would get from each slit independently. Each eigenstate α is like a "slit" that "illuminates" the device and the important point is that the "slits" have no phase coherence. That is why we calculate the device density matrix for each "slit" a independently and add them up.

Now that we have been through this exercise once, it is convenient to devise the following rule for dealing with the contact and channel wavefunctions

$$\boldsymbol{\Phi}_\mathrm{R}\, \boldsymbol{\Phi}_\mathrm{R}^+ \Rightarrow \int (\mathrm{d}E/2\pi)\, f_0(E-\mu)\, \boldsymbol{A}_\mathrm{R}(E) \tag{1.3.27a}$$

$$\boldsymbol{\psi}\boldsymbol{\psi}^+ \Rightarrow \int (\mathrm{d}E/2\pi)\, \boldsymbol{G}^n(E) \tag{1.3.27b}$$

reflecting the fact that the electrons in the contact are distributed according to the Fermi function $f_0(E-\mu)$ in a continuous distribution of eigenstates described by the spectral function $\boldsymbol{A}_R(E)$. This rule can be used to shorten the algebra considerably. For example, to evaluate the density matrix we first write down the result for a single eigenstate

$$\boldsymbol{\psi} = \boldsymbol{G\tau\Phi}_R \rightarrow \boldsymbol{\psi\psi}^+ = \boldsymbol{G\tau\Phi}_R\boldsymbol{\Phi}_R{}^+\boldsymbol{\tau}^+\boldsymbol{G}^+$$

and then apply Eq. (1. 3. 21) to obtain $\boldsymbol{G}^n = \boldsymbol{G\tau A}_R\boldsymbol{\tau}^+\boldsymbol{G}^+ f_0(E-\mu)$, which reduces to Eq. (1. 3. 25) making use of Eq. (1. 3. 26).

Now we are ready to tackle the actual problem with two contacts. We assume that before connecting to the channel, the electrons in the source and the drain contact have wavefunctions $\boldsymbol{\Phi}_1$, $\boldsymbol{\Phi}_2$ obeying the "Schrödinger" equations for the isolated contacts:

$$(E\boldsymbol{I}-\boldsymbol{H}_1+\mathrm{i}\boldsymbol{\eta})\boldsymbol{\Phi}_1 = \boldsymbol{S}_1 \text{ and } (E\boldsymbol{I}-\boldsymbol{H}_2+\mathrm{i}\boldsymbol{\eta})\boldsymbol{\Phi}_2 = \boldsymbol{S}_2 \tag{1. 3. 28}$$

where $\boldsymbol{H}_1$, $\boldsymbol{H}_2$ are the Hamiltonians for contacts 1 and 2 respectively and we have added a small positive infinitesimal times an identity matrix, $\boldsymbol{\eta} = 0^+\boldsymbol{I}$, to introduce dissipation as before. When we couple the device to the contacts as shown in Fig. 1. 3. 6, these electronic states from the contacts "spill over" giving rise to a wavefunction $\boldsymbol{\psi}$ inside the device which in turn excites scattered waves $\boldsymbol{\chi}_1$ and $\boldsymbol{\chi}_2$ in the source and drain respectively.

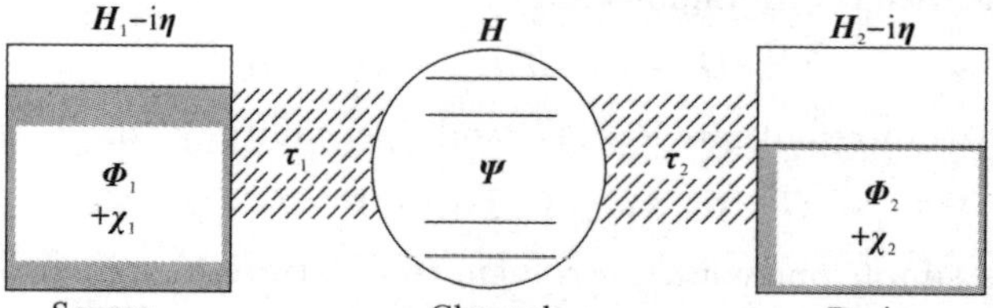

Fig. 1. 3. 6 A channel connected to two contacts

The overall wavefunction will satisfy the composite Schrödinger equation for the composite contact-1-device-contact-2 system which we can write in three blocks:

$$\begin{bmatrix} E\boldsymbol{I}-\boldsymbol{H}_1+\mathrm{i}\boldsymbol{\eta} & -\boldsymbol{\tau}_1{}^+ & \boldsymbol{O} \\ -\boldsymbol{\tau}_1 & E\boldsymbol{I}-\boldsymbol{H} & -\boldsymbol{\tau}_2 \\ \boldsymbol{O} & -\boldsymbol{\tau}_2{}^+ & E\boldsymbol{I}-\boldsymbol{H}_2+\mathrm{i}\boldsymbol{\eta} \end{bmatrix} \begin{bmatrix} \boldsymbol{\Phi}_1+\boldsymbol{\chi}_1 \\ \boldsymbol{\psi} \\ \boldsymbol{\Phi}_2+\boldsymbol{\chi}_2 \end{bmatrix} = \begin{bmatrix} \boldsymbol{S}_1 \\ \boldsymbol{O} \\ \boldsymbol{S}_2 \end{bmatrix} \tag{1. 3. 29}$$

where $\boldsymbol{H}$ is the channel Hamiltonian. Using straightforward matrix algebra we obtain from the first and last equations

$$\boldsymbol{\chi}_1 = \boldsymbol{G}_1\boldsymbol{\tau}_1{}^+\boldsymbol{\psi},\ \boldsymbol{\chi}_2 = \boldsymbol{G}_2\boldsymbol{\tau}_2{}^+\boldsymbol{\psi} \tag{1. 3. 30}$$

where

$$\boldsymbol{G}_1 = (E\boldsymbol{I}-\boldsymbol{H}_1+\mathrm{i}\boldsymbol{\eta})^{-1},\ \boldsymbol{G}_2 = (E\boldsymbol{I}-\boldsymbol{H}_2+\mathrm{i}\boldsymbol{\eta})^{-1} \tag{1. 3. 31}$$

are the Green's functions for the isolated reservoirs. Using Eq. (1. 3. 26) to eliminate $\boldsymbol{\chi}_1$, $\boldsymbol{\chi}_2$ from the middle equation in Eq. (1. 3. 27) we obtain

$$(E\boldsymbol{I}-\boldsymbol{H}-\boldsymbol{\Sigma}_1-\boldsymbol{\Sigma}_2)\boldsymbol{\psi} = \boldsymbol{S} \tag{1. 3. 32}$$

where

$$\boldsymbol{\Sigma}_1 = \boldsymbol{\tau}_1 \boldsymbol{G}_1 \boldsymbol{\tau}_1^+, \boldsymbol{\Sigma}_2 = \boldsymbol{\tau}_2 \boldsymbol{G}_2 \boldsymbol{\tau}_2^+ \tag{1.3.33}$$

are the self-energy matrices that we discussed in Chapter 2. The corresponding broadening matrices(see Eq. (1.3.18)) are given by

$$\boldsymbol{\Gamma}_1 = \boldsymbol{\tau}_1 \boldsymbol{A}_1 \boldsymbol{\tau}_1^+, \boldsymbol{\Gamma}_2 = \boldsymbol{\tau}_2 \boldsymbol{A}_2 \boldsymbol{\tau}_2^+ \tag{1.3.34}$$

where $\boldsymbol{A}_1 = \mathrm{i}(\boldsymbol{G}_1 - \boldsymbol{G}_1^+)$ and $\boldsymbol{A}_2 = \mathrm{i}(\boldsymbol{G}_2 - \boldsymbol{G}_2^+)$ are the spectral functions for the isolated contacts 1 and 2 respectively. Also,

$$\boldsymbol{S} \equiv \boldsymbol{\tau}_1 \boldsymbol{\Phi}_1 + \boldsymbol{\tau}_2 \boldsymbol{\Phi}_2 \tag{1.3.35}$$

is the sum of the source terms $\boldsymbol{\tau}_1 \boldsymbol{\Phi}_1$ (from the source) and $\boldsymbol{\tau}_2 \boldsymbol{\Phi}_2$ (from the drain) as shown in Fig. 1.3.7.

To evaluate the density matrix, we define the channel Green's function

$$\boldsymbol{G} \equiv (E\boldsymbol{I} - \boldsymbol{H} - \boldsymbol{\Sigma}_1 - \boldsymbol{\Sigma}_2)^{-1} \tag{1.3.36}$$

and use it to express the channel wavefunction in terms of the source terms from Eq. (1.3.25):

$$\boldsymbol{\psi} = \boldsymbol{GS} \rightarrow \boldsymbol{\psi}\boldsymbol{\psi}^+ = \boldsymbol{GS}\,\boldsymbol{S}^+\,\boldsymbol{G}^+ \tag{1.3.37}$$

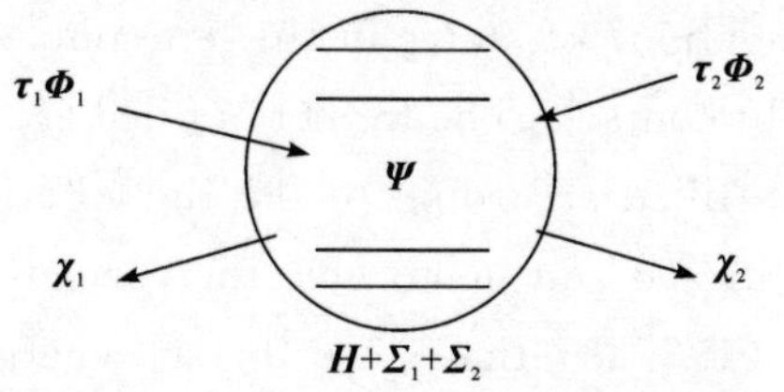

Fig. 1.3.7 Channel excited by $\boldsymbol{\tau}_1 \boldsymbol{\Phi}_1$ (from the source) and $\boldsymbol{\tau}_2 \boldsymbol{\Phi}_2$ (from the drain). The channel response is described by Eq. (1.3.29) and it in turn generates $\boldsymbol{\chi}_1$, $\boldsymbol{\chi}_2$ in the contacts

Note that the cross-terms in the source

$$\boldsymbol{S}\boldsymbol{S}^+ = \boldsymbol{\tau}_1 \boldsymbol{\Phi}_1 \boldsymbol{\Phi}_1^+ \boldsymbol{\tau}_1^+ + \boldsymbol{\tau}_2 \boldsymbol{\Phi}_2 \boldsymbol{\Phi}_2^+ \boldsymbol{\tau}_2^+ + \boldsymbol{\tau}_1 \boldsymbol{\Phi}_1 \boldsymbol{\Phi}_2^+ \boldsymbol{\tau}_2^+ + \boldsymbol{\tau}_2 \boldsymbol{\Phi}_2 \boldsymbol{\Phi}_1^+ \boldsymbol{\tau}_1^+$$

cross-terms=0 are zero since $\boldsymbol{\Phi}_1$ and $\boldsymbol{\Phi}_2$ are the wavefunctions (before connecting to the channel) in the source and drain contacts which are physically disjoint and unconnected. The direct terms are evaluated using the basic principle [see Eq. (1.3.25)] that we formulated earlier with the one-contact problem:

$$\boldsymbol{\Phi}_1 \boldsymbol{\Phi}_1^+ \Rightarrow \int (\mathrm{d}E/2\pi) f_1(E) \boldsymbol{A}_1(E) \tag{1.3.38a}$$

$$\boldsymbol{\Phi}_2 \boldsymbol{\Phi}_2^+ \Rightarrow \int (\mathrm{d}E/2\pi) f_2(E) \boldsymbol{A}_2(E) \tag{1.3.39b}$$

to write down the density matrix from $\boldsymbol{\psi}\boldsymbol{\psi}^+ = \boldsymbol{GS}\,\boldsymbol{S}^+\,\boldsymbol{G}^+$

$$\boldsymbol{\rho} = \int (\mathrm{d}E/2\pi)(\boldsymbol{G}\boldsymbol{\tau}_1 \boldsymbol{A}_1 \boldsymbol{\tau}_1^+ \boldsymbol{G}^+ f_1 + \boldsymbol{G}\boldsymbol{\tau}_2 \boldsymbol{A}_2 \boldsymbol{\tau}_2^+ \boldsymbol{G}^+ f_2)$$

Making use of Eq. (1.3.29) we can simplify this expression to write

$$G^n = G\Sigma^{in} G^+ \tag{1.3.40}$$

and

$$\Sigma^{in} = \Gamma_1 f_1 + \Gamma_2 f_2 \tag{1.3.41}$$

noting that $\rho = \int (dE/2\pi) G^n$ as defined in Eq. (1.3.2a). Just as G^n is obtained from $\psi\psi^+$, Σ^{in} is obtained from SS^+. One could thus view Eq. (1.3.31) as a relation between the "electron density" in the device created by the source term S representing the spill-over of electrons from the contacts.

Partial spectral function: Substituting Eq. (1.3.28) into Eq. (1.3.29) we can write

$$G^n = A_1 f_1 + A_2 f_2 \tag{1.3.42}$$

where

$$A_1 = G\Gamma_1 G^+ , A_2 = G\Gamma_2 G^+$$

Comparing this with the equilibrium result (see Eq. (1.3.2b)), $G^n = A f_0$, it seems natural to think of the total spectral function $A(E)$ as consisting of two parts: $A_1(E)$ arising from the spill-over (or propagation) of states in the left contact and $A_2(E)$ arising from the spill-over of states in the right contact. The former is filled according to the left Fermi function $f_1(E)$ while the latter is filled according to the right Ferm function $f_2(E)$. To show that the two partial spectra indeed add up to give the correct total spectral function, $A = A_1 + A_2$, we note from Eq. (1.3.25) that since the self-energy Σ has two parts Σ_1 and Σ_2 coming from two contacts, $A = G(\Gamma_1 + \Gamma_2)G^+ = A_1 + A_2$ as stated in Eq. (1.3.6).

An important conceptual point before we move on. Our approach is to use the Schrödinger equation to calculate the evolution of a specific eigenstate Φ_α from one of the contacts and then superpose the results from distinct eigenstates to obtain the basic rule stated in Eq. (1.3.25) or Eq. (1.3.26). It may appear that by superposing all these individual fluxes we are ignoring the Pauli exclusion principle. Wouldn't the presence of electrons evolving out of one eigenstate block the flux evolving out of another eigenstate? The answer is no, as long as the evolution of the electrons is coherent. This is easiest to prove in the time domain, by considering two electrons that originate in distinct eigenstates Φ_1 and Φ_2. Initially there is no question of one blocking the other since they are orthogonal: $\Phi_1^+ \Phi_2 = 0$. At later times their wavefunctions can be written as

$$\psi_1(t) = \exp(-iHt/\hbar)\Phi_1$$

$$\psi_2(t) = \exp(-iHt/\hbar)\Phi_2$$

if both states evolve coherently according to the Schrödinger equation:

$$i\hbar\, d\psi/dt = H\psi$$

It is straightforward to show that the overlap between any two states does not change as a result of this evolution: $\psi_1(t)^+ \ \psi_2(t) = \Phi_1^+ \ \Phi_2$. Hence, wavefunctions originating from orthogonal at all times and never "Pauli block" each other. Note, however, that this argument cannot be used when phase-breaking processes (briefly explained in the introduction to this chapter) are involved since the evolution of electrons cannot be described by a one-particle Schrödinger equation.

Now that we have derived the results for the non-equilibrium density matrix [see Eq. (1. 3. 3)- Eq. (1. 3. 6)], let us discuss the current flow at the terminals [see Eq. (1. 3. 7)- Eq. (1. 3. 9)]. As before let us start with the "one-contact" problem shown in Fig. 1. 3. 6.

Consider again the problem of a channel connected to one contact described by:

$$E\begin{bmatrix}\psi\\ \Phi\end{bmatrix}=\begin{bmatrix}H & \tau\\ \tau^+ & H_R - i\eta\end{bmatrix}\begin{bmatrix}\psi\\ \Phi\end{bmatrix}$$

which is the same as Eq. (1. 3. 18) with $\Phi \equiv \Phi_R + \chi$ and S_R dropped for clarity. How can we evaluate the current flowing between the channel and the contact? when discussing the velocity of a band electron, we need to look at the time-dependent version of this equation

$$i\hbar\frac{d}{dt}\begin{bmatrix}\psi\\ \Phi\end{bmatrix}=\begin{bmatrix}H & \tau\\ \tau^+ & H_R - i\eta\end{bmatrix}\begin{bmatrix}\psi\\ \Phi\end{bmatrix}$$

and obtain an expression for the time rate of change in the probability density inside the channel, which is given by

$$\text{Trace}(\psi\psi^+) = \text{Trace}(\psi^+\psi) = \psi^+\psi$$

(note that $\psi^+\psi$ is just a number and so it does not matter if we take the trace or not):

$$I \equiv \frac{d}{dt}\psi^+\psi = \frac{\text{Trace}(\psi^+\tau\Phi - \Phi^+\tau^+\psi)}{i\hbar} \qquad (1.3.43)$$

Noting that $\Phi \equiv \Phi_R + \chi$, we can divide this net current I conceptually into an inflow, proportional to the "incident" wave Φ_R , and an outflow proportional to the "scattered" wave χ :

$$I \equiv \underbrace{\frac{\text{Trace}(\psi^+\tau\Phi_R - \Phi_R^+\tau^+\psi]}{i\hbar}}_{\text{Inflow}} - \underbrace{\frac{\text{Trace}(\chi^+\tau^+\psi - \psi^+\tau\chi)}{i\hbar}}_{\text{Outflow}} \qquad (1.3.44)$$

Making use of Eq. (1. 3. 21) and Eq. (1. 3. 22) we can write the inflow as

Inflow= Trace $(S^+ G^+ S - S^+ G S)/i\hbar$ = Trace $(SS^+A)/\hbar$

Since $i(G - G^+) = A$. To obtain the total inflow we need to sum the inflows due to each contact eigenstate α , all of which as we have seen are taken care of by the replacement (see Eq. (1. 3. 27a))

$$\boldsymbol{\Phi}_R\, \boldsymbol{\Phi}_R^+ \Rightarrow \int \frac{dE}{2\pi} f_0(E-\mu)(\boldsymbol{A}_R(E))$$

Since $\boldsymbol{S}=\boldsymbol{\tau}\boldsymbol{\Phi}_R$, this leads to

$$\boldsymbol{S}\boldsymbol{S}^+ \Rightarrow \int \frac{dE}{2\pi} f_0(E-\mu)\boldsymbol{\tau}\,\boldsymbol{A}_R\,\boldsymbol{\tau}^+ = \int \frac{dE}{2\pi} f_0(E-\mu)\boldsymbol{\Gamma}$$

so that the inflow term becomes

$$\text{Inflow} = \frac{1}{\hbar}\int \frac{dE}{2\pi} f_0(E-\mu)\,\text{Trace}(\boldsymbol{\Gamma}\boldsymbol{A}) \tag{1.3.45a}$$

Similarly, we make use of Eq. (1.3.22) and Eq. (1.3.29) to write the outflow term as

$$\text{Outflow} = \text{Trace}(\boldsymbol{\psi}^+\boldsymbol{\tau}\boldsymbol{G}_R^+\,\boldsymbol{\tau}^+\boldsymbol{\psi} - \boldsymbol{\psi}^+\boldsymbol{\tau}\boldsymbol{G}_R\,\boldsymbol{\tau}^+\boldsymbol{\psi})/\text{i}\hbar = \text{Trace}(\boldsymbol{\psi}\boldsymbol{\psi}^+\boldsymbol{\Gamma})/\hbar$$

On summing overall the eigenstates, $\boldsymbol{\psi}\boldsymbol{\psi}^+ \Rightarrow \int dE\,\boldsymbol{G}^n/2\pi$, so that

$$\text{Outflow} = \frac{1}{\hbar}\int \frac{dE}{2\pi}\text{Trace}(\boldsymbol{\Gamma}\boldsymbol{G}^n) \tag{1.3.46b}$$

It is easy to see that the inflow and outflow are equal at equilibrium, since $\boldsymbol{G}^n = \boldsymbol{A} f_0$.

Now we are ready to calculate the inflow and outflow for the channel with two contacts, We consider one of the interfaces, say the one with the source contact, and write the inflow as [see Eq. (1.3.44)]

$$I_1 = \underbrace{\frac{\text{Trace}(\boldsymbol{\psi}^+\boldsymbol{\tau}\boldsymbol{\Phi}_1 - \boldsymbol{\Phi}_1^+\,\boldsymbol{\tau}_1^+\boldsymbol{\psi})}{\text{i}\hbar}}_{\text{Inflow}} - \underbrace{\frac{\text{Trace}(\boldsymbol{\chi}_1^+\,\boldsymbol{\tau}_1^+\boldsymbol{\psi} - \boldsymbol{\psi}^+\boldsymbol{\tau}\boldsymbol{\chi}_1)}{\text{i}\hbar}}_{\text{Outflow}}$$

Making use of the relations $\boldsymbol{\psi}=\boldsymbol{G}\boldsymbol{S}$ [with $\boldsymbol{S}$ and $\boldsymbol{G}$ defined in Eq. (1.3.35) and Eq. (1.3.36)] and $\boldsymbol{S}_1 \equiv \boldsymbol{\tau}_1\,\boldsymbol{\Phi}_1$, we can write

$$\text{Inflow} = \text{Trace}(\boldsymbol{S}^+\,\boldsymbol{G}^+\,\boldsymbol{S}_1 - \boldsymbol{S}_1^+\boldsymbol{G}\boldsymbol{S})/\text{i}\hbar = \text{Trace}(\boldsymbol{S}_1\,\boldsymbol{S}_1^+\boldsymbol{A})/\hbar$$

Since $\boldsymbol{S}=\boldsymbol{S}_1+\boldsymbol{S}_2$ and $\boldsymbol{S}_1^+\,\boldsymbol{S}_2 = \boldsymbol{S}_2^+\,\boldsymbol{S}_1 = 0$.

Next we sum the inflow due to each contact eigenstate α, all of which is taken care of by the replacement [see Eq. (1.3.27a)]

$$\boldsymbol{\Phi}_1\,\boldsymbol{\Phi}_1^+ \Rightarrow \int \frac{dE}{2\pi} f_1(E)\boldsymbol{A}_1(E)$$

leading to $\boldsymbol{S}_1\,\boldsymbol{S}_1^+ = \boldsymbol{\tau}_1\,\boldsymbol{\Phi}_1\,\boldsymbol{\Phi}_1^+\,\boldsymbol{\tau}_1^+ \Rightarrow \int \frac{dE}{2\pi}(\boldsymbol{\tau}_1\,\boldsymbol{A}_1\,\boldsymbol{\tau}_1^+)f_1(E) = \int \frac{dE}{2\pi}\boldsymbol{\Gamma}_1 f_1(E)$

so that

$$\text{Inflow} = \frac{1}{\hbar}\int \frac{dE}{2\pi} f_1(E)\,\text{Trace}(\boldsymbol{\Gamma}_1\boldsymbol{A}) \tag{1.3.47a}$$

Similarly we make use of Eq. (1.3.29) and Eq. (1.3.30) to write the outflow term as

$$\text{Ouflow} = \text{Trace}(\boldsymbol{\psi}^+\,\boldsymbol{\tau}_1\,\boldsymbol{G}_1^+\,\boldsymbol{\tau}_1^+\boldsymbol{\psi} - \boldsymbol{\psi}^+\,\boldsymbol{\tau}_1\,\boldsymbol{G}_1\,\boldsymbol{\tau}_1^+\boldsymbol{\psi})/\text{i}\hbar = \text{Trace}(\boldsymbol{\psi}\boldsymbol{\psi}^+\,\boldsymbol{\Gamma}_1)/\hbar$$

On summing over all the eigenstates, $\boldsymbol{\psi}\boldsymbol{\psi}^+ \Rightarrow \int (dE/2\pi)\boldsymbol{G}^n$, so that

$$\text{Ouflow} = (1/\hbar)\int \frac{dE}{2\pi}\text{Trace}(\boldsymbol{\Gamma}_1 \boldsymbol{G}^n) \qquad (1.3.47b)$$

The net current I_i at terminal i is given by the difference between the inflow and the outflow (multiplied by the charge $-q$ of an electron) as stated in Eq (1.3.7a)

$$I_i = (-q/\hbar)\int_{-\infty}^{+\infty} \frac{dE}{2\pi}\widetilde{I}_i(E)$$

with

$$\widetilde{I}_i = \text{Trace}(\boldsymbol{\Gamma}_i \boldsymbol{A})f_i - \text{Trace}(\boldsymbol{\Gamma}_i \boldsymbol{G}^n) \qquad (1.3.48)$$

and illustrated in Fig. 1.3.2.

1.4 Equilibrium density matrix

The density matrix is one of the central concepts in statistical mechanics. The reason I am bringing it up in this chapter is that it provides an instructive example of the concept of basis functions. Let me start by briefly explaining what it means. We calculated the electron density, $n(r)$, in multi-electron atoms by summing up the probability densities of each occupied eigenstate α : $n(r) = \sum_{\text{occ}\alpha} |\Phi_\alpha(r)|^2$.

This is true at low temperatures for closed systems having a fixed number of electrons that occupy the lowest available energy levels. In general, however, states can be partially occupied and in general the equilibrium electron density can be written as

$$n(r) = \sum_\alpha f_0(\varepsilon_\alpha - \mu)|\Phi_\alpha(r)|^2 \qquad (1.4.1)$$

where $f_0 E \equiv [1 + \exp(E/k_B T)]^{-1}$ is the Fermi function whose value indicates the extent to which a particular state is occupied: "0" indicates unoccupied states, "1" indicates occupied states, while a value between 0 and l indicates the average occupancy of a state that is sometimes occupied and sometimes unoccupied.

Could we write a "wavefunction" $\Psi(r)$ for this multi-electron system such that its magnitude will give us the electron density $n(r)$? One possibility is to write it as

$$\Psi(r) = \sum_\alpha C_\alpha \Psi_\alpha(r) \qquad (1.4.2)$$

where $|C_\alpha|^2 = f_0(\varepsilon_\alpha - \mu)$. But this is not quite right. If we square the magnitude of this multi-electron "wavefunction" we obtain

$$n(r) = |\Psi(r)|^2 = \sum_\alpha\sum_\beta C_\alpha C_\beta^* \Phi_\alpha(r)\Phi_\beta(r)$$

which is equivalent to Eq. (1.4.1) if and only if,

$$|C_\alpha|^2 = f_0(\varepsilon_\alpha - \mu) \equiv f_\alpha, C_\alpha C_\beta^* = 0, \alpha \neq \beta \qquad (1.4.3)$$

This is impossible if we view the coefficients C_α as ordinary numbers—in that case $C_\alpha C_\beta^*$ must equal $\sqrt{f_\alpha f_\beta}$ and cannot be zero unless both C_α and C_β are zero. If we wish to write the multi-electron wavefunction in the form shown in Eq. (1.4.2) we should view the coefficients C_α as stochastic numbers whose correlation coefficients are given by Eq. (1.4.3).

So instead of writing a wavefunction for multi-electron systems, it is common to write down $\rho(\alpha,\beta)$ indicating the correlation $C_\alpha C_\beta^*$ between every pair of coefficients. This matrix $\boldsymbol{\rho}$ is called the density matrix and in the eigenstate presentation we can write its elements as [see Eq. (1.4.3)]

$$\rho(\alpha,\beta) = f_\alpha \delta_{\alpha\beta} \tag{1.4.4}$$

where $\delta_{\alpha\beta}$ is the Kronecker delta defined as

$$\delta_{\alpha\beta} = \begin{cases} 1, \text{ if } & \alpha = \beta \\ 0, \text{ if } & \alpha \neq \beta \end{cases}$$

We can rewrite Eq. (1.4.1) for the electron density $n(\boldsymbol{r})$ in the form

$$n(r) = \sum_\alpha \sum_\beta \rho(\alpha,\beta)\Phi_\alpha(r)\Phi_\beta^*(r) \tag{1.4.5}$$

which can be generalized to define

$$\tilde{\rho}(r,r') = \sum_\alpha \sum_\beta \rho(\alpha,\beta)\Phi_\alpha(r)\Phi_\beta^*(r') \tag{1.4.6}$$

such that the electron density $n(r)$ is given by its diagonal elements:

$$n(r) = \rho(r,r')_{r'=r} \tag{1.4.7}$$

Now the point I want to make is that Eq. (1.4.6) represents a unitary transformation from an eigenstate basis to a real space basis. This is seen by noting that the trans-formation matrix $\boldsymbol{V}$ is obtained by writing each of the eigenstates (the old basis) as a column vector using the position (the new basis) representation:

$$V_{r,\alpha} = \Phi_\alpha(r)$$

and that this matrix is unitary: $\boldsymbol{V}^{-1} = \boldsymbol{V}^{+}$

$$\Rightarrow V^{-1}{}_{\alpha,r} = V^{+}{}_{\alpha,r} = V_{\alpha,r}^* \Phi_\alpha^*(r)$$

so that Eq. (1.4.6) can be written in the form of a unitary transformation:

$$\tilde{\rho}(r,r') = \sum_\alpha \sum_\beta V(r,\alpha)\rho(\alpha,\beta)V^{+}(\beta,r') \Rightarrow \tilde{\boldsymbol{\rho}} = \boldsymbol{V}\boldsymbol{\rho}\,\boldsymbol{V}^{+}$$

This leads to a very powerful concept: The density matrix $\boldsymbol{\rho}$ at equilibrium can be written as the Fermi function of the Hamiltonian matrix ($\boldsymbol{I}$ is the identity matrix of the same size as $\boldsymbol{H}$):

$$\boldsymbol{\rho} = f_0(\boldsymbol{H} - \mu\boldsymbol{I}) \tag{1.4.8}$$

This is a general matrix relation that is valid in any representation. For example, if we use the eigenstates α of $\boldsymbol{H}$ as a basis then $\boldsymbol{H}$ is a diagonal matrix:

$$\boldsymbol{H}=\begin{bmatrix} \varepsilon_1 & 0 & 0 & 0 & \cdots \\ 0 & \varepsilon_2 & 0 & 0 & \cdots \\ 0 & 0 & \varepsilon_3 & 0 & \cdots \\ \cdots & \cdots & \cdots & \cdots & \varepsilon_m \end{bmatrix}$$

and so is $\boldsymbol{\rho}$:

$$\boldsymbol{\rho}=\begin{bmatrix} f_0(\varepsilon_1-\mu) & 0 & 0 & 0 & \cdots \\ 0 & f_0(\varepsilon_2-\mu) & 0 & 0 & \cdots \\ 0 & 0 & f_0(\varepsilon_3-\mu) & 0 & \cdots \\ \cdots & \cdots & \cdots & \cdots & f_0(\varepsilon_m-\mu) \end{bmatrix}$$

This is exactly what Eq. (1.4.4) tells us. But the point is that the relation given in Eq. (1.4.8) is valid, not just in the eigenstate representation, but in any representation. Given the matrix representation $\boldsymbol{H}$, it takes just three commands in MATLAB to obtain the density matrix:

$$[\mathrm{V,D}]=\mathrm{eig(H)}$$

$$\mathrm{rho}=\frac{1.}{\left(1+\exp\left(\frac{(\mathrm{diag(D)-mu).}}{\mathrm{kT}}\right)\right)}$$

$$\mathrm{rho}=\mathrm{V}*\mathrm{diag(rho)}*\mathrm{V}'$$

The first command calculates a diagonal matrix $\boldsymbol{D}$ whose diagonal elements are the eigenvalues of $\boldsymbol{H}$ and a matrix $\mathbf{V}$ whose columns are the corresponding eigenvectors In other words, $\boldsymbol{D}$ is the Hamiltonian $\boldsymbol{H}$ transformed to the eigenstate basis $\Rightarrow \boldsymbol{D}=\mathbf{V}^{+}\boldsymbol{H}$.

The second command gives us the density matrix in the eigenstate representation, which is easy since in this representation both $\boldsymbol{H}$ and $\boldsymbol{\rho}$ are diagonal. The third command then transforms $\boldsymbol{\rho}$ back to the original representation.

Fig. 1.4.1 shows the equilibrium electron density for a 1D box modeled with a discrete lattice of 100 points spaced by 2Å, with $\mu=0.25$ eV. The Hamiltonian $\boldsymbol{H}$ is a (100 × 100) matrix can be set up. The density matrix is then evaluated as described above and its diagonal elements give us the electron density $n(x)$ (times the lattice constant, a). Note the standing wave patterns in Figs. 1.4.1(c)(d) which are absent when we use periodic boundary conditions [see Figs. 1.4.1(e)(f)]. Figs 1.4.1(e)(f) also show the standing wave patterns in the electron density when a large repulsive potential

$$U_0\delta\left[x-\left(\frac{L}{2}\right)\right]\text{where } U_0=2\ \mathrm{eV/nm}$$

is included at the center of the box.

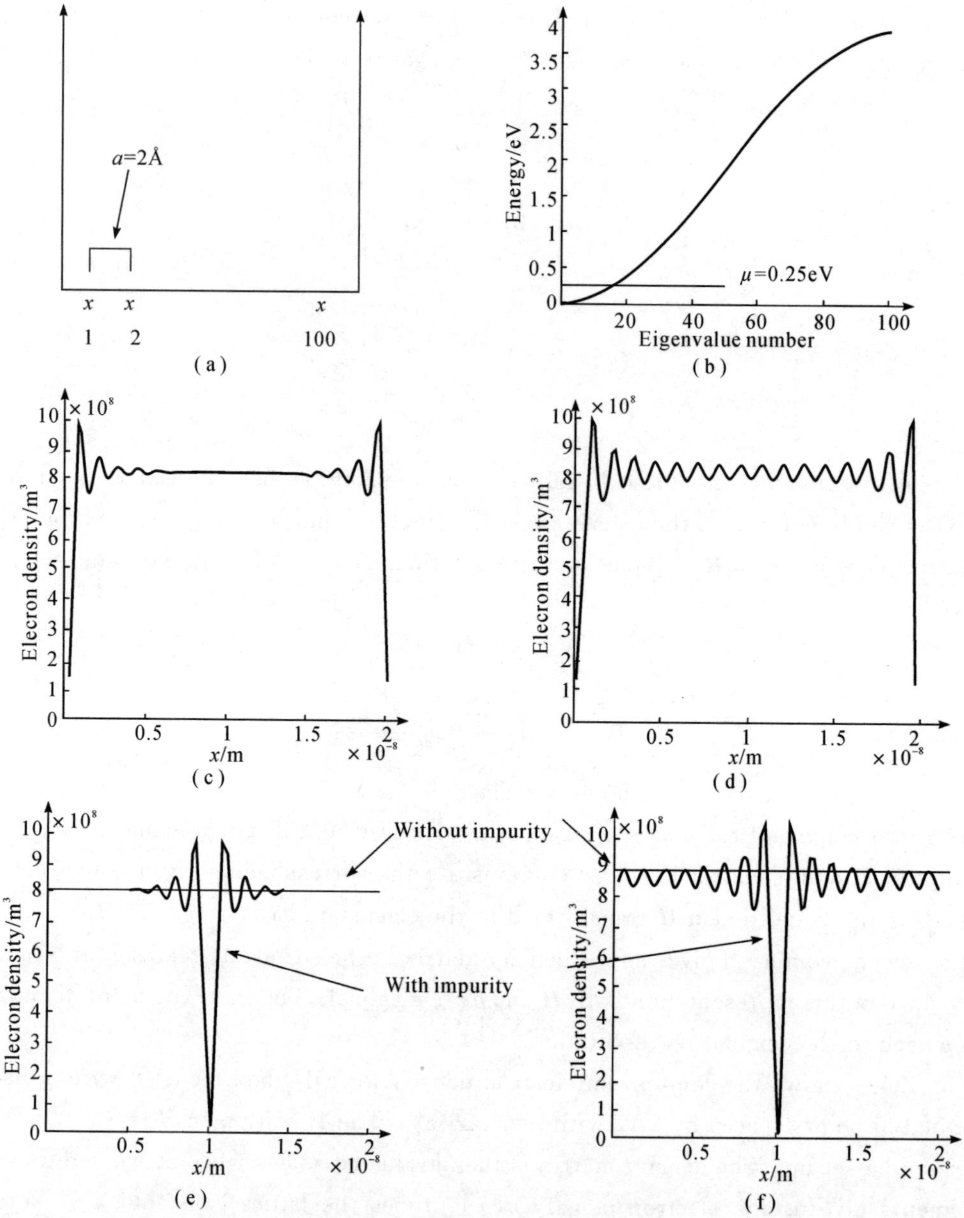

Fig. 1. 4. 1 Equilibrium electron density for a 1D box modeled with a discrete lattice of 100 points spaced by 2 Å

(a) electrons in a 1D box; (b) energy eigenvalues and electrochemical potential μ;

(c) $n(x)$ vs. x, $k_B T = 0.025$ eV; (d) $n(x)$ vs. x, $k_B T = 0.002,5$ eV;

(e) periodic boundary conditions with and without impurity in middle;

(f) boundary conditions with and without impurity in middle

Note that the density matrix can look very different depending on what basis functions we use. In the eigenstate representation it is diagonal since the Hamiltonian is diagonal, but in the real-space lattice representation it has off-diagonal elements. In any basis m , the diagonal elements $\rho(m,m)$ tell us the number of electrons occupying the state m . In a real-space representation, the diagonal elements $\rho(r,r)$ give us the electron density $n(r)$. The trace (sum of diagonal elements) of $\boldsymbol{\rho}$, which is invariant in all representations, gives us the total number of electrons N :

$$N = \text{Trave}\ (\boldsymbol{\rho}) \tag{1.4.9}$$

If we are only interested in the electron density, then the diagonal elements of the density matrix are all we need. But we cannot "throw away" the off-diagonal elements they are needed to ensure that the matrix will transform correctly to another representation. Besides, depending on what we wish to calculate, we may need the off-diagonal elements too.

It is common in quantum mechanics to associate every observable $\mathbf{A}$ with an operator A_{op} for which we can find a matrix representation $\mathbf{A}$ in any basis. The expectation value $\langle \mathbf{A} \rangle$ for this observable (that is, the average value we expect to get in a series of measurements) is given by

$$\langle \mathbf{A} \rangle = \int \mathrm{d}\, r \Psi^*(r) A_{\mathrm{op}} \Psi(r)$$

Substituting for the wavefunction in terms of the basis functions, we can show that

$$\langle \mathbf{A} \rangle = \sum_{\alpha} \sum_{\beta} C_{\alpha} C_{\beta}^* \int \mathrm{d}\, r \Phi_{\beta}^*(r) A_{\mathrm{op}} \Phi_{\alpha}(r)$$

so that

$$\langle \mathbf{A} \rangle = \sum_{\alpha} \sum_{\beta} \rho_{\alpha\beta} A_{\beta\alpha} = \mathrm{Trace}(\boldsymbol{\rho}\mathbf{A})$$

We could use this result to evaluate the expectation value of any quantity, even if the system is out of equilibrium, provided we know the density matrix. But what we have discussed here is the equilibrium density matrix. It is much harder to calculate the non-equilibrium density matrix, as we will discuss later in the book.

Consider now a conductor of length L having just two plane wave (pw) states

$$\Psi_+(x) = \frac{1}{\sqrt{L}} \mathrm{e}^{+ikx}, \quad \Psi_-(x) = \frac{1}{\sqrt{L}} \mathrm{e}^{-ikx}$$

The current operator in this basis is given by

$$\boldsymbol{J}_{\mathrm{op}\pm} = \frac{-q}{L} \begin{bmatrix} \dfrac{\hbar k}{m} & 0 \\ 0 & \dfrac{-\hbar k}{m} \end{bmatrix}$$

and we could write the density matrix as

$$\boldsymbol{\rho}_{\pm} = \begin{bmatrix} f_+ & 0 \\ 0 & f_- \end{bmatrix}$$

where f_+ and f_- are the occupation probabilities for the two states. We wish to transform both these matrices basis using cosl ne and sine states:

$$\Psi_c(x) = \sqrt{\frac{2}{L}}\cos kx \ , \ \Psi_s(x) = \sqrt{\frac{2}{L}}\sin kx$$

It is straightforward to write down the transformation matrix $\mathbf{V}$ whose columns represent the old basis (+, −) in terms of the new basis (c, s)

$$\mathbf{V} = \frac{1}{\sqrt{2}}\begin{bmatrix} 1 & 1 \\ +i & -i \end{bmatrix}$$

so that in the "cs" representation

$$\boldsymbol{\rho}_{cs} = \mathbf{V}\ (\boldsymbol{J}_{op})_{\pm}\ \mathbf{V}^{+} = \begin{array}{c} \text{"c"} \qquad\qquad \text{"s"} \\ \begin{bmatrix} f_+ + f_- & -(f_+ - f_-) \\ -i(f_+ - f_-) & f_+ + f_- \end{bmatrix} \end{array}$$

and

$$(\boldsymbol{J}_{op})_{cs} = \mathbf{V}\ (\boldsymbol{J}_{op})_{\pm}\ \mathbf{V}^{+} = \begin{array}{c} \text{"s"} \qquad \text{"c"} \\ \begin{bmatrix} 0 & -\dfrac{\hbar k}{mL} \\ \dfrac{\hbar k}{mL} & 0 \end{bmatrix} \end{array}$$

It is easy to check that the current $\langle \boldsymbol{J} \rangle = \mathrm{Trace}(\boldsymbol{\rho}\boldsymbol{J}_{op})$ is the same in either representation:

$$\langle \boldsymbol{J} \rangle = \left(\frac{-q}{L}\right)\left(\frac{\hbar k}{m}\right)[f_+ - f_-]$$

This is expected since the trace is invariant under a unitary transformation and thus remains the same no matter which representation we use. But the point to note is that the current in the cosine-sine representation arises from the off-diagonal elements of the current operator and the density matrix, rather than the diagonal elements. The off diagonal elements do not have an intuitive physical meaning like the diagonal elements. As long as the current is carried by the diagonal elements, we can use a semiclassical picture in terms of occupation probabilities. But if the "action" is in the off-diagonal c elements then we need a more general quantum framework.

Suppose we wish to find the energy levels of a hydrogen atom in the presence of an electric field F applied along the z-direction. Let us use the eigenstates 1s, 2s, $2p_x$, $2p_y$ and $2p_z$, as our basis set and write down the Hamiltonian matrix. If the field were absent the matrix would be diagonal:

$$
H_0 = \begin{array}{c} \\ \end{array}
\begin{array}{c}
\begin{array}{ccccc} 1s & 2s & 2p_x & 2p_y & 2p_z \end{array} \\
\begin{bmatrix} E_1 & 0 & 0 & 0 & 0 \\ 0 & E_2 & 0 & 0 & 0 \\ 0 & 0 & E_2 & 0 & 0 \\ 0 & 0 & 0 & E_2 & 0 \\ 0 & 0 & 0 & 0 & E_2 \end{bmatrix}
\end{array}
$$

where $E_0 = 13.6$ eV, $E_1 = -E_0$, and $E_2 = -E_0/4$. The electric field leads to a matrix $\boldsymbol{H}_F$ which has to be added to $\boldsymbol{H}$. Its elements are given by

$$
(\boldsymbol{H}_F)_{nm} = qF\int_0^{\infty} dr\, r^2 \int_0^{\pi} \sin\theta d\theta \int_0^{2\pi} d\phi\, u_n^*(r) r\cos\theta\, u_m(r)
$$

Using the wavefunctions

$$
u_{1s} = \sqrt{\frac{1}{\pi a_0^3}}\, e^{-\frac{r}{a_0}}
$$

$$
u_{2s} = \sqrt{\frac{1}{32\pi a_0^3}} \left(2 - \frac{r}{a_0}\right) e^{\frac{r}{2a_0}}
$$

$$
u_{2p_x} = \sqrt{\frac{1}{16\pi a_0^3}} \left(\frac{r}{a_0}\right) e^{-\frac{r}{2a_0}} \sin\theta\cos\phi
$$

$$
u_{2p_y} = \sqrt{\frac{1}{16\pi a_0^3}} \left(\frac{r}{a_0}\right) e^{\frac{-r}{2a_0}} \sin\theta\sin\phi
$$

$$
u_{2p_z} = \sqrt{\frac{1}{16\pi a_0^3}} \left(\frac{r}{a_0}\right) e^{-\frac{r}{2a_0}} \cos\theta
$$

we can evaluate the integrals straightforwardly to show that

$$
H_F = \begin{array}{c}
\begin{array}{ccccc} 1s & 2s & 2p_x & 2p_y & 2p_z \end{array} \\
\begin{bmatrix} 0 & 0 & 0 & 0 & A \\ 0 & 0 & 0 & 0 & B \\ 0 & 0 & 0 & 0 & 0 \\ 0 & 0 & 0 & 0 & 0 \\ A & B & 0 & 0 & 0 \end{bmatrix}
\end{array}
$$

where A and B are linear functions of the field:

$$
A = \left(\frac{128\sqrt{2}}{243}\right) a_0 F,\ B = -3\, a_0 F
$$

Hence

$$
H_0 + H_F = \begin{array}{c}
\begin{array}{ccccc} 1s & 2s & 2p_x & 2p_y & 2p_z \end{array} \\
\begin{bmatrix} E_1 & 0 & A & 0 & 0 \\ 0 & E_2 & B & 0 & 0 \\ A & B & E_2 & 0 & 0 \\ 0 & 0 & 0 & E_2 & 0 \\ 0 & 0 & 0 & 0 & E_2 \end{bmatrix}
\end{array}
$$

Note that we have relabeled the rows and columns to accentuate the fact that $2p_x$ and $2p_y$ levels are decoupled from the rest of the matrix and are unaffected by the field, while the ls and 2s and $2p_z$ levels will be affected by the field.

If the field were absent we would have one eigenvalue E and four degenerate eigenvalues E_2. How do these eigenvalues change as we turn up the field? As we have mentioned before, the eigenvalues are more or less equal to the diagonal values unless the off-diagonal term $\boldsymbol{H}_{mn}$ becomes comparable to the difference between the corresponding diagonal terms ($\boldsymbol{H}_{mm} - \boldsymbol{H}_{nn}$). This means that whenever we have two degenerate eigenvalues (that is, $\boldsymbol{H}_{mm} - \boldsymbol{H}_{nn} = 0$) even a small off-diagonal element $\boldsymbol{H}_{mn}$ has a significant effect. We thus expect the 2s and $2p_z$ levels to be significantly affected by the field since they are degenerate to start with. We can get a very good approximation for these eigenvalues simply by looking at a subset of the $\boldsymbol{h}$ matrix containing just these levels:

$$\begin{matrix} & 2s & 2p_z \end{matrix}$$
$$\boldsymbol{H}_0 + \boldsymbol{H}_F = \begin{bmatrix} E_2 & B \\ B & E_2 \end{bmatrix}$$

It is easy to show that the eigenvalues are $E = E_2 \pm B$ and the corresponding eigenvectors are

$$|2s\rangle - |2\,p_z\rangle\,,\ |2s\rangle + |2\,p_z\rangle$$

This approximate approach (known as degenerate perturbation theory) describes the exact eigenvalues quite well (see Fig. 1. 4. 2) as long as the off-diagonal elements (like A) coupling these levels to the other levels are much smaller than the energy difference between these levels (like $E_2 - E_1$).

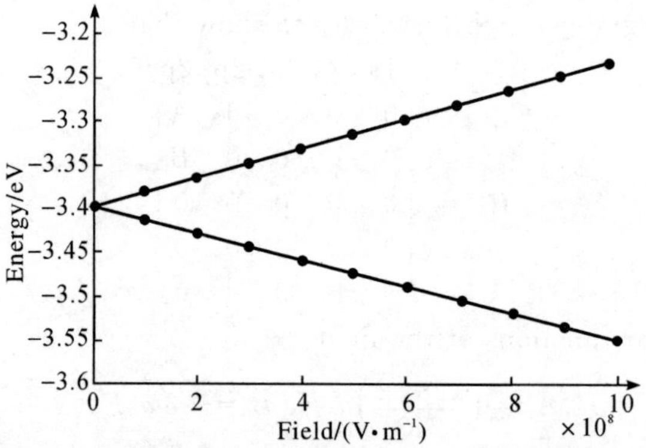

Fig. 1. 4. 2 Energy of 2s-$2p_z$ levels due to an applied electric field F (The solid lines show the results obtained by direct diagonalization while o and x show perturbation theory results $E = E_2 \pm B$)

How is the 1s eigenvalue affected? Since there are no other degenerate levels the effect is much less and to first order one could simply ignore the rest of the matrix:

$$\begin{matrix} 1s \end{matrix}$$
$$|\,\boldsymbol{H}_0 + \boldsymbol{H}_F\,| = E_1$$

and argue that the eigenvalue remains E_1. We could do better by "renormalizing" the matrix as follows. Suppose we partition the $\boldsymbol{H}$ matrix and write

$$\boldsymbol{H\phi} = E\boldsymbol{\phi} \rightarrow \begin{bmatrix} \boldsymbol{H}_{11} & \boldsymbol{H}_{12} \\ \boldsymbol{H}_{21} & \boldsymbol{H}_{22} \end{bmatrix} \begin{bmatrix} \boldsymbol{\phi}_1 \\ \boldsymbol{\phi}_2 \end{bmatrix} = E \begin{bmatrix} \boldsymbol{\phi}_1 \\ \boldsymbol{\phi}_2 \end{bmatrix}$$

where $\boldsymbol{H}_{11}$ denotes the part of the matrix we wish to keep (the 1 s block in this case).

It is easy to eliminate $\boldsymbol{\phi}_2$ to obtain

$$\boldsymbol{H}'\boldsymbol{\phi}_1 = E\boldsymbol{\phi}_1$$

where

$$\boldsymbol{H}' = \boldsymbol{H}_{11} + \boldsymbol{H}_{12}\,(E\boldsymbol{I} - \boldsymbol{H}_{22})^{-1}\,\boldsymbol{H}_{21}$$

$\boldsymbol{I}$ being an identity matrix of the same size as $\boldsymbol{H}_{22}$. We haven't gained much if we still have to invert the matrix $E\boldsymbol{I} - \boldsymbol{H}_{22}$ including its off-diagonal elements. But to lowest order we can simply ignore the off-diagonal elements of $\boldsymbol{H}_{22}$ and write down the inverse by inspection. In the present case, this gives us

$$|\,\boldsymbol{H}'\,| \approx E_1 + [0 \quad A] \begin{bmatrix} \dfrac{1}{(E - E_2)} & 0 \\ 0 & \dfrac{1}{(E - E_2)} \end{bmatrix} \begin{bmatrix} 0 \\ A \end{bmatrix} = E_1 + \frac{A^2}{E - E_2}$$

To lowest order, the eigenvalue E is approximately equal to E_1, so that

$$|\,\boldsymbol{H}'\,| \approx E_1 + \left[\frac{A^2}{(E - E_2)}\right]$$

which shows that the correction to the eigenvalue is quadratic for non-degenerate states, rather than linear as it is for degenerate states. This approximate approach (known as non-degenerate perturbation theory) describes the exact eigenvalues quite well (see Fig. 1. 4. 3).

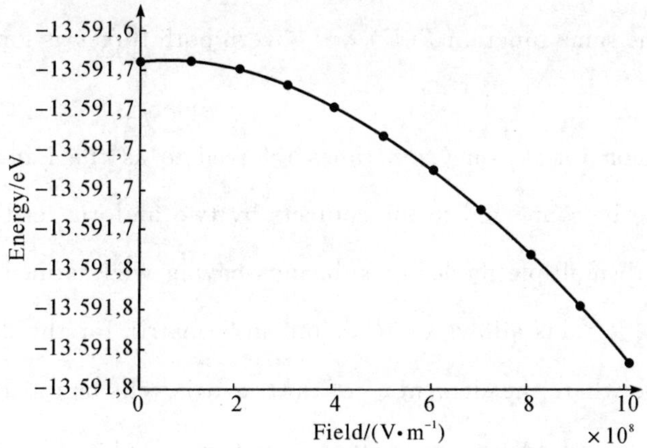

Fig. 1. 4. 3 Energy of 1s level due to an applied electric field F
(The solid curve shows the obtained by direct diagonalization while the crosses denote the perturbation theory results $E = E_1 + [A^2/(E_1 - E_2)]$)

1.5 Transmission

In the last section we obtained expressions for the current at each of the contact which can be expressed as the difference between an inflow and an outflow. In this section we will express the current in a slightly different form that gives a different perspective to the problem of current flow and helps establish a connection with the transmission formalism widely used in the literature. We start by the current as (noting that $2\pi\hbar = h$)

$$I = (q/h)\int_{-\infty}^{+\infty} \mathrm{d}E\, \bar{T}(E)[f_1(E) - f_2(E)] \tag{1.5.1}$$

where

$$\bar{T}(E) \equiv \mathrm{Trace}(\boldsymbol{\Gamma}_1 \boldsymbol{A}_2) = \mathrm{Trace}(\boldsymbol{\Gamma}_2 \boldsymbol{A}_1) = \mathrm{Trace}(\boldsymbol{\Gamma}_1 \boldsymbol{G} \boldsymbol{\Gamma}_2 \boldsymbol{G}^+) = \mathrm{Trace}(\boldsymbol{\Gamma}_2 \boldsymbol{G} \boldsymbol{\Gamma}_1 \boldsymbol{G}^+) \tag{1.5.2}$$

is called the transmission function. Physically we can view the current in Eq. (1.5.1) as the difference between two counterpropagating fluxes, one from the source to the drain and the other from the drain to the source. One could view the device as a "semi-permeable membrane" that separates two reservoirs of electrons (source and drain) and the transmission function $\bar{T}(E)$ as a measure of the permeability of this membrane to electrons with energy E. We will show that the same function $\bar{T}(E)$ will govern both fluxes as long as transport is coherent.

In the transmission formalism (sometimes referred to as the Landauer approach) the channel is assumed to be connected to the contacts by two uniform leads that can be viewed as quantum wires with multiple modes or subbands having well-defined E-k relationships as sketched in Fig. 1.5.1. This allows us to define an $\boldsymbol{S}$-matrix for the device analogous to a microwave waveguide where the element $\boldsymbol{t}_{nm}$ of the $\boldsymbol{t}$-matrix tells us the amplitude for an electron incident in mode m in lead 1 to transmit to a mode n in lead 2. It can then be shown that the current is given by Eq. (1.5.1) with the transmission function given by

$$\bar{T}(E) = \sum_m \sum_n |\boldsymbol{t}_{nm}|^2 = \mathrm{Trace}(\boldsymbol{t}\boldsymbol{t}^+) \tag{1.5.3}$$

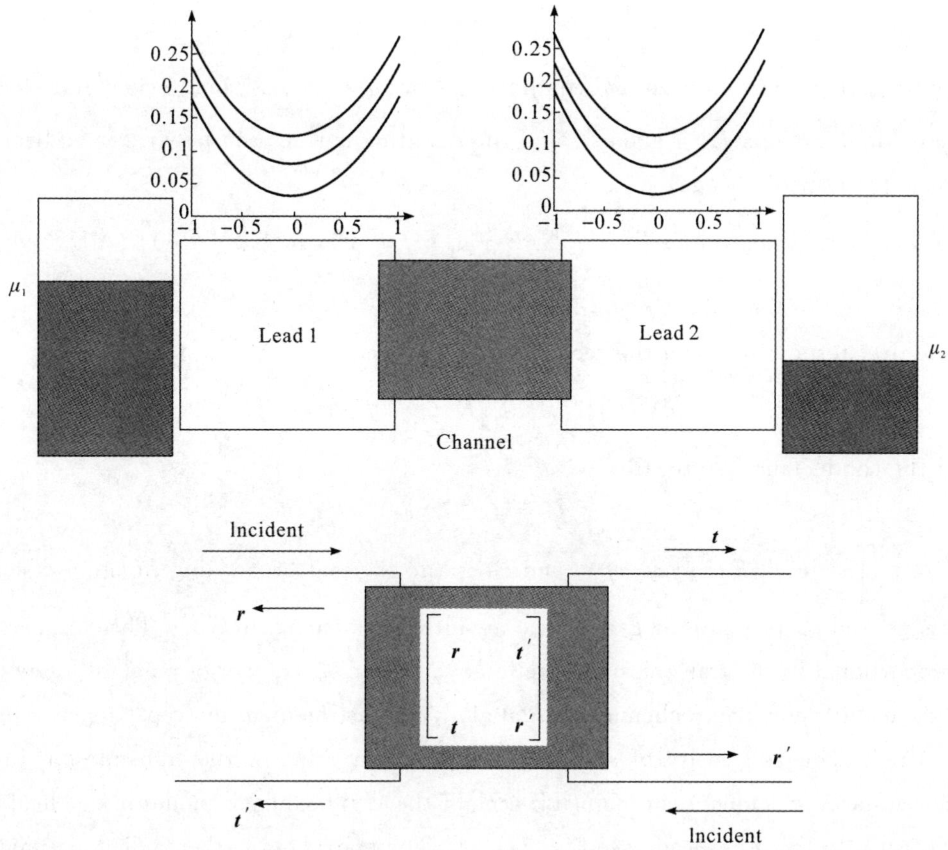

Fig. 1.5.1 In the transmission formalism, the channel is assumed to be connected to the contacts by two uniform leads that can be viewed as quantum wires with multiple subbands having well-defined E-k relationships as shown. This allows us to define an $\boldsymbol{S}$-matrix for the device analogous to a microwave waveguide

This viewpoint, which is very popular, has the advantage of being based on relatively elementary concepts and also allows one to calculate the transmission function by solving a scattering problem. In the next section we will show with simple examples that this approach yields the same result as that obtained from $\bar{T} = \mathrm{Trace}(\boldsymbol{\Gamma}_2 \boldsymbol{G} \boldsymbol{\Gamma}_1 \boldsymbol{G}^+)$ applied to devices with uniform leads.

Landauer pioneered the use of the scattering theory of transport as a conceptual framework for clarifying the meaning of electrical conductance and stressed its fundamental connection to the transmission function: "Conductance is transmission." This basic relation can be seen starting from Eq. (1.5.1)

$$I = (q/h)\int_{-\infty}^{+\infty} \mathrm{d}E\,\bar{T}(E)[f_0(E-\mu_1) - f_0(E-\mu_2)]$$

and noting that the current is zero at equilibrium since $\mu_1 = \mu_2$. A small bias voltage V changes each of the functions $\bar{T}$, μ_1 and μ_2, and the resulting current can be written to first order as (δ denotes a small change)

$$I \approx (q/h)\int_{-\infty}^{+\infty} \mathrm{d}E\delta\,\bar{T}(E)[f_0(E-\mu_1) - f_0(E-\mu_2)] + (q/h)\int_{-\infty}^{+\infty} \mathrm{d}E \times$$
$$\bar{T}(E)\delta[f_0(E-\mu_1) - f_0(E-\mu_2)]$$

The first term is zero and the second can be written as

$$I \approx (q^2V/h)\int_{-\infty}^{+\infty} \mathrm{d}E\,\bar{T}(E)(-\partial f_0(E)/\partial E)_{E=\mu}$$

so that the conductance is given by

$$G = (q^2/h)T_0 \tag{1.5.4}$$

where $T_0 \equiv \int_{-\infty}^{+\infty} \mathrm{d}E\,\bar{T}(E)F_T(E-\mu)$, and F_T is the thermal broadening function discussed, which is peaked sharply around $E = \mu$ with a width proportional to k_BT. The conductance is thus proportional to the transmission function averaged over an energy range of a few k_BT around the equilibrium electrochemical potential μ, just as the quantum capacitance is proportional to the averaged density of states. The maximum value of the transmission function (and hence the conductance) is obtained if each of the M subbands or modes in one lead transmits perfectly to the other lead (see Fig. 1.5.2). The matrix $\boldsymbol{tt}^+$ is then a diagonal matrix of size ($M \times M$) with ones along the diagonal, so that the transmission is equal to M. This suggests that the maximum transmission is equal to the number of modes M in the lcads. But what happens if the device is narrower than the lead and has only N modes, $N < M$[see Fig. 1.5.2(a)] ?

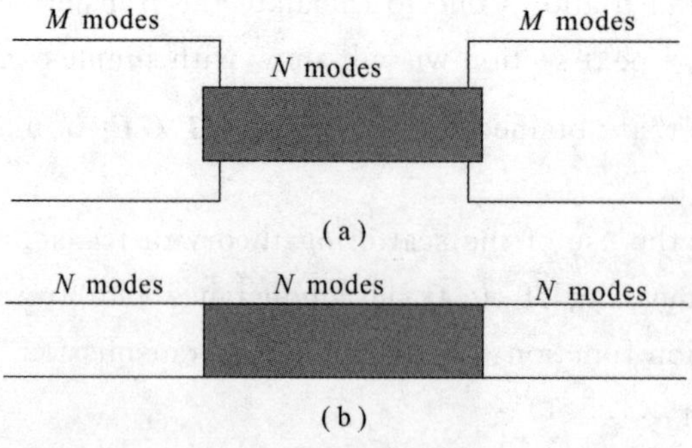

Fig. 1.5.2 M modes and N modes

It can be argued that such a structure could not have a transmission any greater than a structure with the leads the same size as the channel [see Fig. 1. 5. 2(b)] since in either case the electrons have to transmit through the narrow device region(assuming that the device is not so short as to allow direct tunneling). Since this latter structure has a maximum transmission of N that must be true of the first structure as well and detailed calculations do indeed show this to be the case. In general we can expect that the maximum transmission is equal to the number of modes in the narrowest segment. Earlier, we argued that the maximum conductance of a wire with N modes is equal to $(q^2/h)N$ based on the maximum current it could possibly carry.

Conductance measurements are often performed using a four probe structure (see Fig. 1. 5. 3) and their interpretation in small structures was initially unclear till Büttiker came up with an elegant idea (Büttiker, 1988). He suggested that the Landauer formula

$$G = (q^2/h)\widetilde{T} \rightarrow I = (q/h)\widetilde{T}(\mu_1 - \mu_2)$$

be extended to structures with multiple terminals by writing the current I_i at the ith terminal as

$$I_i = (q/h)\sum_j \widetilde{T}_{ij}(\mu_i - \mu_j) \tag{1.5.5}$$

where $\widetilde{T}_{ij}$ is the average transmission from terminal j to i. We know the electrochemical potentials μ at the current terminals (1 and 2) but we do not know them at the voltage terminals, which float to a suitable potential so as to make the current zero. How do we calculate the currents from Eq. (1. 5. 5) since we do not know all the potentials? The point is that of the eight variables (four potentials and four currents), if we know any we can calculate the other four with simple matrix algebra. Actually, there are six independent variables. We can always set one of the potentials to zero, since only potential differences give rise to currents. Also, Kirchhoff's law requires all the currents to add up to zero, so that knowing any three currents we can figure out the fourth. So, it is convenient to set the potential at one terminal (say terminal 2) equal to zero and write Eq. (1. 5. 5) in the form of a (3 × 3) matrix equation:

$$\begin{bmatrix} I_1 \\ I_3 \\ I_4 \end{bmatrix} = \frac{q}{h}\begin{bmatrix} \widetilde{T}_{12}+\widetilde{T}_{13}+\widetilde{T}_{14} & -\widetilde{T}_{13} & -\widetilde{T}_{14} \\ -\widetilde{T}_{31} & \widetilde{T}_{31}+\widetilde{T}_{32}+\widetilde{T}_{34} & \widetilde{T}_{34} \\ -\widetilde{T}_{41} & \widetilde{T}_{43} & \widetilde{T}_{41}+\widetilde{T}_{42}+\widetilde{T}_{43} \end{bmatrix}\begin{bmatrix} \mu_1 \\ \mu_3 \\ \mu_4 \end{bmatrix}$$

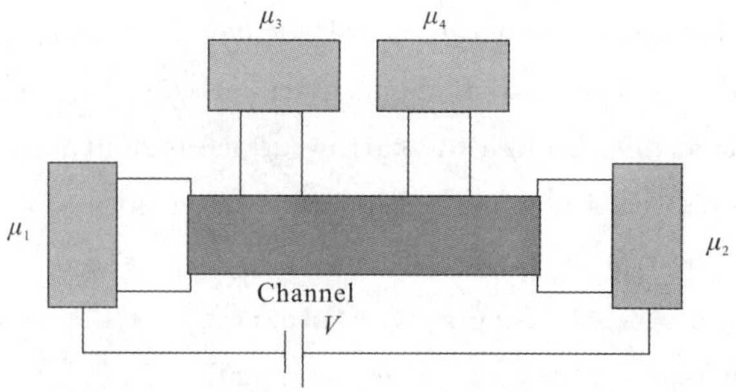

Fig. 1.5.3 Conductance measurements are commonly carried out in a four-probe configuration that can be analyzed using the Büttiker equations

Knowing μ_1 , $I_3 = 0$, $I_4 = 0$, we can calculate I_1 , μ_3 , μ_4 and hence the four-probe conductance:

$$G_{\text{four-probe}} = (\mu_3 - \mu_4)/q I_1$$

We can visualize the Büttiker equations with a simple circuit model if the transmission coefficients are reciprocal, that is, if $\widetilde{T}_{ij} = \widetilde{T}_{ji}$. These equations are then identical to Kirchhoff's law applied to a network of conductors $G_{ij} \propto \widetilde{T}_{ij} = \widetilde{T}_{ji}$. connecting each pair of contacts i and j (see Fig. 1.5.4). However, this picture cannot be used if the transmission coefficients are non-reciprocal: $\widetilde{T}_{ij} \neq \widetilde{T}_{ji}$, as they are in Hall effect measurements where a magnetic field is present, some of the most notable applications of the Büttiker equations, Eq. (1.5.4), are to the interpretation of such measurements.

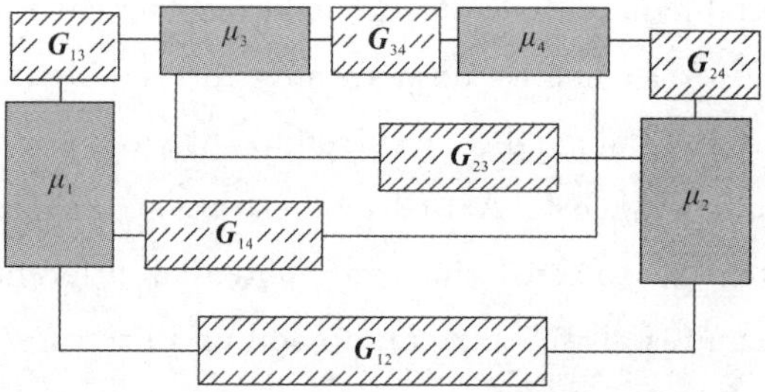

Fig. 1.5.4 The Büttiker equations can be visualized in terms of a conductor network is the transmission between terminals is reciprocal

We mentioned earlier that the scattering theory of transport can only be used if the electrons transmit coherently through the device so that an **S**-matrix can be defined. But floating probes effectively extract electrons from the device and reinject them after phase randomization, thus effectively acting as phase-breaking scatterers. This is a seminal observation due to Büttiker that provides a simple phenomenological technique for including the effects of phase-breaking processes in the calculation of current. We simply connect one or more purely conceptual floating probes to the device and then calculate the net current using the Büttiker equations, which can be applied any number of terminals.

We could even use the general current equation [see Eq. (1.5.1)], rather than the low-bias conductance relation, extended to include multiple floating probes:

$$I_i = (q/h)\int_{-\infty}^{+\infty} \mathrm{d}E\, \bar{I}_i(E) \tag{1.5.6}$$

where

$$\bar{I}_i(E) = \sum_j \bar{T}_{ij}(E)[f_i(E) - f_j(E)] \tag{1.5.7}$$

One could then adjust the potential μ_j to make the current at each energy equal to zero: $\bar{I}_j(E) = 0$. In principle this could result in different values for μ_j at different energies. Alternatively, we could require a single value for μ_j at all energies, adjusted to make the total current at all energies equal to zero, $\int \mathrm{d}E\, \bar{I}_j(E) = 0$. One could then have positive values of $\bar{I}_j(E)$ at certain energies balanced by negative values at other energies making the total come out to zero, indicating a flow of electrons from one energy to another due to the scattering processes that the "probe" is expected to simulate. This makes the detailed implementation more complicated since different energy channels get coupled together.

The transmission coefficients at a given energy are usually calculated from the $\boldsymbol{s}$-matrix for the composite device including the conceptual probes:

$$\bar{T}_{ij} = \mathrm{Trace}\,[s_{ij}(E)\, s_{ij}^{+}(E)] \tag{1.5.8}$$

But we could just as well combine this phenomenological approach with our Green's function method using separate self-energy matrices $\boldsymbol{\Sigma}_i$ to represent different floating probes and then use the expression

$$\bar{T}_{ij}(E) = \mathrm{Trace}(\boldsymbol{\Gamma}_i \boldsymbol{G} \boldsymbol{\Gamma}_j \boldsymbol{G}^{+}) \tag{1.5.9}$$

to evaluate the transmission. This expression can be derived using the same procedure described earlier for two-terminal structures. The current at terminal i is given by the difference between the inflow and outflow:

$$I_i(E) = (1/\hbar)\mathrm{Trace}\,\{\boldsymbol{\Gamma}_i(E)[\boldsymbol{A}(E) f_i - \boldsymbol{G}^n(E)]\}$$

Making use of the relations $\boldsymbol{A} = \sum_j \boldsymbol{G}\boldsymbol{\Gamma}_j \boldsymbol{G}^{+}$ and $\boldsymbol{G}^n = \sum_j \boldsymbol{G}\boldsymbol{\Gamma}_j \boldsymbol{G}^{+} f_j$, we can write

$$I_i(E) = (1/\hbar)\sum_q \mathrm{Trace}(\boldsymbol{\Gamma}_i \boldsymbol{G} \boldsymbol{\Gamma}_j \boldsymbol{G}^{+})(f_i - f_j)$$

so that the current can be written as

$$I_i(E) = (1/\hbar)\sum_j \bar{T}_{ij}(f_i - f_j) \tag{1.5.10}$$

in terms of the transmission function defined above in Eq. (1.5.9).

A very useful result in the scattering theory of transport is the requirement that the sum of the rows or columns of the transmission matrix equals the number of modes:

$$\sum_j \bar{T}_{ij} = \sum_j \bar{T}_{ji} = M_i \tag{1.5.11}$$

where M_i is the number of modes in lead i. One important consequence of this sum rule is that for a two-terminal structure $\bar{T}_{12} = \bar{T}_{21}$, even in a magnetic field, since with a (2×2) $\bar{\boldsymbol{T}}$ matrix:

$$\begin{bmatrix} \bar{T}_{11} & \bar{T}_{12} \\ \bar{T}_{21} & \bar{T}_{22} \end{bmatrix}$$

we have

$$\bar{T}_{11} + \bar{T}_{12} = M_1 = \bar{T}_{11} + \bar{T}_{21} \rightarrow \bar{T}_{12} = \bar{T}_{21}$$

Note that a similar argument would not work with more than two terminals. For example with a three-terminal structure we could show that $\bar{T}_{12} + \bar{T}_{13} = \bar{T}_{21} + \bar{T}_{31}$, but we could not prove that $\bar{T}_{12} = \bar{T}_{21}$ or that $\bar{T}_{13} = \bar{T}_{31}$. The Green's function-based expression for the transmission [see Eq. (1.5.9)] also yields a similar sum rule:

$$\sum_j \bar{T}_{ij} = \sum_j \bar{T}_{ji} = \text{Trace}(\boldsymbol{\Gamma}_i \boldsymbol{A}) \tag{1.5.12}$$

1.6 Non-coherent transport

We discussed a quantum mechanical model that describes the flow of electrons coherently through a channel. All dissipative/phase-breaking processes we reassumed to be limited to the contacts where they act to keep the electrons in local equilibrium. In practice, such processes are present in the channel as well and their role becomes increasingly significant as the channel length is increased. Indeed, prior to the advent of mesoscopic physics, the role of contacts was assumed to be minor and quantum transport theory was essentially focused on the effect of such processes. By contrast, we have taken a "bottom-up" view of the subject and now that we understand how to model a small coherent device, we are ready to discuss dissipative/phase-breaking processes.

Phase-breaking processes arise from the interaction of one electron with the surrounding bath of photons, phonons, and other electrons. Compared to the coherent processes that we have discussed so far, the essential difference is that phase-breaking processes involve a change in the "surroundings". In coherent interactions, the back ground is rigid and the electron interacts elastically with it, somewhat like a ping pong ball bouncing off a truck. The motion of the truck is insignificant. In reality, the background is not quite as rigid as a truck and is set in "motion" by the passage of an electron and this excitation of the background is described in terms of phonons, photons, etc. This is in general a difficult problem with no exact solutions and what we will be describing here is the lowest order approximation, sometimes called the self-consistent Born approximation, which usually provides an adequate description. Within this approximation, these interactions can essentially be viewed as a coupling of the channel from the "standard" configuration with N_ω phonons/photons (in different modes with different frequencies ω) to a neighboring configuration with one less (absorption)or one more (emission) phonon/photon as depicted in Fig. 1. 6. 1.

This coupling to neighboring configurations results in an outflow of electrons from our particular subspace and a subsequent return or inflow back into this subspace. A general model for quantum transport needs to include this inflow and outflow into the coherent transport model, through an additional terminal "s" described by the additional terms $\boldsymbol{\Sigma}_s^{in}$ and $\boldsymbol{\Sigma}_s$ (see Fig. 1. 6. 2). My objective in this chapter is to explain how these additional terms are calculated. We have seen that for the regular contacts, the inscattering is related to the broadening:

$$\boldsymbol{\Sigma}_1^{in} = \boldsymbol{\Gamma}_1 f_1 \text{ and } \boldsymbol{\Sigma}_2^{in} = \boldsymbol{\Gamma}_2 f_2$$

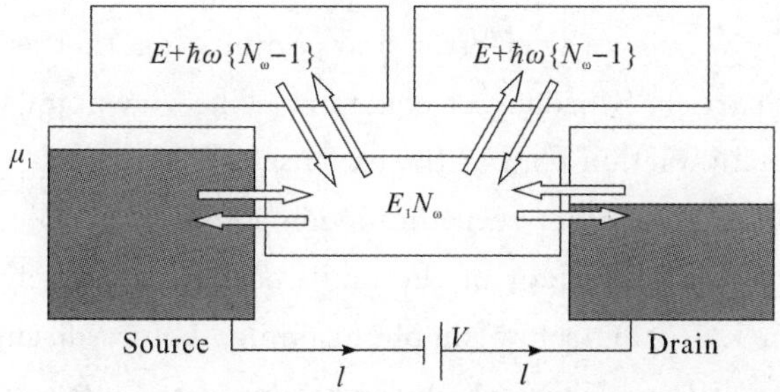

Fig. 1. 6. 1 Phase-breaking introduces a coupling to neighboring configurations having one more or one less number of excitations N_ω than the original

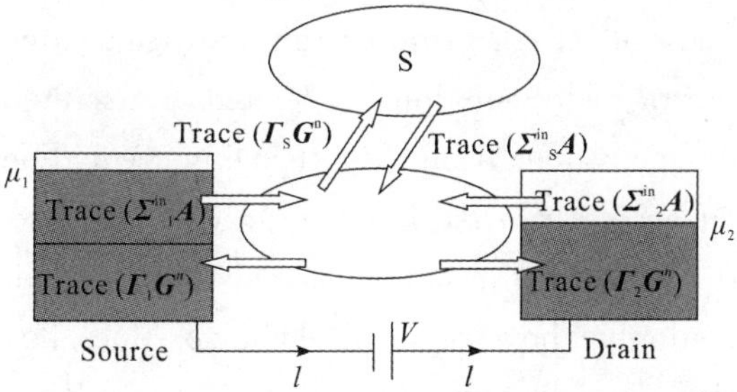

Fig. 1. 6. 2 Non-coherent quantum transport: inflow and outflow

However, for the scattering terminal both $\boldsymbol{\Sigma}_s^{in}$ and $\boldsymbol{\Sigma}_s$ have to be determined separately since there is no Fermi function f_s describing the scattering "terminal" and hence no simple connection between $\boldsymbol{\Sigma}_s^{in}$ and $\boldsymbol{\Sigma}_s$ (or $\boldsymbol{\Gamma}_s$), unlike the contacts. Of course, one could adopt a phenomenological point of view and treat the third terminal like another contact whose chemical potential μ_s is adjusted to ensure zero current at this terminal. That would be in the spirit of the "Büttiker probe" and could well be adequate for many applications. However, I will describe microscopic (rather than phenomenological) models for $\boldsymbol{\Sigma}_s^{in}$ and $\boldsymbol{\Sigma}_s$ that can be used to benchmark any phenomenological models that the reader may choose to use. They can also use these models as a starting Point to include more sophisticated scattering mechanisms as needed.

The inflow and outflow associated with dissipative processes involve subtle conceptual issues beyond what we have encountered so far with coherent transport. We will start in Section 1.6 by explaining two viewpoints that one could use to model the interaction of an electron with its surroundings, say the electromagnetic vibrations or photons. One viewpoint is based on the one-particle picture where we visualize the electron as being affected by its surroundings through a scattering potential U_s. However, as we will see, in order to explain the known equilibrium properties it is necessary to endow this potential U_s with rather special properties that make it difficult to include in the Schrödinger equation. Instead we could adopt a viewpoint whereby we view the electron and photons together as one giant system described by a giant multi-particle Schrödinger equation. This viewpoint leads to a more satisfactory description of the interaction, but at the expense of conceptual complexity. In general it is important to be able to switch between these viewpoints so as to combine the simplicity of the one-particle picture with the rigor of the multi-particle approach. I will illustrate the basic principle in Section 1.6 using a few simple examples before discussing the general expressions for inflow and outflow elaborates on the nature of the lattice vibrations or phonons in common semiconductors for interested readers.

We started this book by noting that the first great success of the Schrödinger equation was to explain the observed optical spectrum of the hydrogen atom. It was found that the light emitted by a hot vapor of hydrogen atoms consisted of discrete frequencies $\omega = 2\pi v$ that were related to the energy eigenvalues from the Schrödinger equation: $\hbar\omega = \varepsilon_n - \varepsilon_m$. This is explained by saying that if an electron is placed in an excited state $|2\rangle$, it relaxes to the ground state $|1\rangle$, and the difference in energy is radiated in the form of light or photons (see Fig. 1.6.3). Interestingly, however, this behavior does not really follow from the Schrödinger equation, unless we add something to it. To see this let us write the time-dependent Schrödinger equation in the form of a matrix equation

$$\mathrm{i}\hbar \frac{\mathrm{d}}{\mathrm{d}t}\boldsymbol{\psi} = \boldsymbol{H}\boldsymbol{\psi} \qquad (1.6.1)$$

using a suitable set of basis functions. If we use the eigenfunctions of $\boldsymbol{H}$ as our basis then this equation has tie form:

$$\mathrm{i}\hbar \frac{\mathrm{d}}{\mathrm{d}t}\begin{bmatrix} \psi_1 \\ \psi_2 \\ \cdots \\ \psi_m \end{bmatrix} = \begin{bmatrix} \varepsilon_1 & 0 & 0 & \cdots \\ 0 & \varepsilon_2 & 0 & \cdots \\ 0 & 0 & \varepsilon_3 & \cdots \\ \cdots & \cdots & \cdots & \varepsilon_m \end{bmatrix}\begin{bmatrix} \psi_1 \\ \psi_2 \\ \cdots \\ \psi_m \end{bmatrix}$$

which decouples neatly into a set of independent equations:

$$\mathrm{i}\hbar \frac{\mathrm{d}}{\mathrm{d}t}\psi_n = \varepsilon_n \psi_n \tag{1.6.2}$$

one for each energy eigenvalue. It is easy to write down the solution to Eq. (1.6.2) for a given set of initial conditions at $t=0$:

$$\psi_n(t) = \psi_n(0)\exp(-\mathrm{i}\varepsilon_n t/\hbar) \tag{1.6.3a}$$

This means that the probability for finding an electron in state does not change with time:

$$P_n(t) = |\psi_n(t)|^2 = |\psi_n(0)|^2 = P_n(0) \tag{1.6.3b}$$

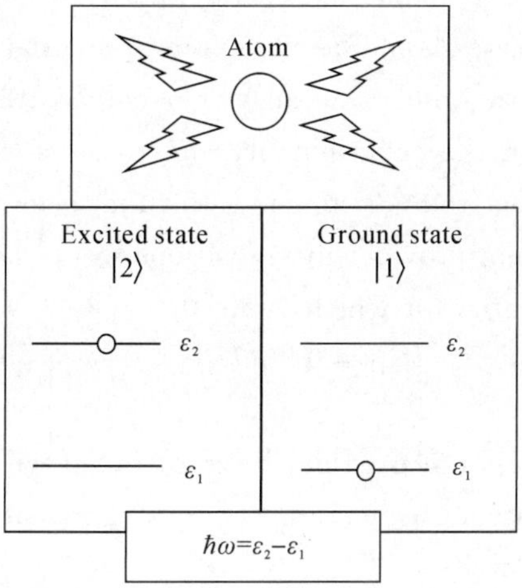

Fig. 1.6.3 If an electron is placed in an excited state |2⟩, it will lose energy by radiating light and relax to the ground state |1⟩. However, this behavior does not follow from the Schrödinger equation, unless we modify it appropriately

According to the Schrödinger equation, an electron placed in an excited state would stay there forever! Whatever it is that causes the excited state to relax is clearly not a part of Eq. (1.6.1) or Eq. (1.6.2). So what is missing? There are two ways to answer this question. Let us look at these one by one.

One-particle viewpoint: This view point says that an electron feels a random external potential U^S due to the photons in the surrounding "box" which causes it to relax to the ground state (see Fig. 1. 6. 4). This potential gives rise to off-diagonal terms in the Hamiltonian that couple the different states together. With just two states we could write

$$i\hbar \frac{d}{dt}\begin{bmatrix} \psi_1 \\ \psi_2 \end{bmatrix} = \begin{bmatrix} \varepsilon_1 & U_{12}^S \\ U_{21}^S & \varepsilon_2 \end{bmatrix}\begin{bmatrix} \psi_1 \\ \psi_2 \end{bmatrix} \tag{1.6.4}$$

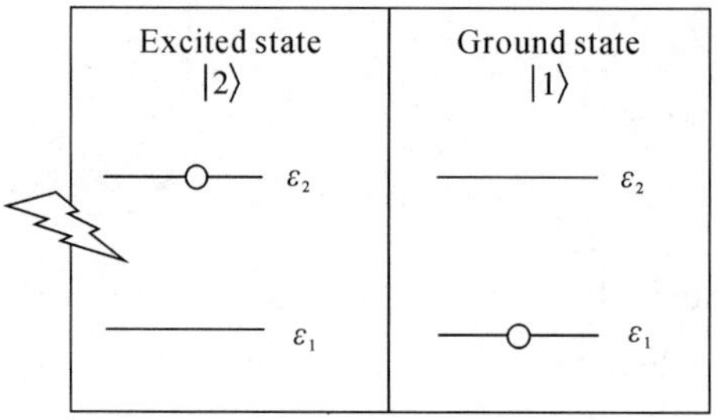

Fig. 1. 6. 4 In the one-particle viewpoint an electron is said to feel an external potential U^S due to the photons in the surrounding "box" which causes it to relax from | 2⟩ to | 1⟩

Without getting into any details it is clear that if the electron is initially in state | 2⟩ , the term U_{12}^S will tend to drive it to state | 1⟩ . But this viewpoint is not really satisfactory. Firstly, one could ask why there should be any exteral potential U^S at zero temperature when all thermal excitations are frozen out. The answer usually is that even at zero temperature there is some noise present in the environment, and these so-called zero-point fluctuations tickle the electron into relaxing from | 2⟩ to | 1⟩ . But that begs the second question: why do these zero-point fluctuations not provide any transitions from | 1⟩ to | 2⟩ ? Somehow we need to postulate a scattering potential for which (note that $\varepsilon_2 > \varepsilon_1$)

$$U_{21}^S = 0 \text{ but } U_{12}^S \neq 0$$

at zero temperature.

For non-zero temperatures, U_{21}^S need not be zero, but it will still have to be much smaller than U_{12}^S , so as to stimulate a greater rate $S(2\to1)$ of transitions from 2 to 1 than from 1 to 2. For example, we could write

$$S(2\to1) = K_{2\to1} f_2(1-f_1) \text{ and } S(1\to2) = K_{1\to2} f_1(1-f_2)$$

where $f_1(1-f_2)$ is the probability for the system to be in state | 1⟩ (level 1 occupied with level 2 empty) and $f_2(1-f_1)$ is the probability for it to be in state | 2⟩ (level 2 occupied with level 1 empty). At equilibrium the two rates must be equal, which requires that

$$\frac{K_{1\to2}}{K_{2\to1}} = \frac{f_2(1-f_1)}{f_1(1-f_2)} = \frac{(1-f_1)/f_1}{(1-f_2)/f_2} \tag{1.6.5}$$

But at equilibrium, the occupation factors f_1 and f_2 are given by the Fermi function:

$$f_n = \frac{1}{1+\exp[(\varepsilon_n-\mu)/k_BT]} \to \frac{1-f_n}{f_n} = \exp(\frac{\varepsilon_n-\mu}{k_BT})$$

Hence from Eq. (1.6.5),

$$\left(\frac{K_{1\to 2}}{K_{2\to 1}}\right)_{\text{equilibrium}} = \exp\left(-\frac{\varepsilon_2-\varepsilon_1}{k_B T}\right) \quad (1.6.6)$$

Clearly at equilibrium, $K_{2\to 1} \gg K_{1\to 2}$, as long as the energy difference

$$(\varepsilon_2-\varepsilon_1) \gg k_B T$$

Early in the twentieth century, Einstein argued that if the number of photons with energy $\hbar\omega$ present in the box is N, then the rate of downward transitions is proportional to $(N+1)$ while the rate of upward transitions is proportional to

$$\left.\begin{aligned} N: K(1\to 2) &= \alpha N && \text{photon absorption} \\ K(2\to 1) &= \alpha(N+1) && \text{photon emission}\end{aligned}\right\} \quad (1.6.7)$$

This ensures that at equilibrium Eq. (1.6.6) is satisfied since the number of photons is given by the Bose-Einstein factor

$$N_{\text{equilibrium}} = \frac{1}{\exp(\hbar\omega/k_B T)-1} \quad (1.6.8)$$

and it is easy to check that

$$\left(\frac{K_{1\to 2}}{K_{2\to 1}}\right)_{\text{equilibrium}} = \left(\frac{N}{N+1}\right)_{\text{equilibrium}} = \exp\left(-\frac{\hbar\omega}{k_B T}\right) \quad (1.6.9)$$

Since $\hbar\omega = \varepsilon_2-\varepsilon_1$, Eq. (1.6.9) and Eq. (1.6.6) are consistent.

What is not clear is why the external potential should stimulate a greater rate of downward transitions($2\to 1$)than upward transitions($1\to 2$), but clearly this must be the case if we are to rationalize the fact that, at equilibrium, lower energy states are more likely to be occupied than higher energy states as predicted by the Fermi function. But there is really no straightforward procedure for incorporating this effect into the Schrödinger equation with an appropriate choice of the scattering potential U^S. Any Hermitian operator U^S will have $U_{12}^S = U_{21}^S$ and thus provide equal rates of upward and downward transitions.

This brings us to the other viewpoint, which provides a natural explanation for the difference between upward and downward transition rates, but is conceptually more complicated. In this viewpoint we picture the electron + photon as one big many-particle system whose dynamics are described by an equation that formally looks just like the Schrödinger equation

$$\mathrm{i}\hbar\frac{\mathrm{d}}{\mathrm{d}t}\psi = \boldsymbol{H}\psi \quad (1.6.10)$$

However, ψ now represents a state vector in a multi-particle Hilbert space, which includes both the electron and the photon systems. The basis functions in this multi-particle space can be written as a product of the electronic and photonic subspaces (see Fig. 1.6.5):

$$|n,N\rangle = |n\rangle \otimes |N\rangle$$

Just as the basis functions in a two-dimensional problem can be written as the product of the basis states of two one-dimensional problems:

$$|k_x, k_y\rangle = |k_x\rangle \otimes |k_y\rangle \sim \exp(ik_x x)\exp(ik_y y)$$

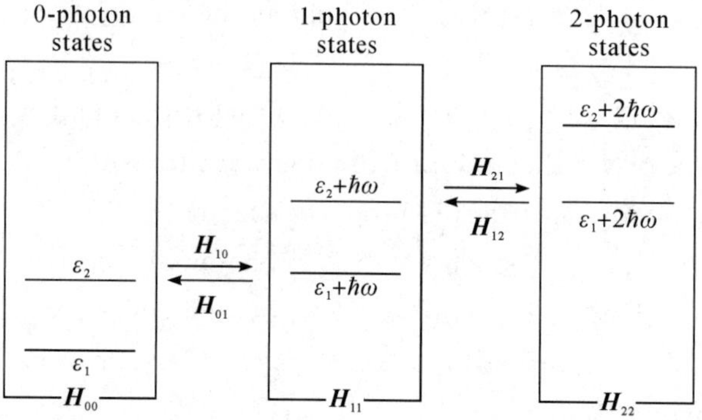

Fig. 1.6.5 In the multi-particle viewpoint, the electron-photon coupling causes transitions between $|2, N\rangle$ and $|1, N-1\rangle$, which are degenerate states of the composite system

We can write Eq. (1.6.10) in the form of a matrix equation:

$$i\hbar \frac{d}{dt}\begin{bmatrix}\psi_0\\ \psi_1\\ \psi_2\end{bmatrix} = \begin{bmatrix}\boldsymbol{H}_{00} & \boldsymbol{H}_{01} & 0\\ \boldsymbol{H}_{10} & \boldsymbol{H}_{11} & \boldsymbol{H}_{12}\\ 0 & \boldsymbol{H}_{21} & \boldsymbol{H}_{22}\end{bmatrix}\begin{bmatrix}\psi_0\\ \psi_1\\ \psi_2\end{bmatrix}$$

where $\boldsymbol{\psi}_N$ represents the N-photon component of the wavefunction. If the electronic subspace is spanned by two states $|1\rangle$ and $|2\rangle$ as shown in Fig. 1.6.5 then $\boldsymbol{\psi}_N$ is a (2×1) column vector

$$\boldsymbol{\psi}_N = \begin{bmatrix}\psi_{1,N}\\ \psi_{2,N}\end{bmatrix}$$

and the matrices $\boldsymbol{H}_{NM}$ are each (2×2) matrices given by

$$\left.\begin{aligned}\boldsymbol{H}_{NN} &= \begin{bmatrix}\varepsilon_1 + N\hbar\omega & 0\\ 0 & \varepsilon_2 + N\hbar\omega\end{bmatrix}\\ \boldsymbol{H}_{N,N+1} &= \begin{bmatrix}0 & K\sqrt{N+1}\\ K^*\sqrt{N+1} & 0\end{bmatrix}\end{aligned}\right\} \tag{1.6.11}$$

with $\boldsymbol{H}_{N+1,N} = \boldsymbol{H}_{N+1,N}^{+}$.

The point is that if we consider the N-photon subspace, it is like an open system that is connected to the $(N+1)$-and $(N-1)$-photon subspaces, just as a device is connected to the source and drain contacts. We saw that the effect of the source or drain contact could be represented by a self-energy matrix $\boldsymbol{\Sigma} = \boldsymbol{\tau g \tau}^{+}$ whose imaginary(more precisely anti-Hermitian) part represents the broadening $\boldsymbol{\Gamma} \equiv i(\boldsymbol{\Sigma} - \boldsymbol{\Sigma}^{+}) = \boldsymbol{\tau a \tau}^{+}$, $\boldsymbol{a} \equiv i(\boldsymbol{g} - \boldsymbol{g}^{+})$ being the spectral function of the isolated reservoir. We could use the same relation to calculate the self-energy function that describes the effect of the rest of the photon reservoir on the N-photon subspace, which we view as the "channel". Actually the details are somewhat more complicated because (unlike coherent interactions) we have to account for the exclusion principle. For the moment, however, let us calculate the broadening (or the outflow) assuming all other states to be "empty" so that there is no exclusion principle to worry about. Also, to keep things simple, let us focus just on the diagonal element of the broadening:

$$[\boldsymbol{\Gamma}_{mm}(E)]_{N,N} = (\boldsymbol{H}_{nm})_{N,N+1}(\boldsymbol{a}_{mm})_{N+1,N+1}(\boldsymbol{H}_{mn})_{N+1,N} + (\boldsymbol{H}_{nm})_{N,N-1}(\boldsymbol{a}_{mm})_{N-1,N-1}(\boldsymbol{H}_{mn})_{N-1,N}$$

Assuming that the coupling from one photon subspace to the next is weak, we can approximate the spectral functions with their unperturbed values:

$$[\boldsymbol{\Gamma}_{mn}(E)]_{N,N} = |K_{mn}^{\mathrm{em}}|^2(N+1)2\pi\delta[E-\varepsilon_m-(N+1)\hbar\omega] + |K_{mn}^{\mathrm{ab}}|^2 N 2\pi\delta \times [E-\varepsilon_m-(N-1)\hbar\omega]$$

where

$$K_{mn}^{\mathrm{em}} \equiv (\boldsymbol{H}_{mn})_{N+1,N} \tag{1.6.12a}$$

and

$$K_{mn}^{\mathrm{ab}} \equiv (\boldsymbol{H}_{mn})_{N-1,N} \tag{1.6.12b}$$

Again with weak coupling between photon subspaces we can assume that the state $|n, N\rangle$ remains an approximate eigenstate with an energy $\varepsilon_n + N\hbar\omega$, so that we can evaluate the broadening at $E = \varepsilon_n + N\hbar\omega$:

$$\boldsymbol{\Gamma}_{mn} = 2\pi|K_{mn}^{\mathrm{em}}|^2(N+1)\delta(\varepsilon_n-\varepsilon_m-\hbar\omega) + 2\pi|K_{mn}^{\mathrm{ab}}|^2 N\delta(\varepsilon_n-\varepsilon_m+\hbar\omega) \tag{1.6.13}$$

The first term arises from the coupling of the N-photon subspace to the $(N+1)$-photon subspace, indicating that it represents a photon emission process. Indeed it is peaked for photon energies $\hbar\omega$ for which $\varepsilon_n - \hbar\omega = \varepsilon_m$, suggesting that we view it as a process in which an electron in state n transits to state m and emits the balance of the energy as a photon. The second term in Eq. (1.6.13) arises from the coupling of the N-photon subspace to the $(N-1)$-photon subspace, indicating that it represents a photon absorption process. Indeed it is peaked for photon energies $\hbar\omega$ for which $\varepsilon_n + \hbar\omega = \varepsilon_m$, suggesting that we view it as a process in which an electron in state n transits to state m and absorbs the balance of the energy from a photon.

How do we write down the coupling constants K appearing in Eq. (1.6.13)? This is where it helps to invoke the one-electron viewpoint (see Fig. 1.6.4). The entire problem then amounts to writing down the "potential" U_S that an electron feels due to one photon or phonon occupying a particular mode with a frequency ω in the form:

$$U_S(\boldsymbol{r},t) = U^{\mathrm{ab}}(\boldsymbol{r})\exp(-\mathrm{i}\omega t) + U^{\mathrm{em}}(\boldsymbol{r})\exp(+\mathrm{i}\omega t) \tag{1.6.14}$$

where $U^{ab}(\boldsymbol{r}) = U^{em}(\boldsymbol{r})$.

Once we have identified this "interaction potential", the coupling constants for emission and absorption can be evaluated simply from the matrix elements of U^{em} and U^{ab} :

$$K_{mn}^{em} = \int d\boldsymbol{r}\phi_m^*(\boldsymbol{r})U^{em}\phi_n(\boldsymbol{r}) \equiv \langle m \mid U^{em} \mid n\rangle$$

and

$$K_{mn}^{ab} = \int d\boldsymbol{r}\phi_m^*(\boldsymbol{r})U^{ab}\phi_n(\boldsymbol{r}) \equiv \langle m \mid U^{ab} \mid n\rangle \tag{1.6.15}$$

where ϕ_m and ϕ_n are the wavefunctions for levels m and n respectively.

I will try to elaborate on the meaning of phonons. But for the moment we can simply view them as representing the vibrations of the lattice of atoms, just as photons represent electromagnetic vibrations. To write down the interaction potential for phonons, we need to write down the atomic displacement or the strain due to the presence of a single phonon in a mode with frequency ω and then multiply it by the change D in the electronic energy per unit displacement or strain. The quantity D, called the deformation potential, is known experimentally for most bulk materials of interest and one could possibly use the same parameter unless dealing with very small structures. Indeed relatively little work has been done on phonon modes in nanostructures and it is common to assume plane wave modes labeled by a wave vector $\boldsymbol{\beta}$, which is appropriate for bulk materials. The presence of a (longitudinal) phonon in such a plane wave mode gives rise to a strain (ρ is the mass density and Ω is the normalization volume)

$$S = \boldsymbol{\beta}\sqrt{2\hbar/\rho\omega\Omega}\cos(\boldsymbol{\beta}\cdot\boldsymbol{r} - \omega(\boldsymbol{\beta})t) \tag{1.6.16}$$

so that the interaction potentials in Eq. (1.6.14) are given by

$$U_{\boldsymbol{\beta}}^{ab}(\boldsymbol{r}) \equiv (U_{\boldsymbol{\beta}}/2)\exp(i\boldsymbol{\beta}\cdot\boldsymbol{r}), \quad U_{\boldsymbol{\beta}}^{em}(\boldsymbol{r}) = U_{\boldsymbol{\beta}}^{ab}(\boldsymbol{r})^* \tag{1.6.17}$$

where $U_{\boldsymbol{\beta}} = D\boldsymbol{\beta}\sqrt{2\hbar\omega\Omega}$.

The basic principle for writing down the electron—photon coupling coefficient is similar: we need to write down the interaction potential that an electron feels due to the presence of a single photon in a particular mode. However, the details are complicated by the fact that the effect of an electromagnetic field enters the Schrödinger equation through the vector potential rather than a scalar potential.

First, we write down the electric field due to a single photon in mode $(\boldsymbol{\beta}\cdot\boldsymbol{r})$ in the form $\boldsymbol{E} = \hat{v}E_0\sin(\boldsymbol{\beta}\cdot\boldsymbol{r} - \omega(\boldsymbol{\beta})t)$, whose amplitude E_0 is evaluated by equating the associated energy to $\hbar\omega$ (Ω is the volume of the "box"): $\frac{\varepsilon E_0^2\Omega}{2} = \hbar\omega \rightarrow |E_0| = \sqrt{\frac{2\hbar\omega}{\varepsilon\Omega}}$, The corresponding vector potential $\boldsymbol{A}$ is written as (noting that for electromagnetic waves $\boldsymbol{E} = -\partial\boldsymbol{A}/\partial t$):

$$\mathbf{A} = \hat{v}\mathbf{A}_0 \cos(\boldsymbol{\beta} \cdot \boldsymbol{r} - \omega(\boldsymbol{\beta})t) \text{ with } |\mathbf{A}_0| = \sqrt{2\hbar/\omega\varepsilon\Omega} \qquad (1.6.18)$$

Next we separate the vector potential due to one photon into two parts:

$$\mathbf{A}(\boldsymbol{r},t) = \mathbf{A}^{ab}(\boldsymbol{r})\exp(-i\omega(\boldsymbol{\beta})t) + \mathbf{A}^{em}(\boldsymbol{r})\exp(+i\omega(\boldsymbol{\beta})t) \qquad (1.6.19)$$

where $\mathbf{A}^{ab}(\boldsymbol{r}) \equiv \hat{v}(\mathbf{A}_0/2)\exp(i\boldsymbol{\beta} \cdot \boldsymbol{r})$ and $\mathbf{A}^{em}(\boldsymbol{r}) = \mathbf{A}^{ab}(\boldsymbol{r})^*$

The coupling coefficient for absorption processes is given by the matrix element for $\left(\frac{q}{m}\right)\mathbf{A}^{ab}(\boldsymbol{r}) \cdot \boldsymbol{p}$, while the coupling coefficient for emission processes is given by the matrix element for $\left(\frac{q}{m}\right)\mathbf{A}^{em}(\boldsymbol{r}) \cdot \boldsymbol{p}$, so that $(\boldsymbol{p} \equiv -i\hbar\boldsymbol{\nabla})$

$$K_{mn}(\boldsymbol{\beta},\hat{v}) = (q\mathbf{A}_0/2m)\langle m|\rangle \exp(i\boldsymbol{\beta} \cdot \boldsymbol{r})\, \boldsymbol{p} \cdot \hat{v}\,|\,n\rangle \text{ Absorption} \qquad (1.6.20a)$$

$$K_{mn}(\boldsymbol{\beta},\hat{v}) = (q\mathbf{A}_0/2m)\langle m|\rangle \exp(-i\boldsymbol{\beta} \cdot \boldsymbol{r})\, \boldsymbol{p} \cdot \hat{v}\,|\,n\rangle \text{ Emission} \qquad (1.6.20b)$$

Note that the m appearing here stands for mass and is different from the index m we are using to catalog basis functions.

Eqs. (1.6.20a, b) require a slightly extended justification since we have not had much occasion to deal with the vector potential. We know that the scalar potential $\phi(\boldsymbol{r})$ enters the Schrödinger equation additively: $p^2/2m \rightarrow (p^2/2m) - q\Phi(\boldsymbol{r})$ and if the photon could be represented by scalar potentials the coupling coefficients would simply be given by the matrix elements of $(-q)\,\Phi^{ab}(\boldsymbol{r})$ and $(-q)\,\Phi^{em}(\boldsymbol{r})$ for absorption and emission respectively as we did in writing down the electron-phonon coupling. But photons require a vector potential which enters the Schrödinger equation as $p^2/2m \rightarrow (\boldsymbol{p}+q\mathbf{A}) \cdot (\boldsymbol{p}+q\mathbf{A})/2m$ so that the change due to the photon is given by $(q/2m)(\mathbf{A} \cdot \boldsymbol{p} + \boldsymbol{p} \cdot \mathbf{A}) + (q^2/2m)\,\mathbf{A} \cdot \mathbf{A} \approx (q/2m)(\mathbf{A} \cdot \boldsymbol{p} + \boldsymbol{p} \cdot \mathbf{A})$ assuming that the vector potential is small enough that the quadratic term is negligible. Finally we note that for any scalar function $\Phi(\boldsymbol{r})$, $\boldsymbol{p} \cdot (\mathbf{A}\Phi) = \mathbf{A} \cdot (\boldsymbol{p}\Phi) + \Phi(\boldsymbol{p} \cdot \mathbf{A})$ so that we can write, $(q/2m)(\mathbf{A} \cdot \boldsymbol{p} + \boldsymbol{p} \cdot \mathbf{A}) = \left(\frac{q}{m}\right)\mathbf{A} \cdot \boldsymbol{p}$, as long as $\boldsymbol{p} \cdot \mathbf{A} = 0$. It can be checked that this is indeed true of the photon vector potential given in Eq. (1.6.19) because of the transverse nature of electromagnetic waves which requires that the wavevector and the polarization be orthogonal to each other. This allows us to obtain the coupling coefficient from the matrix element for $(q/m)\,\mathbf{A}(\boldsymbol{r}) \cdot \boldsymbol{p}$ using $\mathbf{A} \rightarrow \mathbf{A}^{ab}(\boldsymbol{r})$ for absorption processes and $\mathbf{A} \rightarrow \mathbf{A}^{em}(\boldsymbol{r})$ for emission processes.

In this part, I will go through a few examples to illustrate the basic approach for describing incoherent interactions. I will take up the more general case of inflow and outflow, but in this part I will assume all other states to be "empty" so that there is no exclusion principle to worry about and I will calculate the broadening (or the outflow), which can be identified with $\hbar/\tau$, τ being the lifetime of the state. This will include: ① the photon-induced (radiative) lifetime due to atomic transitions; ② the radiative lifetime due to interband transitions in semiconductors; and ③ the phonon-induced(non-radiative) lifetime due to intraband transitions in semiconductors (see Fig. 1. 6. 6). The basic approach is to write down the interaction potential(see Eq. (1. 6. 14)), evaluate the coupling constants(see Eq. (1. 6. 15)), and obtain the broadening and hence the lifetime from Eq. (1. 6. 13).

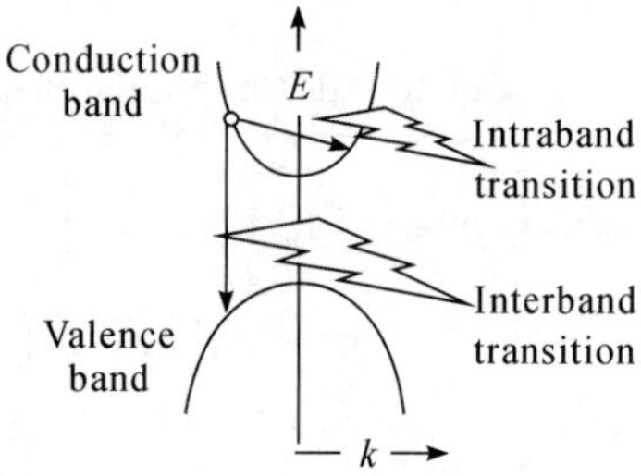

Fig. 1. 6. 6 Electronic transitions in semiconductors can be classified as interband and intraband. The former are associated primarily with electron-photon interactions while the latter are associated primarily with electron-phonon interactions

From Eq. (1. 6. 13)it is apparent that the broadening is large when the argument of the delta function vanishes. How large it is at that point depends on the value of 0^+ that we choose to broaden each reservoir state. As we have seen, the precise value of 0^+ usually does not matter as long as the system is coupled to a continuous distribution of reservoir states. This is true in this case, because we usually do not have a single photon mode with energy $\hbar\omega$. Instead, we have a continuous distribution of photons with different wavevectors $\boldsymbol{\beta}$ with energiesgiven by

$$\hbar\omega(\boldsymbol{\beta}) = \hbar\bar{c}\boldsymbol{\beta} \tag{1.6.21}$$

where $\bar{c}$ is the velocity of light in the solid. Consequently the states in a particular subspace do not look discrete as shown in Fig. 1. 6. 3, but look more like as shown in Fig. 1. 6. 7.

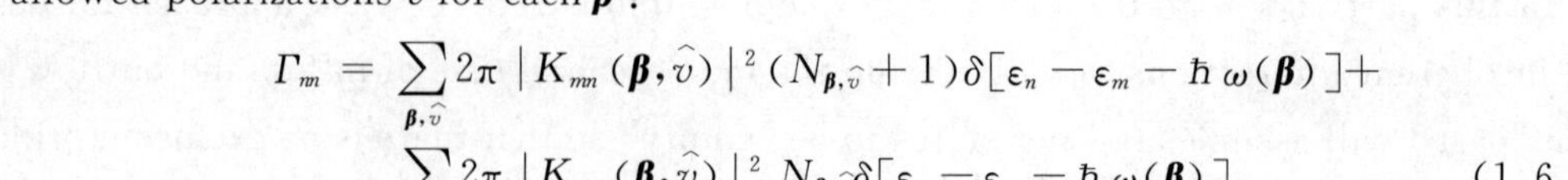

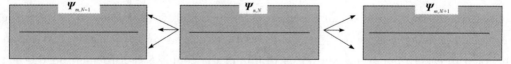

Fig. 1. 6. 7 The continuous distribution of reservoir states

The broadening is obtained from Eq. (1. 6. 13) after summing over all modes $\boldsymbol{\beta}$ and the two allowed polarizations $\hat{v}$ for each $\boldsymbol{\beta}$:

$$\Gamma_{nn} = \sum_{\boldsymbol{\beta},\hat{v}} 2\pi \left| K_{mn}(\boldsymbol{\beta},\hat{v}) \right|^2 (N_{\boldsymbol{\beta},\hat{v}}+1)\delta[\varepsilon_n - \varepsilon_m - \hbar\omega(\boldsymbol{\beta})] + \sum_{\boldsymbol{\beta},\hat{v}} 2\pi \left| K_{mn}(\boldsymbol{\beta},\hat{v}) \right|^2 N_{\boldsymbol{\beta},\hat{v}}\delta[\varepsilon_n - \varepsilon_m - \hbar\omega(\boldsymbol{\beta})] \tag{1.6.22}$$

If the photon reservoir is in equilibrium then the number of photons in mode $\boldsymbol{\beta}$ and polarization $\hat{v}$ is given by the Bose-Einstein factor [see Eq. (1. 6. 8)]:

$$N_{\boldsymbol{\beta},\hat{v}} = \frac{1}{\exp\left[\frac{\hbar\,\omega(\boldsymbol{\beta})}{k_B T}\right] - 1} \tag{1. 6. 23}$$

If we consider transitions involving energies far in excess of $k_B T$, then we can set $N_{\boldsymbol{\beta},\hat{v}}$ equal to zero, so that the broadening (which is proportional to the inverse radiative lifetime τ_r) is given by

$$\Gamma_{mn} = \left(\frac{\hbar}{\tau_r}\right)_n = \sum_{\boldsymbol{\beta},\hat{v}} 2\pi \,|\, K_{mn}(\boldsymbol{\beta},\hat{v}) \,|^2 \delta(\varepsilon_n - \varepsilon_m - \hbar\,\omega_{\boldsymbol{\beta}}) \tag{1. 6. 24a}$$

which is evaluated by converting the summation into an integral assuming periodic boundary conditions for the photon modes. We follow the same prescription that we have used in the past for electronic states, namely, $\sum_{\boldsymbol{\beta}} \rightarrow \frac{\Omega}{8\,\pi^3}\int_0^{\infty} \mathrm{d}\boldsymbol{\beta}\boldsymbol{\beta}^2 \int_0^{+\pi} \mathrm{d}\theta \sin\theta \int_0^{2\pi} \mathrm{d}\Phi$

$$\Gamma_{mn} = \frac{\Omega}{8\,\pi^3}\sum_{v}\int_0^{\infty} \mathrm{d}\boldsymbol{\beta}\boldsymbol{\beta}^2 \int_{-\pi}^{+\pi} \mathrm{d}\theta \sin\theta \int_0^{2\pi} \mathrm{d}\Phi 2\pi \,|\, K_{mn}(\boldsymbol{\beta},\hat{v}) \,|^2 \delta(\varepsilon_n - \varepsilon_m - \hbar\,\omega_{\boldsymbol{\beta}}) \tag{1. 6. 24b}$$

To proceed further, we need to insert the electron-photon coupling coefficients K from Eqs. (1. 6. 24a, b). For atomic wavefunctions that are localized to extremely short dimensions (much shorter than an optical wavelength) we can neglect the factor exp($\boldsymbol{\beta}$, $\hat{v}$) and write from Eqs. (1. 6. 24a, b)

$$K_{mn}(\boldsymbol{\beta},\hat{v}) = \frac{q}{m}\sqrt{\frac{\hbar}{2\varepsilon\omega\Omega}} \,|\, \boldsymbol{P} \,|\, \sin\theta \tag{1. 6. 25}$$

where $\boldsymbol{P} \equiv \langle m \,|\, \boldsymbol{p} \,|\, n\rangle$ and θ is the complement of the angle between the dipole moment of the transition and the polarization of the photon (see Fig. 1. 6. 8).

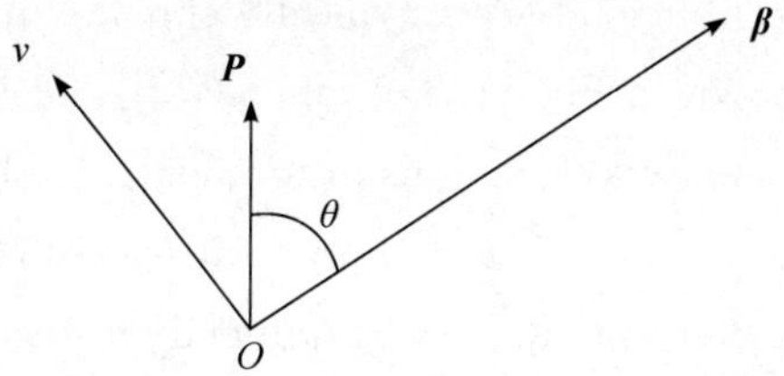

Fig. 1. 6. 8 Emission of photon with wavevector $\boldsymbol{\beta}$ and polarization $\hat{v}$ by an atomic transition with an equivalent dipole moment $\boldsymbol{P}$ (where Ω is again the volume of the "box") to obtain

Using Eq. (1. 6. 10) and Eq. (1. 6. 6) we can find the radiative lifetime from Eq. (1. 6. 24b):

$$\boldsymbol{\Gamma} = \frac{\Omega}{8\,\pi^3}\int_0^{\infty} \frac{\mathrm{d}\omega\,\omega^2}{\bar{c}^3}\int_{-\pi}^{\pi} \mathrm{d}\theta \sin^3\theta \int_0^{2\pi} \mathrm{d}\Phi \frac{q^2\,\hbar}{2\,m^2\,\varepsilon\omega\Omega}\boldsymbol{P}^2 2\pi\delta(\varepsilon_n - \varepsilon_m - \hbar\omega) \text{ so that}$$

$$\frac{1}{\tau_r} = \left|\frac{\boldsymbol{\Gamma}}{\hbar}\right| = \frac{q^2}{4\pi\varepsilon\,\hbar\,\bar{c}}\,\frac{2(\varepsilon_n - \varepsilon_m)}{3\,\hbar\,m\,\bar{c}^2}\left|\frac{2\,\boldsymbol{P}^2}{m}\right| \tag{1. 6. 26}$$

Note that the answer is obtained without having to worry about the precise height of the delta function (which is determined by the value of 0^+). But if the photon modes do not form a quasi-continuous spectrum (as in small nanostructures) then it is conceivable that there will be reversible effects that are affected by the precise values of 0^+.

We can calculate the amount of power radiated per electron (note that $\hbar\omega = \varepsilon_n - \varepsilon_m$) from:

$$W = \frac{\hbar\,\omega}{\tau} = \frac{q^2}{4\pi\varepsilon\,\hbar\,\bar{c}}\,\frac{2(\varepsilon_n - \varepsilon_m)}{3\,\hbar\,m\,\bar{c}^2}\left|\frac{2\,\mathbf{P}^2}{m}\right| = \frac{\omega^2}{12\pi\varepsilon\,\bar{c}^3}\left|\left(\frac{2q\mathbf{P}}{m}\right)^2\right| \tag{1.6.27}$$

It is interesting to note that the power radiated from a classical dipole antenna of length d carrying a current $I\cos\omega t$ is given by $W = \frac{\omega^2}{12\pi\varepsilon\,\bar{c}^3}(Id)^2$, suggesting that an atomic radiator behaves like a classical dipole with

$$Id = |\,2q\mathbf{P}/m\,| = (2q/m)\,|\,\langle m\,|\,\mathbf{p}\,|\,n\rangle\,| \tag{1.6.28}$$

Indeed it is not just the total power, even the polarization and angular distribution of the radiation are the same for a classical antenna and an atomic radiator. The light is polarized in the plane containing the direction of observation and $\mathbf{P}$, and its strength is proportional to $\sin^2\theta$, θ being the angle between the direction of observation and the dipole as shown in Fig. 1.6.8.

The basic rule stated in Eqs. (1.6.20a, b) for the coupling coefficients can be applied to delocalized electronic states too, but we can no longer neglect the factor $\exp(\boldsymbol{\beta}, \hat{v})$ as we did when going to Eq. (1.6.5). For example, in semiconductors (see Fig. 1.6.6), the conduction (c) and valence (v) band electronic states are typically spread out over the entire solid consisting of many unit cells as shown in Fig. 1.6.9 where $|\,c\rangle_n$ and $|\,v\rangle_n$ are the atomic parts of the conduction and valence band wavefunctions in unit cell n. These functions depend on the wavevector $\boldsymbol{k}_c$ or $\boldsymbol{k}_v$, but are the same in each unit cell, except for the spatial shift. This allows us to write the coupling elements for absorption and emission from Eqs. (1.6.20a, b) in the form $\langle v\,|\,\mathbf{p}\cdot\hat{v}\,|\,c\rangle\sum_n \frac{1}{N}\exp[\mathrm{i}(\boldsymbol{k}_c \pm \boldsymbol{\beta} - \boldsymbol{k}_v)\cdot\boldsymbol{r}_n]$, where we take the upper sign (+) for absorption and the lower sign (−) for emission, $\langle\cdots\rangle$ denotes an integral over a unit cell, and we have neglected the variation of the factor $\exp(\boldsymbol{\beta}, \hat{v})$ across a unit cell.

This leads to non-zero values only if

$$\boldsymbol{k}_v = \boldsymbol{k}_c \pm \boldsymbol{\beta} \tag{1.6.29}$$

Eq. (1.6.29) can be viewed as a rule for momentum conservation, if we identify $\boldsymbol{k}$ as the electron momentum and $\boldsymbol{\beta}$ as the photon momentum. The final electronic momentum, the photon wavevector is typically very small compared to the electronic wavevector, so that radiative transitions are nearly "vertical" with $\boldsymbol{k}_v = \boldsymbol{k}_c \pm \boldsymbol{\beta} \approx \boldsymbol{k}_c$. This is easy to see if we note that the range of k extends over a Brillouin zone which is $\sim 2\pi$ divided by an atomic distance, while the photon wavevector is equal to 2π divided by the optical wavelength which is thousands of atomic distances. Assuming that the momentum conservation rule in Eq. (1. 6. 29) is satisfied, we can write the coupling coefficients from Eqs. (1. 6. 24a, b)as

$$K_{mn}(\boldsymbol{\beta},\hat{v}) = \frac{q}{m}\sqrt{\frac{\hbar}{2\varepsilon\omega\Omega}}\,|\,\boldsymbol{P}\,|\,\sin\theta\ ,\ \text{where}\ \boldsymbol{P} \equiv \langle v\,|\,\boldsymbol{p}\,|\,c\rangle \qquad (1.6.30)$$

showing that "vertical" radiative transitions in semiconductors can be understood in much the same way as atomic transitions using the atomic parts of the conduction and valence band wavefunctions. For example, if we put the numbers characteristic of conduction-valence band transitions in a typical semiconductor like GaAs into Eq. (1. 6. 30), $\varepsilon_n - \varepsilon_m = 1.5$ eV , $2P^2/m = 20$ eV , $\varepsilon = 10\varepsilon_0$, we obtain $\tau_r = 0.7$ ns for the radiative lifetime.

We know that the electronic states at the bottom of the conduction band near the Γ-point are isotropic or s-type denoted by $|S\rangle$. If the states at the top of the valence band were purely P_x , P_y and P_z types denoted by $|X\rangle$, $|Y\rangle$, $|Z\rangle$, then we could view the system as being composed of three independent antennas with their dipoles pointing along x- , y- and z-directions since $\langle s\,|\,\boldsymbol{p}\,|\,X\rangle = \hat{x}\boldsymbol{P}$, $\langle s\,|\,\boldsymbol{p}\,|\,Y\rangle = \hat{y}\boldsymbol{P}$, $\langle s\,|\,\boldsymbol{p}\,|\,Z\rangle = \hat{z}\boldsymbol{P}$. The resulting radiation can then be shown to be unpolarized and isotropic. In reality, however, the top of the valence band is composed of light hole and heavy hole bands which are mixtures of up-and down-spin states and the equivalent dipole moments for each of the conduction-valence band pairs can be written as shown in Table 1. 6. 1.

We thus have eight independent antennas, one corresponding to each conduction band-valence band pair. If the C state is occupied, then the first row of four antennas is active. If we look at the radiation coming out in the z- direction then we will see the radiation from the C-HH and C-$\overline{LH}$ transitions. The C-HH transition will emit right circularly polarized(RCP) light, which will be three times as strong as the left circularly polarized(LCP) light from the C-$\overline{LH}$ transition. If the $\overline{C}$ state is also occupied then the $\overline{C}$-$\overline{HH}$ transition would yield three times as much LCP light as the RCP light from the $\overline{C}$-LH transition. Overall there would be just as much LCP as RCP light. But if only the C state is occupied then there would be thrice as much RCP light as LCP light. Indeed the degree of circular polarization of the emission is often used as a measure of the degree of spin polarization that has been achieved in a given experiment.

Table 1.6.1 Optical matrix elements for conduction-valence band transitions

	HH	$\overline{HH}$	LH	$\overline{LH}$
	$\begin{bmatrix} \frac{\lvert X\rangle + i\lvert Y\rangle}{\sqrt{2}} \\ 0 \end{bmatrix}$	$\begin{bmatrix} 0 \\ \frac{\lvert X\rangle - i\lvert Y\rangle}{\sqrt{2}} \end{bmatrix}$	$\begin{bmatrix} -\sqrt{\frac{2}{3}}\lvert Z\rangle \\ \frac{\lvert X\rangle + i\lvert Y\rangle}{\sqrt{6}} \end{bmatrix}$	$\begin{bmatrix} \frac{\lvert X\rangle - i\lvert Y\rangle}{\sqrt{6}} \\ \sqrt{\frac{2}{3}}\lvert Z\rangle \end{bmatrix}$
$C\begin{bmatrix} \lvert S\rangle \\ 0 \end{bmatrix}$	$\frac{\hat{x}+i\hat{y}}{\sqrt{2}}P$	0	$-\sqrt{\frac{2}{3}}\hat{z}P$	$\frac{\hat{x}-i\hat{y}}{\sqrt{6}}P$
$\overline{C}\begin{bmatrix} 0 \\ \lvert S\rangle \end{bmatrix}$	0	$\frac{\hat{x}-i\hat{y}}{\sqrt{2}}P$	$\frac{\hat{x}+i\hat{y}}{\sqrt{6}}P$	$\sqrt{\frac{2}{3}}\hat{z}P$

We have discussed the radiation of light due to interband transitions in semiconductors. But what about intraband transitions? Can they lead to the emission of light? We will show that the simultaneous momentum and energy conservation requirements prevent any radiation of light unless the electron velocity exceeds the velocity of light: $\hbar k/m > c$. This is impossible in vacuum, but could happen in a solid and such Cerenkov radiation of light by fast-moving electrons has indeed been observed. However, this is usually not very relevant to the operation of solid-state devices because typical electronic velocities are about a thousandth of the speed of light. What is more relevant is the Cerenkov emission of acoustic waves or phonons that are five orders of magnitude slower than light. Electron velocities are typically well in excess of the phonon velocity leading to extensive Cerenkov emission (and absorption) of phonons, somewhat like the sonic booms generated by supersonic jets.

For intraband transitions, both the final and initial states have the same atomic wavefunctions and, for clarity, I will not write them down explicitly. Instead I will write the initial and final states in the form of plane waves as if we are dealing with electrons in vacuum:

$$\lvert \boldsymbol{k}\rangle \equiv (1/\sqrt{\Omega})\exp(i\,\boldsymbol{k}\cdot\boldsymbol{r}) \text{ and } \lvert \boldsymbol{k}_f\rangle \equiv (1/\sqrt{\Omega})\exp(i\,\boldsymbol{k}_f\cdot\boldsymbol{r}) \qquad (1.6.31)$$

From Eqs. (1.6.20a, b) we find that the radiative coupling constant is equal to

$$K(\boldsymbol{k}_f, \boldsymbol{k}, \boldsymbol{\beta}, \hat{v}) = \left(\frac{q\boldsymbol{A}_0}{2m}\right)\hbar\,\boldsymbol{k}\cdot\hat{v} \qquad (1.6.32)$$

if

$$\boldsymbol{k}_f = \boldsymbol{k} - \boldsymbol{\beta} \qquad \boldsymbol{k}_f = \boldsymbol{k} + \boldsymbol{\beta} \qquad (1.6.33)$$

and is zero otherwise. Like Eq. (1.6.29), Eq. (1.6.33) too can be interpreted as a condition for momentum conservation. Energy conservation, on the other hand, is enforced by the delta function in Eq. (1.6.13):

$$\varepsilon(\boldsymbol{k}_f) = \varepsilon(\boldsymbol{k}) - \hbar\,\omega(\boldsymbol{\beta}) \qquad \varepsilon(\boldsymbol{k}_f) = \varepsilon(\boldsymbol{k}) + \hbar\,\omega(\boldsymbol{\beta}) \tag{1.6.34}$$

Emission　　　　Absorption

From Eq. (1.6.33) and Eq. (1.6.34) we obtain for emission processes

$$\varepsilon(\boldsymbol{k}_f) - \varepsilon(\boldsymbol{k}) + \hbar\omega(\boldsymbol{\beta}) = 0 = \frac{\hbar^2}{2m}[(\boldsymbol{k}-\boldsymbol{\beta})\cdot(\boldsymbol{k}-\boldsymbol{\beta}) - \boldsymbol{k}^2] + \hbar\,\bar{c}\boldsymbol{\beta}$$

so that

$$\frac{\hbar^2}{2m}(-2k\,|\,\boldsymbol{\beta}\,|\cos\theta + |\,\boldsymbol{\beta}^2\,|) + \hbar\,\bar{c}\,|\,\boldsymbol{\beta}\,| = 0$$

Which yields

$$\cos\theta = \frac{\bar{c}}{\hbar\,k/m} + \frac{|\,\boldsymbol{\beta}\,|}{2k} \tag{1.6.35}$$

The point is that, since $\cos\theta$ must be smaller than one, Cerenkov emission cannot take place unless the electron velocity $\hbar k/m$ exceeds the velocity of light $\bar{c}$. As mentioned above, this is impossible in vacuum, but possible in a solid and Cerenkov radiation has indeed been observed. The emitted light forms a cone around the electronic wavevector $\boldsymbol{k}$ with a maximum angle $\theta_{\max} = \cos^{-1}(m\bar{c}/\hbar\,|\,\boldsymbol{k}\,|)$ (see Fig. 1.6.9).

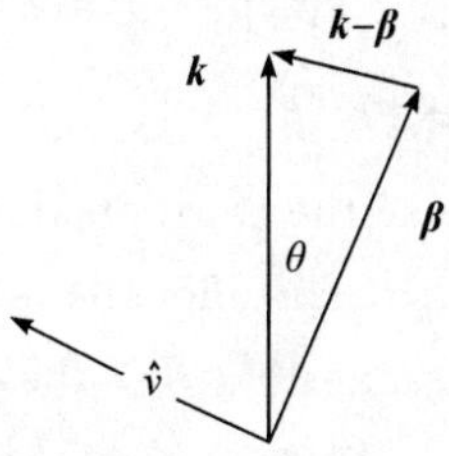

Fig. 1.6.9 Cerenkov emission: initial state $\boldsymbol{k}$, photon wavevector $\boldsymbol{\beta}$, and final state $\boldsymbol{k}-\boldsymbol{\beta}$. The photon polarization is along $\hat{v}$

As stated before, Cerenkov emission of light is not very relevant to the operation of solid-state devices. But the emission (and absorption) of sound waves or acoustic phonons is quite relevant. Acoustic waves are five orders of magnitude slower than light and the velocity of electrons routinely exceeds the sound velocity. Since acoustic phonons typically have energies less than k_BT they are usually present in copious numbers at equilibrium:

$$N_{\boldsymbol{\beta}} + 1 - N_{\boldsymbol{\beta}} = \{\exp[\hbar\,\omega(\boldsymbol{\beta})/k_BT] - 1\}^{-1} \approx k_BT/\hbar\,\omega(\boldsymbol{\beta}) \qquad \text{(acoustic phonons)}$$

Both terms in Eq. (1.6.13) now contribute to the broadening or inverse lifetime: Cerenkov absorption ("ab") is just as important as Cerenkov emission ("em"):

$$(\boldsymbol{\Gamma})_{\boldsymbol{k},\mathrm{em}} = \sum_{\boldsymbol{\beta}} 2\pi \frac{k_BT}{\hbar\,\omega(\boldsymbol{\beta})}\,|K(\boldsymbol{\beta})|^2\delta[\varepsilon(\boldsymbol{k}) - \varepsilon(\boldsymbol{k}-\boldsymbol{\beta}) - \hbar\,\omega(\boldsymbol{\beta})] \tag{1.6.36a}$$

$$(\boldsymbol{\Gamma})_{\boldsymbol{k},\mathrm{ab}} = \sum_{\boldsymbol{\beta}} 2\pi \frac{k_BT}{\hbar\,\omega(\boldsymbol{\beta})}\,|K(\boldsymbol{\beta})|^2\delta[\varepsilon(\boldsymbol{k}) - \varepsilon(\boldsymbol{k}+\boldsymbol{\beta}) + \hbar\,\omega(\boldsymbol{\beta})] \tag{1.6.36b}$$

The coupling element $K(\boldsymbol{\beta})$ is proportional to the potential that an electron feels due to the presence of a single phonon as discussed earlier [see Eq. (1.6.18)]. Without getting into a detailed evaluation of Eqs. (1.6.36a, b), it is easy to relate the angle of emission θ to the magnitude of the phonon wavevector $\boldsymbol{\beta}$ by setting the arguments of the delta functions to zero and proceeding as we did in deriving Eq. (1.6.35):

$$\left.\begin{aligned} \cos\theta &= \frac{c_s}{\hbar k/m} + \left|\frac{\boldsymbol{\beta}}{2\boldsymbol{k}}\right| \quad \text{(emission)} \\ \cos\theta &= \frac{c_s}{\hbar k/m} - \left|\frac{\boldsymbol{\beta}}{2\boldsymbol{k}}\right| \quad \text{(absorption)} \end{aligned}\right\} \tag{1.6.37}$$

where c_s is the velocity of sound waves: $\omega = c_s \mid \boldsymbol{\beta} \mid$. The detailed evaluation of the electron lifetime due to acoustic phonon emission and absorption from Eqs. (1.6.36a, b).

In addition to acoustic phonons, there are optical phonons whose frequency is nearly constant $\omega = \omega_0$, where the phonon energy $\hbar\omega_0$ is typically a few tens of millivolt, So that the number of such phonons present at equilibrium at room temperature is of order one: $N_{\boldsymbol{\beta}} = [\exp(\hbar\omega_0 / k_B T) - 1]^{-1} \equiv N_0$ (optical phonons), From Eq. (1.6.13) we now obtain for the emission and absorption rates:

$$(\boldsymbol{\Gamma})_{\boldsymbol{k},\text{em}} = (N_0 + 1)\sum_{\boldsymbol{\beta}} 2\pi \mid K(\boldsymbol{\beta}) \mid^2 \delta[\varepsilon(\boldsymbol{k}) - \varepsilon(\boldsymbol{k} - \boldsymbol{\beta}) - \hbar\omega(\boldsymbol{\beta})] \tag{1.6.38a}$$

$$(\boldsymbol{\Gamma})_{\boldsymbol{k},\text{ab}} = N_0 \sum_{\boldsymbol{\beta}} 2\pi \mid K(\boldsymbol{\beta}) \mid^2 \delta[\varepsilon(\boldsymbol{k}) - \varepsilon(\boldsymbol{k} + \boldsymbol{\beta}) + \hbar\omega_0] \tag{1.6.38b}$$

Let me now explain how we can use the concepts developed in this chapter to write down the new terms $\boldsymbol{\Sigma}_s^{\text{in}}$ and $\boldsymbol{\Gamma}_s$. To start with, consider the inflow and outflow into a specific energy level ε due to transitions from levels a and b— one above it and one below it separated by an energy of $\hbar\omega$: $\varepsilon_b - \varepsilon = \varepsilon - \varepsilon_a = \hbar\omega$. We saw that the rate constant (K^{ab}) for absoption processes is proportional to the number of phonons present (N_ω) while the rate constant (K^{em}) for emission processes is proportional to ($N_\omega + 1$). Let's assume the temperature is very low compared to $\hbar\omega$, so that $N_\omega \ll 1$ and we need only worry about emission. Using N, N_a and N_b to denote the number of electrons in each of these levels we can write for the level ε: Inflow $= K^{\text{em}}(1 - N)\, N_b$ and Outflow$= K^{\text{em}} N(1 - N_a)$. We can now write the inflow term as a difference between two terms Inflow$= K^{\text{em}} N_b - K^{\text{em}} N N_b$, where the second term represents the part of the inflow that is blocked by the exclusion principle. A non-obvious result that comes out of the advanced formalism is that this part of the inflow is not really blocked. Rather the outflow is increased by this amount($K^{\text{em}} N N_b$):

$$\text{Inflow} = K^{\text{em}} N_b \text{ and Outflow} = K^{\text{em}} N(1 - N_a + N_b) \tag{1.6.39a}$$

The difference between inflow and outflow is not changed from what we had guessed earlier. But the distinction is not academic, since the outflow(not the difference) determines the broadening of the level. We can compare this with the inflow and outflow between a contact and a discrete level that we discussed earlier:

$$\text{Inflow} = \gamma f \quad \text{and} \quad \text{Outflow} = \gamma N \tag{1.6.39b}$$

The inflow looks similar with K^{em} playing the role of γ, but the outflow involves an extra factor($1 - N_a + N_b$)that reduces to one if $N_a = N_b$, but not in general Eq. (1.6.39) was earlier generalized to the form: Inflow= Trace $(\boldsymbol{\Gamma A})f$ = Trace $\boldsymbol{\Sigma}^{in}\boldsymbol{A}$ and Outflow = Trace $(\boldsymbol{\Gamma G}^n)$. Similarly in the present case we can generalize Eq. (1.6.38a) to Inflow= Trace $\boldsymbol{\Sigma}_s^{in}\boldsymbol{A}$ and Outflow=Trace $(\boldsymbol{\Gamma}_s\boldsymbol{G}^n)$, where the expressions for $\boldsymbol{\Sigma}_s^{in}$ and $\boldsymbol{\Gamma}_s$ can be discovered heuristically by analogy. Let us do this one by one.

Scattering processes are very similar to ordinary contacts. The basic difference is that regular contacts are maintained in equilibrium by external sources, while for scattering processes the "contact" is the device itself (see Fig. 1.6.10) and as such can deviate significantly from equilibrium.

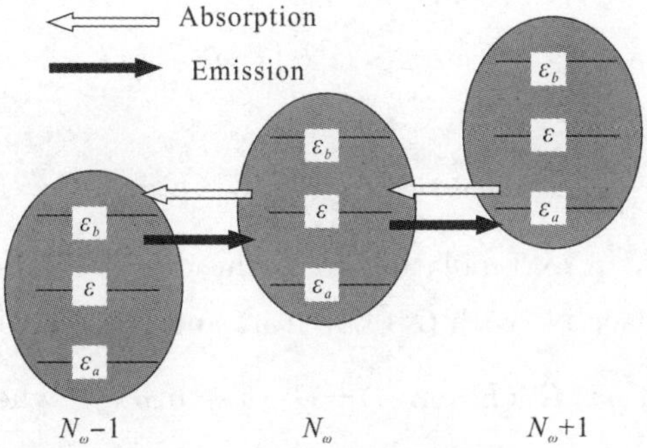

Fig. 1.6.10 An idealized device with three levels used to write down the inflow and outflow terms due to the absorption and emission of photons and phonons

For regular contacts we saw that the inscattering function is given by $\boldsymbol{\Sigma}^{in}(E) = (\boldsymbol{\tau A \tau}^+)f = \boldsymbol{\tau G}^n\boldsymbol{\tau}^+$, where $\boldsymbol{G}^n = \boldsymbol{A}f$ is the correlation function in the contact: the contact correlation function at energy E causes inscattering into the device at an energy E. Now for emission processes, the device correlation function at $E + \hbar\omega$ causes inscattering into the device at energy E, suggesting that we write: $\boldsymbol{\Sigma}_s^{in}(E) = \sum_{\boldsymbol{\beta}}(N_{\boldsymbol{\beta}} + 1)[\boldsymbol{U}_{\boldsymbol{\beta}}^{em}\boldsymbol{G}^n(E+\hbar\omega(\boldsymbol{\beta}))(\boldsymbol{U}_{\boldsymbol{\beta}}^{em})^+]$, where U^{em} is the emission component of the interaction potential(see Eq. (1.6.17)).

Writing out the matrix multiplication in detail we have

$$\boldsymbol{\Sigma}_s^{in}(i,j;E) = \sum_{p',q',\boldsymbol{\beta}}(N_{\boldsymbol{\beta}}+1)\{U_{\boldsymbol{\beta}}^{em}(i,r)\boldsymbol{G}^n[r,s;E+\hbar\omega(\boldsymbol{\beta})]U_{\boldsymbol{\beta}}^{em}(j,s)^*\} =$$

$$\int_0^\infty \frac{d(\hbar\omega)}{2\pi}\sum_{p',q'} D^{em}(i,r;j,s;\hbar\omega)\,\boldsymbol{G}^n(r,s;E+\hbar\omega) \tag{1.6.40}$$

where $D^{em}(i,r;j,s;\hbar\omega) \equiv 2\pi\sum_{\boldsymbol{\beta}}(N_\omega+1)\delta[\hbar\omega - \hbar\omega(\boldsymbol{\beta})]\,U_{\boldsymbol{\beta}}^{em}(i,r)U_{\boldsymbol{\beta}}^{em}(j,s)^*$ (1.6.41)

If we include absorption processes we obtain

$$\boldsymbol{\Sigma}_{s}^{in}(i,j;E)=\int_{0}^{\infty}\frac{d(\hbar\omega)}{2\pi}(D^{em}(i,r;j,s;\hbar\omega)\boldsymbol{G}^{n}(r,s;E+\hbar\omega)+$$

$$D^{ab}(i,r;j,s;\hbar\omega)\,\boldsymbol{G}^{n}(r,s;E-\hbar\omega)) \tag{1.6.42}$$

where the absorption term is given by an expression similar to the emission term with N instead of $(N+1)$:

$$D^{ab}(i,r;j,s;\hbar\omega)=\sum_{\boldsymbol{\beta}}N_{\omega}\delta(\hbar\omega-\hbar\omega(\boldsymbol{\beta}))\,U_{\boldsymbol{\beta}}^{ab}(i,r)U_{\boldsymbol{\beta}}^{ab}(j,s)^{*} \tag{1.6.43}$$

In general, the emission and absorption functions D^{em} and D^{ab} are fourth-rank tensors, but from hereon let me simplify by treating these as scalar quantities. The full tensorial results along with a detailed derivation. We write the simplified emission and absorption functions as

$$D^{em}(\hbar\omega)\equiv(N_{\omega}+1)\,D_{0}(\hbar\omega) \tag{1.6.44a}$$

$$D^{ab}(\hbar\omega)\equiv N_{\omega}D_{0}(\hbar\omega) \tag{1.6.44b}$$

So that

$$\boldsymbol{\Sigma}_{s}^{in}(E)=\int_{0}^{\infty}\frac{d(\hbar\omega)}{2\pi}D_{0}(\hbar\omega)\,((N_{\omega}+1)\boldsymbol{G}^{n}(E+\hbar\omega)+N_{\omega}\boldsymbol{G}^{n}(E-\hbar\omega))$$

For the outflow term, in extrapolating from the discrete version in Eq. (1.6.39a) to a continuous version we replace N_{b} with $\boldsymbol{G}^{n}(E+\hbar\omega)$ and $1-N_{a}$ with $\boldsymbol{G}^{p}(E-\hbar\omega)$ to obtain $\boldsymbol{\Gamma}_{s}(E)=\int\frac{d(\hbar\omega)}{2\pi}D^{em}(\hbar\omega)[\boldsymbol{G}^{p}(E-\hbar\omega)+\boldsymbol{G}^{n}(E+\hbar\omega)]$, where I have defined a new quantity $\boldsymbol{G}^{p}\equiv\boldsymbol{A}-\boldsymbol{G}^{n}$ that tells us the density (eV) of empty states or holes, just as $\boldsymbol{G}^{n}$ tells us the density of filled states or electrons. The sum of the two is equal to the spectral function $\boldsymbol{A}$ which represents the density of states. We can extend this result as before to include absorption terms as before to yield.

$$\boldsymbol{\Gamma}_{s}(E)=\int\frac{d(\hbar\omega)}{2\pi}D_{0}(\hbar\omega)\{(N_{\omega}+1)[\boldsymbol{G}^{p}(E-\hbar\omega)+\boldsymbol{G}^{n}(E+\hbar\omega)]+$$

$$N_{\omega}[\boldsymbol{G}^{n}(E-\hbar\omega)+\boldsymbol{G}^{p}(E+\hbar\omega)]\} \tag{1.6.45}$$

Eq. (1.6.43)-Eq. (1.6.45) are the expressions for the scattering functions that we are looking for. The self-energy function $\boldsymbol{\Sigma}_{S}$ can be written as $\mathrm{Re}(\boldsymbol{\Sigma}_{s})+i\boldsymbol{\Gamma}_{s}/2$ where the real part can be obtained from the Hilbert transform of the imaginary part given in Eq. (1.6.45).

If the electronic distribution can be described by an equilibrium Fermi function $\boldsymbol{G}^{n}(E')\equiv F(E')\boldsymbol{A}(E')$ and $\boldsymbol{G}^{p}(E')\equiv(1-f(E'))\boldsymbol{A}(E')$. Then $\boldsymbol{\Gamma}_{s}(E)=\int d(\hbar\omega)\int d\,E'\,D_{0}(\hbar\omega)\cdot\boldsymbol{A}(E')\times(N_{\omega}+1)-f(E'))\delta(E-E'-\hbar\omega)+(N_{\omega}+f(E'))\delta(E-E'+\hbar\omega)$, which is the expression commonly found in the literature on electron-phonon scattering in metals where it is referred to as Migdal's "theorem". Close to equilibrium, a separate equation for the inscattering function $\boldsymbol{\Sigma}_s^{in}(E)$ is not needed, since it is simply equal to $f(E)\boldsymbol{\Gamma}_s(E)$, just like an ordinary contact.

Note that the expressions simplify considerably for low-energy scattering processes($\hbar\omega \rightarrow 0$) for which we can set $E+\hbar\omega \approx E \approx E-\hbar\omega$:

$$\boldsymbol{\Sigma}_s^{in}(E)=\int_0^{\infty} d(\hbar\omega)[D^{em}(\hbar\omega)+D^{ab}(\hbar\omega)]\boldsymbol{G}^n(E) \qquad (1.6.46)$$

$$\boldsymbol{\Gamma}_s(E)=\int_0^{\infty} d(\hbar\omega)[D^{em}(\hbar\omega)+D^{ab}(\hbar\omega)]\boldsymbol{A}(E) \qquad (1.6.47)$$

Indeed, in this case the real part of the self-energy $\boldsymbol{\Sigma}_S$ does not require a separate Hilbert transform. We can simply write

$$\boldsymbol{\Sigma}_{in}(E)=\int_0^{\infty} d(\hbar\omega)(D^{em}(\hbar\omega)+D^{ab}(\hbar\omega))\boldsymbol{G}(E) \qquad (1.6.48)$$

since Re($\boldsymbol{\Sigma}_s$) is related to $\boldsymbol{\Gamma}_S$ in exactly the same way as Re($\boldsymbol{G}$) is related to $\boldsymbol{A}$, namely through a Hilbert transform.

The complete equations of dissipative quantum transport including the inflow/outflow terms discussed in this section are summarized along with illustrative examples.

Chapter 2 Green's Functions

2.1 Perturbation theory in Green's functions

Among the problems concerning quantum theory, a great amount is difficult to figure out by directly solving the eigenenergies and eigenfunctions of the eigenequations. It is often the case, however, that the Hamiltonian can be divided into two parts: one is strictly solvable and the other can be considered as a perturbation. In such a case, perturbation theory is helpful to solve the problem. For one-body Green's functions, the perturbation method is easy to handle. While for many-body systems, the numerical computation methods such as self-consistent field method can be employed.

By use of the perturbation theory, the poles, or the eigenenergies, of the Green's functions can be calculated conveniently. The imaginary part of a Green's function gives the DOS of the system. The perturbation theory of the Green's function has standard formalism. In the third part of this book, the diagram technique of the many-body Green's functions is just the perturbation theory of the many-body system. The uniform technique enables the Green's function method being applied extensively. For one-body Green's function, the perturbation theory has an additional merit that the formalism is simple and convenient in employings.

The Hamiltonian $\boldsymbol{H}$ is divided into unperturbed part $\boldsymbol{H}_0$ and perturbation part $\boldsymbol{H}_1$:

$$\boldsymbol{H} = \boldsymbol{H}_0 + \boldsymbol{H}_1 \tag{2.1.1}$$

One can easily work out the eigenspectrum of $\boldsymbol{H}_0$, so as to obtain the corresponding Green's function $\boldsymbol{G}_0$. The problem is to find out the Green's function $\boldsymbol{G}$ corresponding to $\boldsymbol{H}$. The procedure used usually is as follows: ①Determine first $\boldsymbol{G}_0$ associated with the unperturbed part $\boldsymbol{H}_0$; ②Express $\boldsymbol{G}$ associated with $\boldsymbol{H}$ in terms of $\boldsymbol{G}_0$ and $\boldsymbol{H}_1$; ③Obtain the information about the eigenspectrum and DOS of $\boldsymbol{H}$ from $\boldsymbol{G}$.

$$\boldsymbol{G}_0(z) = \frac{1}{z - \boldsymbol{H}_0} \tag{2.1.2}$$

and

$$\boldsymbol{G}(z) = \frac{1}{z - \boldsymbol{H}} \tag{2.1.3}$$

Substituting Eq. (2.1.2) into Eq. (2.1.3), we obtain

$$G(z) = (z - H_0 - H_1)^{-1} = \{(z - H_0)[1 - (z - H_0)^{-1} H_1]\}^{-1} =$$

$$\frac{1}{1 - (z - H_0)^{-1} H_1}(z - H_0)^{-1} = [1 - G_0(z) H_1]^{-1} G_0(z) \tag{2.1.4}$$

Here attention should be paid about the rule in calculating the inverse of the products of operators: $(AB)^{-1} = B^{-1} A^{-1}$. Multiplying by $1 - G_0(z) H_1$ on both sides of Eq. (2.1.4), we have

$$G = G_0 + G_0 H_1 G_0 \tag{2.1 5}$$

If the second equal mark of Eq. (2.1.4) is put into another form: $\{[1 - H_1 (z - H_0)^{-1}](z - H_0)\}^{-1}$, then the result will be

$$G = G_0 + G H_1 G_0 \tag{2.1.6}$$

There is no difference between Eq. (2.1.5) and Eq. (2.1.6). Iterating Eq. (2.1.5) or Eq. (2.1.6) repeatedly, we get

$$G = G_0 \sum_{n=0}^{\infty} (H_1 G_0)^n = \sum_{n=0}^{\infty} (G_0 H_1)^n G_0 \tag{2.1.7}$$

Another way to achieve Eq. (2.1.4) to Eq. (2.1.6) is to expand $[1 - G_0(z) H_1]^{-1}$ in Eq. (2.1.4) in power series just as one does for $(1 - x)^{-1}$ when $x < 1$.

Here we should point out that although the above formalism is named as perturbation theory, there is discrepancy between Eq. (2.1.5) and Eq. (2.1.7). In Eq. (2.1.5), the "perturbation Hamiltonian" H_1 is in fact not necessarily small, since equation is derived without any approximation. Therefore, in applying Eq. (2.1.5), the so-called "perturbation" H_1, may not be small, where H_1 can even be infinity. While if in some cases one has to resort to Eq. (2.1.7), the expansion series, to compute the Green's function, then the perturbation H_1 should be comparatively small in order to guarantee the convergence of the series.

In the r-representation, Eq. (2.1.5) becomes

$$G(r, r'; z) = G_0(r, r'; z) + \int \mathrm{d} r_1 \mathrm{d} r_2 \, G_0(r, r_1; z) H_1(r_1, r_2) G(r_2, r'; z) \tag{2.1.8}$$

In some cases

$$H_1(r_1, r_2) = V(r_1) \delta(r_1 - r_2) \tag{2.1.9}$$

Then Eq. (2.1.8) is simplified into

$$G(r, r'; z) = G_0(r, r'; z) + \int \mathrm{d} r_1 \, G_0(r, r_1; z) V(r_1) G(r_1, r'; z) \tag{2.1.10}$$

In the k-representation, Eq. (2.1.5) is written as

$$G(k, k'; z) = G_0(k, k'; z) + \sum_{k_1, k_2} G_0(k, k_1; z) H(k_1, k_2) G(k_2, k'; z) \tag{2.1.11}$$

where $\sum_k |k\rangle\langle k| = 1$ is inserted and $\langle k | G_0(z) | k' \rangle = G_0(k, k'; z)$. Considering that the wave function of a free-particle in the $\boldsymbol{r}$-representation is just a plane wave, then Eq. (2.1.10) and Eq. (2.1.11) are Fourier transformations of each other.

Now we introduce a $\boldsymbol{T}$ matrix, or scattering matrix, $\boldsymbol{T}(z)$ which is of central importance in scattering theory:

$$\boldsymbol{T}(z) = \boldsymbol{H}_1 \boldsymbol{G}(z)(z - \boldsymbol{H}_0) \tag{2.1.12}$$

This equation stands when $z \neq E_n$ where E_n are the eigenvalues of $\boldsymbol{H}$, for there are poles of $\boldsymbol{G}$ at the eigenvalues. If $z = E$ coincides with the continuous spectrum of $\boldsymbol{H}$, the side limits can be defined as

$$\boldsymbol{T}^{\pm}(E) = \boldsymbol{H}_1 \boldsymbol{G}^{\pm}(E)(E - \boldsymbol{H}_0) \tag{2.1.13}$$

For $z = E_n$, the discrete eigenvalues of $\boldsymbol{H}$, the $\boldsymbol{T}$ matrix can be defined only in such case that E_n, happens to be also the eigenvalue of $\boldsymbol{H}_0$. This is because in the denominator of Eq. (2.1.12) $z - \boldsymbol{H} \to E_n - \boldsymbol{H}_0$ which can be cancelled with the numerator $E_n - \boldsymbol{H}_0$, and $\boldsymbol{T}(z)$ is analytic around E_n.

The analytical structure of $\boldsymbol{T}(z)$ is quite similar to that of $\boldsymbol{G}(z)$: it is analytic in the complex z plane except the real axis; it has singularities on the real axis. The positions of the poles of $\boldsymbol{T}(z)$ give new discrete eigenvalues of $\boldsymbol{H}$ other than those of $\boldsymbol{H}_0$, and vice versa. The continuous spectrum of $\boldsymbol{H}$ produces a branch cut of $\boldsymbol{T}(z)$. In this sense, $\boldsymbol{T}$ can be regarded as equivalent to $\boldsymbol{G}$. $\boldsymbol{T}(z)$ cannot be analytically continued across the branch cut, for this continuation will produce a new singularity on the complex z plane.

Substituting Eq. (2.1.7) into Eq. (2.1.12), we have

$$\boldsymbol{T}(z) = \sum_{n=0}^{\infty} (\boldsymbol{H}_1 \boldsymbol{G}_0)^n \boldsymbol{H}_1 = \boldsymbol{H}_1 \sum_{n=0}^{\infty} (\boldsymbol{G}_0 \boldsymbol{H}_1)^n \tag{2.1.14}$$

This equation leads to following results:

$$\boldsymbol{T}(z) = \boldsymbol{H}_1 + \boldsymbol{T}\boldsymbol{G}_0 \boldsymbol{H}_1 = \boldsymbol{H}_1 + \boldsymbol{H}_1 \boldsymbol{G}_0 \boldsymbol{T} = \boldsymbol{H}_1 + \boldsymbol{H}_1 \boldsymbol{G} \boldsymbol{H}_1 \tag{2.1.15}$$

And consequently,

$$\boldsymbol{H}_1 \boldsymbol{G} = \boldsymbol{T}\boldsymbol{G}_0 \text{ and } \boldsymbol{G}\boldsymbol{H}_1 = \boldsymbol{G}_0 \boldsymbol{T} \tag{2.1.16}$$

Using Eq. (2.1.7) and Eq. (2.1.14), we obtain

$$\boldsymbol{G}(z) = \boldsymbol{G}_0(z) + \boldsymbol{G}_0(z)\boldsymbol{T}\boldsymbol{G}_0(z) \tag{2.1.17}$$

This equation shows that $\boldsymbol{G}$ will be determined as soon as $\boldsymbol{T}(z)$ and $\boldsymbol{G}_0$ are available.

The formulas above can be put down in the $\boldsymbol{r}$-or $\boldsymbol{k}$-representation. For instance, in $\boldsymbol{k}$-representation

$$\boldsymbol{T}(\boldsymbol{k},\boldsymbol{k}';z) = \boldsymbol{H}_1(\boldsymbol{k},\boldsymbol{k}') + \sum_{\boldsymbol{k}_1,\boldsymbol{k}_2} \boldsymbol{H}_1(\boldsymbol{k},\boldsymbol{k}_1)\boldsymbol{G}_0(\boldsymbol{k}_1,\boldsymbol{k}_2;z)\boldsymbol{T}(\boldsymbol{k}_2,\boldsymbol{k}';z) \tag{2.1.18}$$

where

$$\boldsymbol{H}_1(\boldsymbol{k},\boldsymbol{k}') = \langle \boldsymbol{k} \mid \boldsymbol{H}_1 \mid \boldsymbol{k}' \rangle = \frac{1}{\boldsymbol{V}}\int \mathrm{d}r\mathrm{d}\,\boldsymbol{r}'\, \mathrm{e}^{-\mathrm{i}kr+\mathrm{i}k'r'}\, \boldsymbol{H}_1(\boldsymbol{r},\boldsymbol{r}') \tag{2.1.19}$$

$$G_0(k_1, k_2; z) = \langle k_1 \mid G_0(z) \mid k_2 \rangle = \frac{1}{V}\int \mathrm{d}\, r_1 \mathrm{d}\, r_2\, \mathrm{e}^{-\mathrm{i}k_1 r_1 + \mathrm{i}k_2 r_2}\, G_0(r_1, r_2; z) \quad (2.1.20)$$

and

$$T(k, k'; z) = \langle k_1 \mid G_0(z) \mid k_2 \rangle = \frac{1}{V}\int \mathrm{d}r \mathrm{d}\, r'\, \mathrm{e}^{-\mathrm{i}kr + \mathrm{i}k'r'}\, T(r, r'; z) \quad (2.1.21)$$

In the simplified case Eq. (2.1.9), $H_1(k, k')$ becomes

$$H_1(k, k') = V(k - k')/V \quad (2.1.22)$$

where $V(q)$ is the Fourier component of $V(r)$:

$$V(q) = \int \mathrm{d}r V(r)\, \mathrm{e}^{-\mathrm{i}qr} \quad (2.1.23)$$

It was mentioned at the end of chapter 1 that in a uniform space $G_0(r, r'; z)$ is a function of $r_1 - r_2$. Subsequently Eq. (2.1.20) is simplified into

$$G_0(k_1, k_2; z) = \delta_{k_1 k_2}\, G_0(k_1; z) \quad (2.1.24)$$

where $G_0(k; z)$ is the Fourier component of $G_0(k_1, k_2; z)$ with respect to variable $p = r_1 - r_2$. Also, Eq. (2.1.18) is simplified as

$$T'(k, k'; z) = V(k - k') + \int \frac{\mathrm{d}\, k_1}{(2\pi)^d} V(k - k')\, G_0(k_1; z)\, T'(k_1, k'; z) \quad (2.1.25)$$

As long as the Green's function G is known, its simple poles determine the discrete eigenenergies of H, and its imaginary part permits us to obtain the DOS. The eigenstates associated with the continuous spectrum of H can be solved. To do so let us rebuild Schröinger equation as

$$(E - H_0) \mid \psi\rangle = H_1 \mid \psi\rangle \quad (2.1.26)$$

Here E belongs to the continuous spectrum of H. The solution of the inhomogeneous equation helps us to put the solution of Eq. (2.1.26):

$$\mid \psi^{\pm}\rangle = \mid \varphi\rangle + G_0^{\pm}(E)\, H_1 \mid \psi^{\pm}\rangle \quad (2.1.27)$$

where $\mid \varphi\rangle$ is an eigenfunction of H_0: $(E - H_0) \mid \psi\rangle = 0$. Here we consider that the continuous spectrum H is also possibly possessed by H_0, but may be associated with different eigenfunctions. This is often the case in one-body Green's functions. The superscripts $\pm$ of ψ in Eq. (2.1.27) distinguish the solutions associated with G_0^+ and G_0^-. Eq. (2.1.27) is an integral equation for the unknown functions $\mid \psi^{\pm}\rangle$, and can be rewritten in the r-representation as

$$\psi^{\pm}(r) = \varphi(r) + \int \mathrm{d}\, r_1 \mathrm{d}\, r_2\, G_0^{\pm}(r, r_1; E)\, H_1(r_1, r_2)\, \psi^{\pm}(r_2) \quad (2.1.28)$$

If Eq. (2.1.9) is valid,

$$\psi^{\pm}(r) = \varphi(r) + \int \mathrm{d}\, r_1\, G_0^{\pm}(r, r_1; E) V(r_1)\, \psi^{\pm}(r_1) \quad (2.1.29)$$

In the case that E does not belong to the eigenenergies of H_0, one should take $\varphi(r) = 0$.

$$\psi(\boldsymbol{r}) = \int \mathrm{d}\boldsymbol{r}_1 \, \boldsymbol{G}_0(\boldsymbol{r},\boldsymbol{r}_1;E)\boldsymbol{V}(\boldsymbol{r}_1)\psi(\boldsymbol{r}_1) \tag{2.1.30}$$

Usually, $\boldsymbol{H}_0$ and $\boldsymbol{H}_1$ are chosen in such a way that the continuous spectra of $\boldsymbol{H}$ and $\boldsymbol{H}_0$ coincide. Then Eq. (2.1.28) and Eq. (2.1.29) are appropriate for finding the eigenfunctions of $\boldsymbol{H}$ associated with the continuous spectrum. Eq. (2.1.30) facilitates to solve the eigenfunctions of $\boldsymbol{H}$ associated with discrete eigenenergies which can be obtained by finding the poles of $\boldsymbol{G}$ or $\boldsymbol{T}$ matrix.

Iterating Eq. (2.1.27) repeatedly and using Eq. (2.1.14), one obtains

$$|\psi^{\pm}\rangle = |\varphi\rangle + \boldsymbol{G}_0^{\pm}\sum_{n=0}^{\infty}\boldsymbol{H}_1(\boldsymbol{G}_0^{\pm}\boldsymbol{H}_1)^n|\varphi\rangle = |\varphi\rangle + \boldsymbol{G}_0^{\pm}\boldsymbol{T}^{\pm}|\varphi\rangle \tag{2.1.31}$$

Substituting Eq. (2.1.16) into Eq. (2.1.31), we have

$$|\psi^{\pm}\rangle = |\varphi\rangle + \boldsymbol{G}^{\pm}\boldsymbol{H}_1|\varphi\rangle \tag{2.1.32}$$

We can conclude that the basic task is to find the $\boldsymbol{T}$ matrix. As soon as $\boldsymbol{T}$ matrix is available, important quantities such as the Green's function $\boldsymbol{G}$ and the eigenenergies and corresponding eigenfunctions of $\boldsymbol{H}$ can be computed.

Comparison of Eq. (2.1.31) and Eq. (2.1.27) results in

$$\boldsymbol{T}^{\pm}|\varphi\rangle = \boldsymbol{H}_1|\psi^{\pm}\rangle \tag{2.1.33}$$

The time-dependent Schröinger equation is written in the form of

$$\left(i\hbar\frac{\partial}{\partial t} - \boldsymbol{H}_0\right)|\psi(t)\rangle = \boldsymbol{H}_1(t)|\psi(t)\rangle \tag{2.1.34}$$

Here the perturbation Hamiltonian $\boldsymbol{H}_1$ can be time-dependent. Following the solution, we achieve the solution of Eq. (2.2.34) as

$$|\psi^{\pm}(t)\rangle = |\varphi(t)\rangle + \int_{-\infty}^{\infty}\mathrm{d}t'\,\boldsymbol{G}_0^{\pm}(t-t')\boldsymbol{H}_1(t')|\psi^{\pm}(t')\rangle \tag{2.1.35}$$

According to the discussion of causality reflected by Eq. (2.1.21), we know that only the solution $|\psi^{+}(t)\rangle$ is of physical significance. That is to say, we discuss the system's behavior after $\boldsymbol{H}_1$ begins to work. Eq. (2.1.35) can be iterated to give

$$|\psi^{+}(t)\rangle = |\varphi(t)\rangle + \int_{-\infty}^{t}\mathrm{d}t_1\,\boldsymbol{G}_0^{+}(t-t_1)\boldsymbol{H}_1(t_1)|\varphi(t_1)\rangle + \int_{-\infty}^{t}\mathrm{d}t_1\mathrm{d}t_2 \times$$
$$\boldsymbol{G}_0^{+}(t-t_1)\boldsymbol{H}_1(t_1)\boldsymbol{G}_0^{+}(t_1-t_2)\boldsymbol{H}_1(t_2)|\varphi(t_2)\rangle + \cdots \tag{2.1.36}$$

Let use assume that $\boldsymbol{H}_1(t) = 0$ for $t < t_0$ and that $|\varphi(t_0)\rangle$ is an eigenfunction of $\boldsymbol{H}_0$, say, φ_n. By use of the definition of the time-evolution operator Eq. (2.1.13) to Eq. (2.1.15), the state of $\boldsymbol{H}_0$ at time t should be

$$|\varphi(t)\rangle = \mathrm{i}\hbar\widetilde{\boldsymbol{G}}_0(t-t_0)|\varphi_n\rangle = \mathrm{e}^{-\mathrm{i}\boldsymbol{H}_0(t-t_0)/\hbar}|\varphi_n\rangle = \mathrm{e}^{-\mathrm{i}E_n(t-t_0)/\hbar}|\varphi_n\rangle \tag{2.1.37}$$

When this condition is employed in every term of Eq. (2.1.36), we have

$$| \psi^+(t)\rangle = \boldsymbol{A}(t,t_0) \mid \varphi_n\rangle \tag{2.1.38}$$

where

$$\boldsymbol{A}(t,t_0) = \mathrm{i}\hbar\widetilde{\boldsymbol{G}}_0(t-t_0) + \mathrm{i}\hbar\int_{t_0}^{t} \mathrm{d}t_1\, \widetilde{\boldsymbol{G}}_0(t-t_1)\boldsymbol{H}_1(t_1)\widetilde{\boldsymbol{G}}_0(t_1-t_0) + \mathrm{i}\hbar\int_{t_0}^{t} \mathrm{d}t_1\mathrm{d}t_2 \times$$
$$\widetilde{\boldsymbol{G}}_0(t-t_1)\boldsymbol{H}_1(t_1)\widetilde{\boldsymbol{G}}_0(t_1-t_2)\boldsymbol{H}_1(t_2)\widetilde{\boldsymbol{G}}_0(t_2-t_0) + \cdots \tag{2.1.39}$$

The identity $\boldsymbol{G}^+(\tau) = \theta(\tau)\widetilde{\boldsymbol{G}}(\tau)$ is used.

At time t the probability amplitude for the system to stay at the state φ_m is

$$\langle\varphi_m \mid \psi^+(t)\rangle = \langle\varphi_m \mid \boldsymbol{A}(t,t_0) \mid \varphi_n\rangle = \mathrm{e}^{(-\mathrm{i}E_m t + \mathrm{i}E_n t_0)/\hbar}[\langle\varphi_m \mid \varphi_n\rangle +$$
$$\frac{-\mathrm{i}}{\hbar}\int_{t_0}^{t} \mathrm{d}t_1 \langle\varphi_m \mid \mathrm{e}^{\frac{\mathrm{i}H_0 t_1}{\hbar}} \boldsymbol{H}_1(t_1)\, \mathrm{e}^{\frac{-\mathrm{i}H_0 t_1}{\hbar}} \mid \varphi_n\rangle + \frac{-\mathrm{i}}{\hbar}\int_{t_0}^{t} \mathrm{d}t_1\mathrm{d}t_2 \langle\varphi_m \mid \times$$
$$\mathrm{e}^{\frac{\mathrm{i}H_0 t_1}{\hbar}} \boldsymbol{H}_1(t_1)\, \boldsymbol{G}_0^+(t_1-t_2)\, \boldsymbol{H}_1(t_2)\, \mathrm{e}^{\frac{-\mathrm{i}H_0 t_2}{\hbar}} \mid \varphi_n\rangle + \cdots] \tag{2.1.40}$$

In order to get rid of the unimportant phase factor in Eq. (2.1.40), we define an operator $\boldsymbol{S}(t,t_0)$ as follows:

$$\boldsymbol{S}(t,t_0) = \mathrm{e}^{\mathrm{i}H_0 t/\hbar}\boldsymbol{A}(t,t_0)\,\mathrm{e}^{-\mathrm{i}H_0 t_0/\hbar} \tag{2.1.41}$$

The series in the square parentheses in Eq. (2.1.40) is just the matrix element $\langle\varphi_m \mid \boldsymbol{S}(t,t_0) \mid \varphi_n\rangle$, which can be written as

$$\varphi_m \mid \boldsymbol{S}(t,t_0) \mid \varphi_n\rangle = \delta_{mn} + \frac{-\mathrm{i}}{\hbar}\int_{t_0}^{t} \mathrm{d}t_1\, \mathrm{e}^{\mathrm{i}\omega_{mn}t_1} \langle\varphi_m \mid \boldsymbol{H}_1(t_1) \mid \varphi_n\rangle +$$
$$\frac{-\mathrm{i}}{\hbar}\int_{t_0}^{t} \mathrm{d}t_1\mathrm{d}t_2 \int \frac{\mathrm{d}\omega}{2\pi}\, \mathrm{e}^{\mathrm{i}(\omega_m-\omega)t_1}\, \mathrm{e}^{\mathrm{i}(\omega-\omega_n)} \times$$
$$\langle\varphi_m \mid \boldsymbol{H}_1(t_1)\, \boldsymbol{G}_0^+(\hbar\omega)\, \boldsymbol{H}_1(t_2) \mid \varphi_n\rangle + \cdots \tag{2.1.42}$$

where $\omega_n = E_n/\hbar$, $\omega_m = E_m/\hbar$, $\omega_{mn} = \omega_m - \omega_n$ and the time Fourier transformation of the Green's function is made. Now extending the upper and lower limit of all the integrals in Eq. (2.1.42) to $-\infty$ and $+\infty$, we define an $\boldsymbol{S}$ matrix:

$$\boldsymbol{S} = \lim_{\substack{t\to+\infty \\ t_0\to-\infty}} \boldsymbol{S}(t,t_0) \tag{2.1.43}$$

For the particular case where $\boldsymbol{H}_1$ is independent of time,

$$\langle\varphi_m \mid \boldsymbol{S} \mid \varphi_n\rangle = \delta_{mn} + \langle\varphi_m \mid \boldsymbol{H}_1 \mid \varphi_n\rangle \frac{-\mathrm{i}}{\hbar}\int_{-\infty}^{\infty} \mathrm{d}t_1\, \mathrm{e}^{\mathrm{i}\omega_{mn}t_1} +$$
$$\frac{-\mathrm{i}}{\hbar}\int \frac{\mathrm{d}\omega}{2\pi} \langle\varphi_m \mid \boldsymbol{H}_1(t_1)\, \boldsymbol{G}_0^+(\hbar\omega)\, \boldsymbol{H}_1 \mid \varphi_n\rangle \times$$
$$\int_{-\infty}^{\infty} \mathrm{d}t_1 \int_{-\infty}^{\infty} \mathrm{d}t_2\, \mathrm{e}^{\mathrm{i}(\omega_m-\omega)t_1}\, \mathrm{e}^{\mathrm{i}(\omega-\omega_n)t_2} + \cdots =$$
$$\delta_{mn} - 2\pi\mathrm{i}\delta(E_n - E_m)\langle\varphi_m \mid \boldsymbol{T}^+(E_n) \mid \varphi_n\rangle \tag{2.1.44}$$

Here we have employed the relation

$$\delta(E) = \frac{1}{2\pi\hbar}\int_{-\infty}^{\infty} \mathrm{e}^{\mathrm{i}Et/\hbar}\,\mathrm{d}t \tag{2.1.45}$$

The transition probability between the states $|\varphi_n\rangle$ and $|\varphi_m\rangle$ is $|\langle\varphi_m|S|\varphi_n\rangle|^2 = (2\pi)^2\delta(E_n - E_m)\delta(E_n - E_m)\,|\langle\varphi_m|T^+(E_n)|\varphi_n\rangle|^2$, where the first δ_{mn} is dropped for it is much less than the δ function term as $m = n$. One of the two δ functions is expanded in terms of Eq. (2.1.45). Since the other $\delta(E_n - E_m)$ guarantees that only when $E_n = E_m$ the final result is nonzero, we are able to let $\omega_{mn} = 0$ in the integrand, so as to obtain the transition probability:

$$|\langle\varphi_m|S|\varphi_n\rangle|^2 = \frac{2\pi}{\hbar}\delta(E_n - E_m)\,|\langle\varphi_m|T^+(E_n)|\varphi_n\rangle|^2\int_{-\infty}^{\infty}\mathrm{d}t \tag{2.1.46}$$

The transition probability per unit time is

$$W_{mn} = \frac{2\pi}{\hbar}\,|\langle\varphi_m|T^+(E_n)|\varphi_n\rangle|^2\delta(E_n - E_m) \tag{2.1.47}$$

Eq. (2.1.47) is termed as Fermi's "golden rule". It is of important physical significance: a system is at an eigenstate φ_n of H_0 at initial time; when a perturbation H_1 is applied, it may transit to another eigenstate, say, φ_m, then the transition probability from φ_n to φ_m is expressed by Eq. (2.1.47). Please note that actually the system is at the state $|\psi^+\rangle$ which is the superposition of all the eigenstates of H_0.

Both $|\varphi_n\rangle$ and $|\psi^+\rangle$ are assumed normalized, then

$$1 = \langle\psi^+(t)|\psi^+(t)\rangle = \langle\varphi_n|A^+(t,t_0)A(t,t_0)|\varphi_n\rangle \tag{2.1.48}$$

This equation shows that $A(t,t_0)$ is unitary. By Eq. (2.1.42), $S(t,t_0)$ is unitary too, and S matrix as well.

$$S^+S = SS^+ = 1 \tag{2.1.49}$$

We are going to recast S matrix into a more compact form. In order to do so we write the S matrix, by Eq. (2.1.42) and Eq. (2.1.43) as

$$S = 1 + \frac{-\mathrm{i}}{\hbar}\int_{-\infty}^{\infty}\mathrm{d}t_1\,H_1^{\mathrm{i}}(t_1) + \left(\frac{-\mathrm{i}}{\hbar}\right)^2\int_{-\infty}^{\infty}\mathrm{d}t_1\,H_1^{\mathrm{i}}(t_1)\int_{-\infty}^{t_1}\mathrm{d}t_2\,H_1^{\mathrm{i}}(t_2) + \cdots \tag{2.1.50}$$

where

$$H_1^{\mathrm{i}}(t) = \mathrm{e}^{\frac{\mathrm{i}H_0 t}{\hbar}}\,H_1(t)\,\mathrm{e}^{-\mathrm{i}H_0 t/\hbar} \tag{2.1.51}$$

is defined as Eq. (2.1.41) and Eq. (2.1.42) are employed:

$$G_0^+(t_1 - t_2) = \theta(t_1 - t_2)\tilde{G}_0(t_1 - t_2) = \theta(t_1 - t_2)\frac{-\mathrm{i}}{\hbar}\mathrm{e}^{-\mathrm{i}H_0(t_1 - t_2)/\hbar} \tag{2.1.52}$$

If $H_1^{\mathrm{i}}(t_i)$ is just a number, the upper integral limits in Eq. (2.1.50) can be extended to $+\infty$, and then a factor $1/(n-1)!$ is multiplied to the term with n integrals, remaining the results unchanged. For instances,

$$\left.\begin{aligned}
&\int_{-\infty}^{\infty}\mathrm{d}t_1\int_{-\infty}^{t_1}\mathrm{d}t_2 = \frac{1}{2!}\int_{-\infty}^{\infty}\mathrm{d}t_1\int_{-\infty}^{\infty}\mathrm{d}t_2\\
&\int_{-\infty}^{\infty}\mathrm{d}t_1\int_{-\infty}^{t_1}\mathrm{d}t_2\int_{-\infty}^{t_2}\mathrm{d}t_3 = \frac{1}{3!}\int_{-\infty}^{\infty}\mathrm{d}t_1\int_{-\infty}^{\infty}\mathrm{d}t_2\int_{-\infty}^{\infty}\mathrm{d}t_3\\
&\qquad\vdots
\end{aligned}\right\} \tag{2.1.53}$$

Unfortunately, $H_1^i(t_i)$ defined by Eq. (2.1.51) does not allow the commutation between $H_1^i(t_i)$ and $H_1^i(t_j)$ when $t_i \neq t_j$, although $H_1(t_i)$ does, because H_0 is not commutable with H_1. Thus the factors in the integrands of Eq. (2.1.50) are not allowed to change the product ordering. For example the ordering $t_1 > t_3$ in the fourth term of Eq. (2.1.51) should be retained. Nevertheless, we still intend to make use of Eq. (2.1.52) in order to put down Eq. (2.1.53) in a compact form. For this purpose we introduce a so-called chronological operator T_t, which arranges the product operators in such a way that the earlier time appears to the right. For instance,

$$T_t[H_1^i(t_1)H_1^i(t_2)] = H_1^i(t_1)H_1^i(t_2), t_1 > t_2 = H_1^i(t_2)H_1^i(t_1), t_2 > t_1 \quad (2.1.54)$$

The integration region associated with Eq. (2.1.54) is plotted in Fig. 2.1.1 The product $T_t[H_1^i(t_1)H_1^i(t_2)H_1^i(t_3)]$ can be expanded to six terms each having unique time ordering with earlier time appearing to the right. A product containing n $H_1^i(t_i)$ is expanded to $n!$ terms after T_t is acted, each having equal contribution to the integral. Because of the introduction of chronological operator, the integral limits are extended to infinity, so that Eq. (2.1.53) can be employed. Eq. (2.1.49) is rewritten as

$$S = \sum_{n=0}^{\infty}\left(\frac{-\mathrm{i}}{\hbar}\right)^n \frac{1}{n!}\int_{-\infty}^{\infty} \mathrm{d}t_1 \mathrm{d}t_2 \cdots \mathrm{d}t_n T_t[H_1^i(t_1)\cdots H_1^i(t_n)] =$$
$$T_t \exp\left[\frac{-\mathrm{i}}{\hbar}\int_{-\infty}^{\infty} \mathrm{d}t\, H_1^i(t)\right] \quad (2.1.55)$$

In this way S matrix is indeed written in a simple form. However, one should keep in mind that Eq. (2.1.55) is just formally written, and cannot consider S matrix to be an exponential function. If one computes the S matrix, he must expand the exponent of Eq. (2.1.55), and apply T_t, to each term, thus finally he still does the computation following Eq. (2.1.52). However, this by no means shows that the introduction of the chronological operator is meaningless. As we shall see in the third part of this book, use of T_t facilitates dealing with problems.

By Eq. (2.1.48) and Eq. (2.1.52) we have

$$\langle \varphi_n \mid T^+(E_n) \mid \varphi_l \rangle - \langle \varphi_n \mid T^-(E_n) \mid \varphi_l \rangle =$$
$$-2\pi i \sum_m \langle \varphi_n \mid T^-(E_n) \mid \varphi_m \rangle \langle \varphi_m \mid T^+(E_n) \mid \varphi_l \rangle \delta(E_m - E_n) \quad (2.1.56)$$

This equation can also be obtained in following way. From $T = H_1 + H_1 G H_1$ one get $T^+ - T^- = H_1(G^+ - G^-)H_1 = -2\pi\mathrm{i}\delta(E - H)H_1$, and makes matrix elements on both sides, inserts the summation of $|\psi^+\rangle\langle\psi^+|$, and then using Eq. (2.1.33) finally one gets Eq. (2.1.32). We will see in the next section that Eq. (2.1.48) is equivalent to the optical theorem in scattering theory. In other words, the optical theorem stems from the unitary of S matrix.

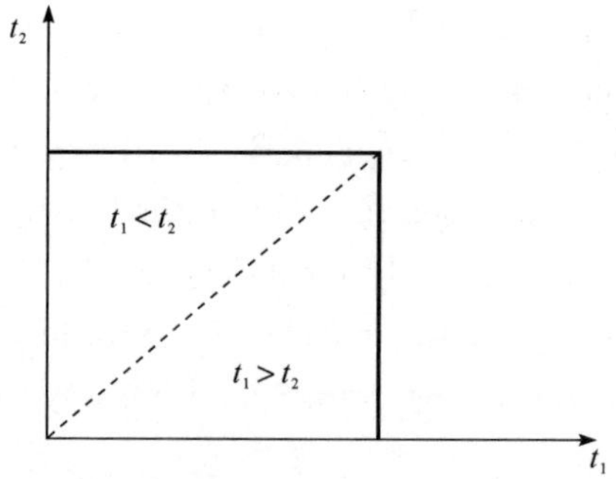

Fig. 2. 1. 1 The integration region associated with Eq. (2. 1. 54)

Here we take $\boldsymbol{H}_0 = \boldsymbol{P}^2/2m = \hbar^2\nabla^2/2\mathrm{m}$; the perturbation $\boldsymbol{H}_1(\boldsymbol{r}-\boldsymbol{r}')$ is of the form $\delta(\boldsymbol{r}-\boldsymbol{r}')\boldsymbol{V}(\boldsymbol{r})$, where $\boldsymbol{V}(\boldsymbol{r})$ is of finite extent (i. e, $\boldsymbol{V}(\boldsymbol{r})$ decays fast enough as $\boldsymbol{r}\to\infty$). Here $\boldsymbol{H}_0$ has a continuous spectrum that extends from zero to $+\infty$. The continuous spectrum of $\boldsymbol{H}$ coincides with that of $\boldsymbol{H}_0$. $\boldsymbol{H}$, however, may develop discrete levels of negative eigenenergies if $\boldsymbol{V}(\boldsymbol{r})$ is negative in some regions.

In this section we deal with the scattering problem for $E>0$: an incident particle with energy $E=\hbar^2 k^2/2m$, described by unperturbed wave function, comes under the influence of the perturbation $\boldsymbol{V}(\boldsymbol{r})$, usually called as scattering center, and then gets out of the region. Through this process, its wave function is changed, or modified. The question is to find how it is modified.

It is rather difficult to deal with the problem in the region where $\boldsymbol{V}(\boldsymbol{r})$ is of finite value, since one has to solve the Schrödinger equation, $\frac{\hbar^2}{2m}\nabla 2\psi(\boldsymbol{r})+\boldsymbol{V}(\boldsymbol{r})\psi(\boldsymbol{r})=E\psi(\boldsymbol{r})$. If $\boldsymbol{V}(\boldsymbol{r})$ is spherically symmetric, the solution will be Bessel functions. The incident plane wave has to be expanded by Bessel functions to determine the expansion coefficients. At the boundaries of $\boldsymbol{V}(\boldsymbol{r})$, the wave function and its derivative should match. This procedure is complicated and is often difficult to work out. In the scope of condensed matter physics, we actually do not need the information about such regions.

It is comparatively much simpler to determine the modified wave function of particle when it leaves the region that the influence of $\boldsymbol{V}(\boldsymbol{r})$ can be omitted. The information about the behavior of the particle when it is far away from the scattering center is needed in scattering experiments. Here we explore the asymptotic behavior of the wave function as $\boldsymbol{r}\to\infty$. Firstly, the wave function of the particle leaving the scattering center can be divided into two parts: incident and scattering parts. The former is the same as the plane wave, while the latter is the scattered part covering all directions of the space as $\boldsymbol{r}\to\infty$, being of the form of the spherical wave $\frac{1}{\boldsymbol{r}}\mathrm{e}^{i\boldsymbol{k}\cdot\boldsymbol{r}}$ with the amplitude to be determined. Secondly, $\boldsymbol{V}(\boldsymbol{r})$ is zero as $\boldsymbol{r}\to$

∞, and thus the energy of the particle is conserved. This results in that the wave vector $\boldsymbol{k}'$ of the scattered wave must be the same as $\boldsymbol{k}$, although heir directions can be different.

We only consider the 3-d case. Eq. (2.1.31) is written in the $\boldsymbol{r}$-presentation,

$$\psi^{\pm}(r)=\frac{1}{\sqrt{V}}e^{i\boldsymbol{k}\cdot\boldsymbol{r}}+\frac{1}{\sqrt{V}}\int d\boldsymbol{r}_1 d\boldsymbol{r}_2\, \boldsymbol{G}_0^{\pm}(\boldsymbol{r},\boldsymbol{r}')\,\boldsymbol{T}^{\pm}(\boldsymbol{r}_1,\boldsymbol{r}_2)\,e^{i\boldsymbol{k}\cdot\boldsymbol{r}_2} \tag{2.1.57}$$

The expressions of $\boldsymbol{G}_0^{\pm}$ are,

$$\sqrt{\boldsymbol{V}}\,\psi^{\pm}(\boldsymbol{r})=e^{i\boldsymbol{k}\cdot\boldsymbol{r}}-\frac{m}{2\pi\hbar^2}\int d\boldsymbol{r}_1 d\boldsymbol{r}_2\,\frac{e^{\pm ik|\boldsymbol{r}-\boldsymbol{r}_1|}}{|\boldsymbol{r}-\boldsymbol{r}_1|}\boldsymbol{T}^{\pm}(\boldsymbol{r}_1,\boldsymbol{r}_2)\,e^{i\boldsymbol{k}\cdot\boldsymbol{r}_2} \tag{2.1.58}$$

where $\boldsymbol{k}=\sqrt{2mE/\hbar^2}$. Because we merely consider the asymptotic behavior when $\boldsymbol{r}\to\infty$, $\boldsymbol{T}^{\pm}(\boldsymbol{r}_1,\boldsymbol{r}_2)\propto \boldsymbol{H}_1(\boldsymbol{r}_1,\boldsymbol{r}_2)=\delta(\boldsymbol{r}_1-\boldsymbol{r}_2)\boldsymbol{V}(\boldsymbol{r})$, so that the integral over $d\boldsymbol{r}_1$ and $d\boldsymbol{r}_2$ is nontrivial within the region where $\boldsymbol{V}(\boldsymbol{r})$ is finite. Therefore, as $\boldsymbol{r}\gg\boldsymbol{r}_1$ the $\boldsymbol{r}_1$ in the denominator of Eq. (2.1.58) can be neglected. In the exponential the simplification is made: $|\boldsymbol{r}-\boldsymbol{r}_1|\approx \boldsymbol{r}-\boldsymbol{r}_1\cos\theta$, and thus $\boldsymbol{k}|\boldsymbol{r}-\boldsymbol{r}_1|\approx \boldsymbol{kr}-\boldsymbol{kr}_1\cos\theta=\boldsymbol{kr}-\boldsymbol{k}_f\cdot\boldsymbol{r}_1$, where θ is the angle between $\boldsymbol{r}_1$ and $\boldsymbol{r}$ and $\boldsymbol{k}_f$ is the scattering wave vector along $\boldsymbol{r}$ direction and with the value $\boldsymbol{k}$. Thus for sufficiently large $\boldsymbol{r}$, Eq. (2.1.58) is recast as

$$\sqrt{\boldsymbol{V}}\,\psi^{\pm}(\boldsymbol{r})\xrightarrow{\boldsymbol{r}\to\infty} e^{i\boldsymbol{k}\cdot\boldsymbol{r}}-\frac{m}{2\pi\hbar^2}\frac{e^{\pm ikr}}{\boldsymbol{r}}\int d\boldsymbol{r}_1 d\boldsymbol{r}_2\, e^{\mp i\boldsymbol{k}_f\cdot\boldsymbol{r}_1}\langle \boldsymbol{r}_1\mid \boldsymbol{T}^{\pm}(E)\mid \boldsymbol{r}_2\rangle e^{i\boldsymbol{k}\cdot\boldsymbol{r}_2}=$$

$$e^{i\boldsymbol{k}\cdot\boldsymbol{r}}-\frac{m}{2\pi\hbar^2}\frac{e^{\pm ikr}}{\boldsymbol{r}}\langle\pm\boldsymbol{k}_f\mid \boldsymbol{T}'^{\pm}(E)\mid \boldsymbol{k}\rangle \tag{2.1.59}$$

To obtain the equal mark in Eq. (2.1.59), we have employed the relation $\langle \boldsymbol{r}\mid\boldsymbol{k}\rangle = e^{i\boldsymbol{k}\cdot\boldsymbol{r}}/\sqrt{\boldsymbol{V}}$, $\boldsymbol{T}'=\boldsymbol{VT}$ and completeness.

The solution ψ^{-} in Eq. (2.1.59) should be dropped since it means that the scattered spherical wave propagates from outside toward the centre, an unphysical picture. It is well known from the scattering theory that the physically significant wave function should be of the following form as $\boldsymbol{r}\to\infty$.

$$\psi(\boldsymbol{r})\xrightarrow{\boldsymbol{r}\to\infty}\text{const.}\times\left[e^{i\boldsymbol{k}\cdot\boldsymbol{r}}+f(\boldsymbol{k}_f,\boldsymbol{k})\frac{e^{ikr}}{\boldsymbol{r}}\right] \tag{2.1.60}$$

Here $f(\boldsymbol{k}_f,\boldsymbol{k})$ is the amplitude of the spherical wave, which is a function of the wave vector $\boldsymbol{k}$ of incident wave (initial state) and $\boldsymbol{k}_f$ of scattered wave (final state). Comparing Eq. (2.1.59) with Eq. (2.1.60), we obtain for the scattering amplitude

$$f(\boldsymbol{k}_f,\boldsymbol{k})=-\frac{m}{2\pi\hbar^2}\langle \boldsymbol{k}_f\mid \boldsymbol{T}'^{+}(E)\mid \boldsymbol{k}\rangle \tag{2.1.61}$$

where $E=\hbar^2\boldsymbol{k}_f^2/2m=\hbar^2\boldsymbol{k}^2/2m$, Eq. (2.1.61) shows that the $\boldsymbol{T}$ matrix element between initial and final states in $\boldsymbol{k}$-representation is proportional to the scattering amplitude, which is directly related to the differential cross section $d\sigma/d\boldsymbol{V}$:

$$\frac{d\sigma}{d\mathbf{V}} = |f|^2 = \frac{m^2}{4\pi^2 \hbar^4} |\langle \boldsymbol{k}_f | \boldsymbol{T}'^{+}(E) | \boldsymbol{k} \rangle|^2 \tag{2.1.62}$$

Substituting Eq. (2.1.61) into Eq. (2.1.25), one obtains

$$f(\boldsymbol{k}_f, \boldsymbol{k}) = -\frac{m}{2\pi\hbar^2} \langle \boldsymbol{k}_f | \boldsymbol{T}'^{+}(E) | \boldsymbol{k} \rangle \mathbf{V}(\boldsymbol{k}_f - \boldsymbol{k}) + \int \frac{d\boldsymbol{k}_1}{(2\pi)^3} \frac{\mathbf{V}(\boldsymbol{k}_f - \boldsymbol{k})}{E - \hbar^2 \boldsymbol{k}_1^2/2m + i0^+} f(\boldsymbol{k}_1, \boldsymbol{k}) \tag{2.1.63}$$

This is the integral equation with respect to the scattering amplitude. To the first ordering the scattering potential,

$$f(\boldsymbol{k}_f, \boldsymbol{k}) \approx -\frac{m}{2\pi\hbar^2} \mathbf{V}(\boldsymbol{k}_f - \boldsymbol{k}) \tag{2.1.64}$$

where $\mathbf{V}(\boldsymbol{q})$ is the Fourier transformation of $\mathbf{V}(\boldsymbol{r})$, see Eq. (2.1.22). Eq. (2.1.64) is the Born approximation for scattering amplitude.

Eq. (2.1.62) can also be derived in an alternative way. The differential cross section is defined as the probability per unit time for the transition $\boldsymbol{k} \rightarrow \boldsymbol{k}_f$, $W_{\boldsymbol{k}_f \boldsymbol{k}}$ times the number of final states divided by the solid angle 4π, and by the flux $j = v/\mathbf{V}$, of the incoming particle:

$$\frac{d\sigma}{d\mathbf{V}} = \frac{\mathbf{V}}{4\pi v} \int dE_f N(E_f) W_{\boldsymbol{k}_f \boldsymbol{k}} \tag{2.1.65}$$

Substituting $W_{\boldsymbol{k}_f \boldsymbol{k}}$ and $N(E_f)$

The total cross section

$$\sigma = \int d\mathbf{V} \frac{d\sigma}{d\mathbf{V}} \tag{2.1.66}$$

can be written in view as

$$\begin{aligned}
\sigma &= \frac{\mathbf{V}}{v} \sum_{\boldsymbol{k}_f} W_{\boldsymbol{k}_f \boldsymbol{k}} = \frac{\mathbf{V}}{v} \frac{2\pi}{\hbar} \sum_{\boldsymbol{k}_f} |\langle \boldsymbol{k}_f | T^{+}(E) | \boldsymbol{k} \rangle|^2 \delta(E_f - E) = \\
&\frac{2\pi}{\hbar v} \mathbf{V} \sum_{\boldsymbol{k}_f} \langle \boldsymbol{k} | \boldsymbol{T}^{-}(E) | \boldsymbol{k}_f \rangle \langle \boldsymbol{k}_f | \boldsymbol{T}^{+}(E) | \boldsymbol{k} \rangle \delta(E_f - E) = \\
&\frac{2\pi}{\hbar v} \mathbf{V} \frac{i}{2\pi} [\langle \boldsymbol{k} | \boldsymbol{T}^{+}(E) | \boldsymbol{k}_f \rangle - \langle \boldsymbol{k}_f | \boldsymbol{T}^{-}(E) | \boldsymbol{k} \rangle] = \frac{2\mathbf{V}}{\hbar v} \mathrm{Im} \langle \boldsymbol{k} | \boldsymbol{T}^{+}(E) | \boldsymbol{k} \rangle
\end{aligned} \tag{2.1.67}$$

we can rewrite Eq. (2.1.67) as

$$\sigma = \frac{2\pi}{\boldsymbol{k}} \mathrm{Im}[f(\boldsymbol{k}, \boldsymbol{k})] \tag{2.1.68}$$

This equation is so-called optical theorem, which connects the total cross section with the forward scattering amplitude.

The scattering amplitude f for positive energies is directly related to the differential cross section which is an observable quantity of great importance. The behavior of f for negative energies is also of physical significance because the poles of $f(E)$, which coincide with the poles of $T(E)$, give the discrete energies of the system. In other words, if the scattering problems has been solve and f vs. E has been obtained, we only need to find the position of the poles f to find the discrete levels of the system. Of course these poles are on the negative E-semiaxis. We should also mention that f vs. E, or T^{+} vs. E, may exhibit sharp peaks at certain positive energies. The states associated with such peaks in f are called resonance.

An elementary example exhibiting the negative discrete energies is provided by the case where $V(r)$ is an attractive Coulomb potential

$$V(r)=-\frac{e^2}{4\pi\varepsilon_0 r} \tag{2.1.69}$$

The scattering amplitude is :

$$f=\frac{-t\Gamma(1-it)}{\Gamma(1+it)}\frac{1}{2k\sin^2(\theta/2)}e^{2it\ln\sin(\theta/2)} \tag{2.1.70}$$

where

$$k=\sqrt{2mE/\hbar^2},\mathrm{Im}k\geqslant 0 \tag{2.1.71}$$

$$t=\frac{m e^2}{4\pi\varepsilon_0\hbar^2 k} \tag{2.1.72}$$

and θ is the angle between k and k_f. The poles of f occur when the argument of $\Gamma(1-it)$ is a nonnegative integer, i. e, $-(1-it)=p=0,1,2,\cdots$, and thus

$$it=1+\rho=n, n=1,2,3,\cdots \tag{2.1.73}$$

We obtain the discrete eigenenergies of the attractive Coulomb potential,

$$E_n=-\frac{m e^4}{2(4\pi\varepsilon_0)^2\hbar^2}\frac{1}{n^2}, n=1,2,3,\cdots \tag{2.1.74}$$

which are the energy levels of a hydrogen atom, as it should be. The problem now is to determine the wave function at finite distance r, not the asymptotic behavior as $r\to\infty$. If the potential were repulsive, the scattering amplitude would be given with i replaced by $-i$. As a result, the argument of Γ function in the numerator of Eq. (2.1.73) could not be negative, and f has no poles, which means that there is no discrete level. This is expected result since the potential is repulsive.

We assume that

$$V(r)=-V_0, r\in\Omega_0 \tag{2.1.75a}$$

$$V(r)=0, r\notin\Omega_0 \tag{2.1.75b}$$

Here Ω_0 is a finite region in real space and V_0 is positive and very small: $V_0\to 0^{+}$. We are interested in finding whether or not a discrete level E_0 appears and how it varies with V_0 when a particle moves around the region Ω_0. To answer the question, it is enough to find the position of the pole of $G(E)$. The Green's function G_0 of the unperturbed Hamiltonian H_0 has been given in chapter 3. The perturbed Hamiltonian is Eq. (2.1.75). We apply Eq. (2.1.7) to put down the total Green's function in $\boldsymbol{r}$-representation:

$$\boldsymbol{G}(\boldsymbol{r},\boldsymbol{r}';z)=\boldsymbol{G}_0(\boldsymbol{r},\boldsymbol{r}';z)-\boldsymbol{V}_0\int\Omega_0\mathrm{d}\boldsymbol{r}_1\,\boldsymbol{G}_0(\boldsymbol{r}_1,\boldsymbol{r}';z)+\boldsymbol{V}_0{}^2\int\Omega_0\mathrm{d}\boldsymbol{r}_1\times$$

$$\int\Omega_0\mathrm{d}\boldsymbol{r}_2\,\boldsymbol{G}_0(\boldsymbol{r},\boldsymbol{r}_1;z)\boldsymbol{G}_0(\boldsymbol{r}_1,\boldsymbol{r}_2;z)\boldsymbol{G}_0(\boldsymbol{r}_2,\boldsymbol{r}';z)+\cdots \qquad (2.1.76)$$

In order to find its possible positions of the poles, we substitute $\boldsymbol{G}_0(E)$ of space with three kinds of dimensionality with $E<0$ into Eq. (2.1.76).

2.2 Tight-binding Hamiltonians in Green's functions

Tight Binding Model (TBM) is an ideal model attracted from the real crystals. Each atom in a crystal is regarded as a geometric point, called as lattice site. The details of the inside of the atom are neglected. Each site plays a role of producing a potential well(barrier). Electrons can transit from one site to another.

The features of the TBM are: ①It is a solvable model. The system has a continuous energy spectrum, which is similar to the case of a free-particle in an infinite space. However, we have to mention that in obtaining the wave function of the free-particle $\mathrm{e}^{\mathrm{i}\boldsymbol{k}\cdot\boldsymbol{r}}/\sqrt{\boldsymbol{V}}$, we have used the condition of box-normalization: regarding the space as a cube with the volume $\boldsymbol{V}=L^3$. There are two kinds of boundary conditions: one is that the wave function value is zero at the surface of $\boldsymbol{V}$, which results in the standing wave in $\boldsymbol{V}$; another is so-called periodic boundary condition, which results in the propagating wave. Usually the latter is adapted since it coincides with the reality better. In this case, the wave vector component along the i-th Cartesian coordinate is $\boldsymbol{k}_i=2\pi n_i/L$, with n_i being natural numbers. The wave vector is substantially discrete. Nevertheless, since the edge L of $\boldsymbol{V}$ is very large, the distribution $\boldsymbol{k}$ points are quite dense, which can be considered as physically continuous. Then the summation over $\boldsymbol{k}$ can be replaced by integral. In a crystal lattice the thing is slightly different: there are N_i periods, each being of length a_i, along the i-th Cartesian coordinate, so that the wave vector can be written as $\boldsymbol{k}_i=2\pi n_i/N_i a_i$. Due to the periodic boundary condition, the wave function with vector $\boldsymbol{k}'_i$, $\boldsymbol{k}'_i=\boldsymbol{k}_i+2\pi m/a_i$, m being an integer, has the same value as that with $\boldsymbol{k}$. Thus it suffices to discuss the wave function and energy band with $\boldsymbol{k}$ being in the region confined by $m=0$, so-called the first Brillouin zone (1BZ). Here the energy band is considered as continuous which has both a lower and an upper bounds. Conventionally, the 1BZ is assigned as the region with $\boldsymbol{k}_i\in[-\pi/a_i,\pi/a_i]$, which is of center symmetry. The two features of the periodic boundary condition and the energy band within finite scope bring more abundant information compared to the free space. Since information is originated from the periodic lattice, it is helpful for us to understand the various features of solid state crystals; ②As impurities are introduced in the model, the system can be treated by means of the perturbation theory. This will help us to know the physics of the crystal when doped.

In the lattice, the lattice sites array uniformly, and the motion of the electrons is simply considered the transition between the sites. Since the information other than the transition is neglected, the space is similar to a vacuum. Hence the lattice sites can be taken as the coordinates of the wave function. Such kind of representation is named as site $\boldsymbol{r}$-representation, which is equivalent to $\boldsymbol{r}$-representation. We will see below how to convert the $\boldsymbol{r}$-representation in a crystal to the site-representation.

The eigenfunctions of lattice Hamiltonian are so-called Bloch functions,

$$\psi_{nk}(\boldsymbol{r}) = \frac{1}{\sqrt{V}}\mathrm{e}^{\mathrm{i}\boldsymbol{k}\cdot\boldsymbol{r}}u_{nk}(\boldsymbol{r}) \tag{2.2.1}$$

which is similar to the plane wave in a free space, except that the amplitude is modified. The modifying function $u_{nk}(\boldsymbol{r})$ is periodic, with the periodicity just the same as the potential, where n is the energy band index and $\boldsymbol{k}$ is restricted within the 1BZ. The Bloch function has a feature that

$$\psi_{nk}(\boldsymbol{r}+\boldsymbol{l}) = \mathrm{e}^{\mathrm{i}\boldsymbol{k}\cdot\boldsymbol{l}}\psi_{nk}(\boldsymbol{r}) \tag{2.2.2}$$

where $\boldsymbol{l}$ is a lattice vector, i. e., both its starting and end points are site points. If the lattice constant along the i-th direction is a_i,

$$\boldsymbol{l} = \sum_{i=1}^{d} \boldsymbol{l}_i a_i \tag{2.2.3}$$

where $\boldsymbol{l}_i$ is an integer and d is the dimension. Let us now construct so-called Wannier functions $W_n(\boldsymbol{r}-\boldsymbol{l})$ from the Bloch functions:

$$W_n(\boldsymbol{r}-\boldsymbol{l}) = \frac{1}{\sqrt{V}}\sum_{k}\mathrm{e}^{-\mathrm{i}\boldsymbol{k}\cdot\boldsymbol{l}}\psi_{nk}(\boldsymbol{r}) \tag{2.2.4}$$

where n is the total number of cell in the crystal. After the summation over $\boldsymbol{k}$ points within the 1BZ, only band index is left in $W_n(\boldsymbol{r}-\boldsymbol{l})$. Usually the function $W_n(\boldsymbol{r}-\boldsymbol{l})$ is of the largest absolute value at the site $\boldsymbol{l}$ and decays rapidly as $\boldsymbol{r}$ is away from $\boldsymbol{l}$. It can be evaluated approximately: suppose that $u_{nk}(\boldsymbol{r})$ is a constant, then, by use of Eq. (2.2.1),

$$W_n(\boldsymbol{r}-\boldsymbol{l}) = \frac{1}{\sqrt{N}}\frac{1}{\sqrt{V}}\sum_{k}\mathrm{e}^{-\mathrm{i}\boldsymbol{k}\cdot\boldsymbol{l}+\mathrm{i}\boldsymbol{k}\cdot\boldsymbol{r}}u_{nk}(\boldsymbol{r}) =$$

$$\frac{1}{\sqrt{V}}u_n\frac{1}{\sqrt{N}}\sum_{k}\mathrm{e}^{\mathrm{i}\boldsymbol{k}\cdot(\boldsymbol{r}-\boldsymbol{l})} = \frac{1}{\sqrt{V}}u_n\delta(\boldsymbol{r}-\boldsymbol{l}) \tag{2.2.5}$$

This evaluation shows that an electron moves around a site $\boldsymbol{l}$ if it is bounded to or localized at the site, so that the Wannier function $W_n(\boldsymbol{r}-\boldsymbol{l})$ is called the bounded or localized state at site $\boldsymbol{l}$. It is the tight binding wave functions, which compose a complete set and are equivalent to Bloch wave functions. The transformation of the two complete sets is expressed by Eq. (2.2.5). The matrix element of any operator, such as the crystal Hamiltonian, can be computed by the Wannier representation. In the following we assume there is only one energy band, called single-band model, thus dropping the index n.

Wannier function $W_n(\boldsymbol{r}-\boldsymbol{l})$ is closely related to site $\boldsymbol{l}$. It is in fact the form of $\boldsymbol{r}$-representation,

$$W(\boldsymbol{r}-\boldsymbol{l}) = \langle \boldsymbol{r} \mid \boldsymbol{l} \rangle \tag{2.2.6}$$

where $\mid \boldsymbol{l} \rangle$ is the wave function in site-representation. In this representation, the Hamiltonian matrix elements are

$$\langle \boldsymbol{l} \mid \boldsymbol{H} \mid m \rangle = \varepsilon_l \delta_{lm} + \boldsymbol{V}_{lm} \tag{2.2.7}$$

The two terms of the element are of physical significance: ε_l is the binding energy, or on-site energy, at site $\boldsymbol{l}$, which plays a role of a potential well, and $\boldsymbol{V}_{lm}$ is the transition probability between sites $\boldsymbol{l}$ and m. We can evaluate from Eq. (2.2.5) that Wannier functions of different sites overlap each other slightly. As an approximation, we consider they are orthogonal to each other:

$$\langle \boldsymbol{l} \mid m \rangle = \delta_{lm} \tag{2.2.8}$$

The site wave functions compose a complete set:

$$\sum_l \mid \boldsymbol{l} \rangle \langle \boldsymbol{l} \mid = 1 \tag{2.2.9}$$

In this way we transform the crystal wave functions, after some reasonable approximation, into site wave functions in the site-representation. The discrete coordinate $\boldsymbol{l}$ is analogous to the continuous coordinate $\boldsymbol{r}$ in real space. The approximation used above is termed as tight binding approximation (TBA), and the model as tight binding model. The Hamiltonian associated with the model is the tight binding Hamiltonian (TBH) which is of the form of

$$\boldsymbol{H} = \sum_l \mid \boldsymbol{l} \rangle \varepsilon_l \mid \boldsymbol{l} \rangle + \sum_{l \neq m} \mid \boldsymbol{l} \rangle \boldsymbol{V}_{lm} \langle m \mid \tag{2.2.10}$$

Its elements have been shown by Eq. (2.2.7), with the diagonal and off-diagonal ones being ε_l and $\boldsymbol{V}_{lm}$, respectively.

Although the Hamiltonian Eq. (2.2.10) is quite simple, a further approximation is desirable in order to make use of it. Since all sites in the lattice are identical, so-called Bravais lattice, their on-site energies should be the same:

$$\varepsilon_l = \varepsilon_0 \tag{2.2.11}$$

For the transitions between sites, only those between the nearest neighbour sites are considered and they are the same to each other. Thus the simplified Hamiltonian becomes

$$H = \varepsilon_0 \sum_l | l\rangle\langle l | + V_t \sum_l \sum_{m(nn)}{}' | l\rangle\langle m | \tag{2.2.12}$$

where $\sum_{m(nn)}{}'$ means that the summation over m merely covers the nearest neighbour sites of $\boldsymbol{l}$. Of course, we may also include the second and the third nearest neighbours and so on if necessary.

If an electron is indeed entirely localized at $\boldsymbol{l}$, then the project of its wave function is 1 on site $\boldsymbol{l}$, and is zero on other sites. The energy associated with this wave function is ε_l. In general, a wave function has an amplitude distribution among the sites. The site wave functions are orthonormalized and complete, as shown by Eq. (2.2.8) and Eq. (2.2.9).

For the simplified Hamiltonian Eq. (2.2.12) where only Bravais lattice and transition between nearest neighbor sites are considered, it is easy to solve its eigenfunctions and corresponding eigenenergies. The eigenfunctions are:

$$| \boldsymbol{k}\rangle = \frac{1}{\sqrt{N}} \sum_l e^{i\boldsymbol{k}\cdot\boldsymbol{l}} | \boldsymbol{l}\rangle \tag{2.2.13}$$

This represents a wave with wave vector $\boldsymbol{k}$ propagating in the lattice, with the amplitudes on all the sites being equal and phase $\varphi_l = \boldsymbol{k} \cdot \boldsymbol{l}$. The eigenenergy corresponding to $|\boldsymbol{k}\rangle$ is

$$E(\boldsymbol{k}) = \varepsilon_0 + V_t \sum_{l(nn)} e^{i\boldsymbol{k}\cdot\boldsymbol{l}} \tag{2.2.14}$$

Since all the sites are identical as mentioned above, we take the on-site energy of the origin, and the summation in Eq. (2.2.14) over $\boldsymbol{l}$ covers the nearest neighbor sites of the origin. The dependence of energy E on the wave vector $\boldsymbol{k}$ is named as dispersion relationship. For the simplest lattices in three kinds spaces, the dispersion relationships are as follows:

1-d simple lattice, $E(\boldsymbol{k}) = \varepsilon_0 + 2V_t \cos ka$; (2.2.15)

2-d square lattice, $E(\boldsymbol{k}) = \varepsilon_0 + 2V_t(\cos k_1 a + \cos k_2 a)$; (2.2.16)

3-d simple cubic lattice, $E(\boldsymbol{k}) = \varepsilon_0 + 2V_t(\cos k_1 a + \cos k_2 a + \cos k_3 a)$; (2.2.17)

Here k_1, k_2 and k_3 represent the components of $\boldsymbol{k}$ along x^-, y^- and z^- directions, respectively, and a is the lattice constant.

In Fig. 2.2.1 the dispersion curve of 1-d case (2.2.15) has been plotted, where $\boldsymbol{k}$ is restricted within 1BZ$[-\pi/a, \pi/a]$. At the center of the 1BZ, $\boldsymbol{k}=0$, energy $E(\boldsymbol{k})$ reaches its maximum $E_{max} = \varepsilon_0 + 2V_t$ which is the upper band edge. At the boundary of the 1BZ $\boldsymbol{k} = \pm\pi/a$, $E(\boldsymbol{k})$ reaches its minimum $E_{min} = \varepsilon_0 - 2V_t$, the lower band edge. The energy band is continuous within $[\varepsilon_0 - 2V_t, \varepsilon_0 + 2V_t]$ and the bandwidth is $4V_t$. From Eq. (2.2.15) to Eq. (2.2.17) one can easily show that the simplest lattices produce a single band extending from $\varepsilon_0 - ZV_t$ to $\varepsilon_0 + ZV_t$, where Z is the nearest neighbour number, and the half bandwidth is $B =$

ZV_t. In the next section we will compute the Green's functions associated with Eq. (2.2.15) to Eq. (2.2.17), respectively.

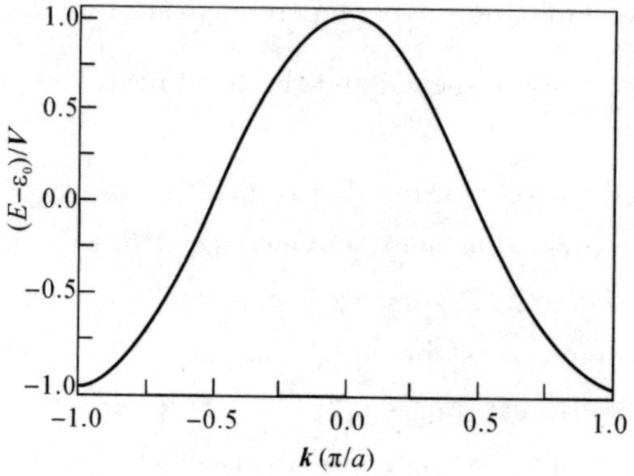

Fig. 2.2.1 $E(\boldsymbol{k})$ vs. $\boldsymbol{k}$ for Eq. (2.2.15)

The Green's function can be computed by

$$\boldsymbol{G}(z) = \sum_{\boldsymbol{k}} \frac{|\boldsymbol{k})(\boldsymbol{k}|}{z - E(\boldsymbol{k})} \tag{2.2.18}$$

where $|\boldsymbol{k})$ is given by Eq. (2.2.13) and $E(\boldsymbol{k})$ by (2.2.14). In the lattice representation the matrix elements of $\boldsymbol{G}(z)$ are

$$\boldsymbol{G}(\boldsymbol{l},\boldsymbol{m};z) = \langle \boldsymbol{l} \mid \boldsymbol{G}(z) \mid \boldsymbol{m}\rangle = \sum_{\boldsymbol{k}} \frac{\langle \boldsymbol{l} \mid \boldsymbol{k})(\boldsymbol{k} \mid \boldsymbol{m}\rangle}{z - E(\boldsymbol{k})} =$$

$$\frac{\boldsymbol{V}}{N\,(2\pi)^d}\int_{1\mathrm{BZ}} \frac{\mathrm{e}^{\mathrm{i}\boldsymbol{k}\cdot(\boldsymbol{l}-\boldsymbol{m})}}{z - E(\boldsymbol{k})}\mathrm{d}\boldsymbol{k} \tag{2.2.19}$$

where the subscript 1BZ means the integration must be restricted within the 1BZ. In particular, the diagonal matrix element $\boldsymbol{G}(\boldsymbol{l},\boldsymbol{l};z)$ is also called site Green's function of site $\boldsymbol{l}$. All the site Green's functions are equal to each other and given by

$$\boldsymbol{G}(\boldsymbol{l},\boldsymbol{l};z) = \frac{\boldsymbol{V}}{N\,(2\pi)^d}\int_{1\mathrm{BZ}} \frac{\mathrm{d}\boldsymbol{k}}{z - E(\boldsymbol{k})} \tag{2.2.20}$$

For sufficiently large z, one can omit $E(\boldsymbol{k})$ in the denominator of the integrand in Eq. (2.2.20) so that $\boldsymbol{G}(\boldsymbol{l},\boldsymbol{l};z) \underset{z\to\infty}{\to} \frac{1}{z}\frac{\boldsymbol{V}}{N\,(2\pi)^d}\int_{1\mathrm{BZ}}\mathrm{d}\boldsymbol{k}$. The volume of the 1BZ equals $(2\pi)^d/\boldsymbol{V}_c$, where $\boldsymbol{V}_c = \boldsymbol{V}/N$ is the volume of the primitive cell of the lattice. Hence,

$$\boldsymbol{G}(\boldsymbol{l},\boldsymbol{l};z) \underset{z\to\infty}{\to} \frac{1}{z} \tag{2.2.21}$$

This behaviour can be understood if one expresses the integration over $\boldsymbol{k}$ in Eq. (2.2.20) in terms of over E with $\mathrm{DOS}_\rho(E)$ as a weighted factor: $\boldsymbol{G}(\boldsymbol{l},\boldsymbol{l};z) = \int \frac{\rho(E)}{z-E}\mathrm{d}E \underset{z\to\infty}{\to} \frac{1}{z}\int \rho(E)\mathrm{d}E$, but $\int \rho(E)\mathrm{d}E = 1$ since there is one state per site. Any Green's function has the asymptotic behaviour Eq. (2.2.21), as can be seen in a general case.

The site DOS can be computed by the site Green's function:

$$\rho_0(E) = \mp \frac{1}{\pi} \mathrm{Im}\, G^{\pm}(l,l;E) \tag{2.2.22}$$

Below we give some explicit results for $d = 1,2,3$.

Substituting Eq. (2.2.15) into Eq. (2.2.19), we obtain

$$G(l,m;z) = \frac{L}{N2\pi} \int_{-\pi/a}^{\pi/a} \frac{e^{ika(l-m)}}{z - \varepsilon_0 - 2V_t \cos ka} dk = \frac{1}{2\pi} \int_{-\pi}^{\pi} \frac{e^{i\varphi(l-m)}}{z - \varepsilon_0 - 2V_t \cos\varphi} d\varphi \tag{2.2.23}$$

We observe first that the integral depends on the absolute value $|l-m|$. Next since $e^{ix} = \cos x + i\sin x$, the odd term in the integrand does not contribute to the integration. Then we let $\omega = e^{i\varphi}$ and $d\omega = i\omega d\varphi$ so that transformation of the intergration over φ into an integral over the complex ω variable along the unit circle is

$$G(l,m;z) = \frac{-1}{2\pi i V_t} \oint d\omega \frac{\omega^{|l-m|}}{\omega^2 - 2x\omega + 1} \tag{2.2.24}$$

where $x = (z - \varepsilon_0)/B$ and $B = 2V_t$. The evaluation of Eq. (2.2.24) depends on the poles in the unit circle. The two roots of $\omega^2 - 2\omega + 1 = (\omega - \rho_1)(\omega - \rho_2) = 0$ are given by

$$\rho_1 = x - \sqrt{x^2 - 1} \tag{2.2.25a}$$

and

$$\rho_2 = x + \sqrt{x^2 - 1} \tag{2.2.25b}$$

Obviously, $\rho_1 \rho_2 = 1$. First we examine the case of $|x| > 1$. When $x > 1$, $|\rho_1| < 1$; when $x < -1$, $|\rho_2| < 1$. That is to say, one of the two roots must be inside the unit circle and the other outside. The result associated with $|\rho_1| < 1$ is

$$G(l,m;z) = \frac{-1}{V_t} \frac{\rho_1^{|l-m|}}{\rho_1 - \rho_2} = -\frac{\rho_1^{|l-m|}}{\sqrt{(z-\varepsilon_0)^2 - B^2}} \tag{2.2.26a}$$

and that with $|\rho_2| < 1$ is

$$G(l,m;z) = \frac{-1}{V_t} \frac{\rho_2^{|l-m|}}{\rho_2 - \rho_1} = -\frac{\rho_2^{|l-m|}}{\sqrt{(z-\varepsilon_0)^2 - B^2}} \tag{2.2.26b}$$

Next we examine the case where x is real and $|x| \leqslant 1$. In this case both the roots are just on the unit circle, and the Green's function is not well defined. Fortunately, we can make use of its side limits:

$$G^{\pm}(l,m;E) = \frac{\pm i}{\sqrt{B^2 - (E - \varepsilon_0)^2}} (x \pm i\sqrt{1 - x^2})^{|l-m|} \tag{2.2.27}$$

where since $|x| \leqslant 1$, the energy E is just within the band

$$\varepsilon_0 - B \leqslant E \leqslant \varepsilon_0 + B \tag{2.2.28}$$

The site DOS is

$$\rho_0(E)=\pm\frac{1}{\pi}\text{Im}\,\boldsymbol{G}^{\pm}(\boldsymbol{l},\boldsymbol{l};E)=\frac{\theta(B-|E-\varepsilon_0|)}{\pi\sqrt{B^2-(E-\varepsilon_0)^2}} \tag{2.2.29}$$

We plot in Fig. 2.2.2 the real and imaginary parts of the site Green's function vs. real x. Within the energy band, there is only imaginary part with no real part, while outside the band, the situation is contrary. Approaching the band edges, DOS blows up in the way of inverse of energy square root. For instance, at the upper band edge, as $E-\varepsilon_0\to B-\delta$, $\frac{1}{\sqrt{B^2-(E-\varepsilon_0)^2}}=\frac{1}{\sqrt{2\delta}}$, where $\delta\to 0$. The real part of $\boldsymbol{G}^{+}(\boldsymbol{l},\boldsymbol{l};E)$ blows up in the way of $1/\delta$ as $E-\varepsilon_0\to B-\delta$, where $\delta\to 0$. This behaviour is the same as in a free space, which is the characteristic of the continuous spectrum associated with 1-d space.

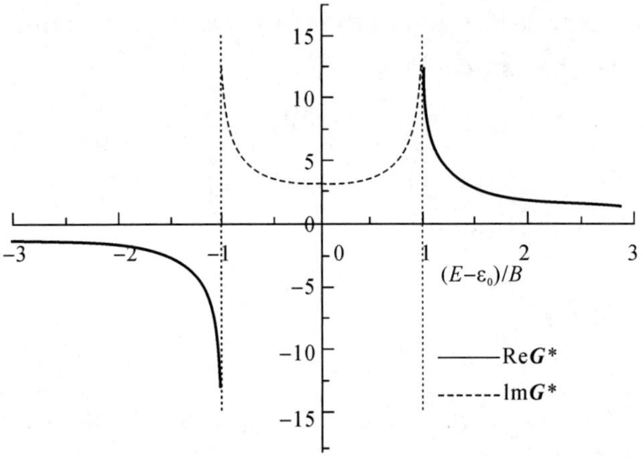

Fig. 2.2.2 The real and imaginary parts of 1-d Green's function Eq. (2.2.10) where $\boldsymbol{l}=m$

The physical significance of the matrix element $\boldsymbol{G}_0(\boldsymbol{l},m;z)$ is the probability amplitude of a particle created at site $\boldsymbol{l}$, with energy z, moved to site m and annihilated, or that of the propagation of a particle from site $\boldsymbol{l}$ to m. It is easily seen from Eq. (2.2.26) that when $|\rho_1|<1$, the energy being outside of the band, this amplitude decreases with the distance between sites. While within the band, $|\rho_1|=1$, this amplitude remains unchanged with the distance, which means that a particle can move to anywhere. We have to point out, however, that the propagation of a particle after it is created is different from the propagating wave solution Eq. (2.2.13).

Substituting Eq. (2.2.16) into Eq. (2.2.19), we obtain:

$$\boldsymbol{G}(\boldsymbol{l},m;z)=\frac{a^2}{(2\pi)^2}\int_{1\text{BZ}}\frac{e^{i\boldsymbol{k}\cdot(\boldsymbol{l}-m)}}{z-\varepsilon_0-2\boldsymbol{V}_t(\cos\boldsymbol{k}_1a+\cos\boldsymbol{k}_2a)}\text{d}\boldsymbol{k} \tag{2.2.30}$$

where

$$\boldsymbol{k}\cdot(\boldsymbol{l}-m) = a[\boldsymbol{k}_1(\boldsymbol{l}_1 - m_1) + \boldsymbol{k}_2(\boldsymbol{l}_2 - m_2)] \tag{2.2.31}$$

$\boldsymbol{l}_1, \boldsymbol{l}_2, m_1$ and m_2 are integers and a is the lattice constant. The 1BZ is the square $-\pi/a \leqslant \boldsymbol{k}_1$, $\boldsymbol{k}_2 \leqslant \pi/a$. Thus Eq. (2.2.30) can be rewritten as

$$\boldsymbol{G}(\boldsymbol{l},m;z) = \frac{1}{(2\pi)^2}\int_{-\pi}^{\pi} \mathrm{d}\varphi_1 \int_{-\pi}^{\pi} \mathrm{d}\varphi_2 \frac{\mathrm{e}^{\mathrm{i}\varphi_1(l_1-m_1)+\mathrm{i}\varphi_2(l_2-m_2)}}{z-\varepsilon_0 - 2\boldsymbol{V}_t(\cos\varphi_1 + \cos\varphi_2)} =$$

$$\frac{1}{(2\pi)^2}\int_{-\pi}^{\pi} \mathrm{d}\varphi_1 \int_{-\pi}^{\pi} \mathrm{d}\varphi_2 \frac{\cos[\varphi_1(\boldsymbol{l}_1 - m_1) + \varphi_2(\boldsymbol{l}_2 - m_2)]}{z-\varepsilon_0 - 2\boldsymbol{V}_t(\cos\varphi_1 + \cos\varphi_2)} = \tag{2.2.32a}$$

$$\frac{1}{(2\pi)^2}\int_{-\pi}^{\pi} \mathrm{d}\varphi_1 \int_{-\pi}^{\pi} \mathrm{d}\varphi_2 \times$$

$$\frac{[\cos(\boldsymbol{l}_1 - m_1 + \boldsymbol{l}_2 - m_2)\varphi_1][\cos(\boldsymbol{l}_1 - m_1 - \boldsymbol{l}_2 + m_2)\varphi_2]}{z-\varepsilon_0 - 4\boldsymbol{V}_t\cos\varphi_1\cos\varphi_2} \tag{2.2.32b}$$

In the last line, we transform the integration over φ_1 and φ_2 into that over $\varphi_1+\varphi_2$ and $\varphi_1-\varphi_2$ (let $\varphi_1 = \alpha+\beta$ and $\varphi_2 = \alpha-\beta$, and then replace α and β by φ_1 and φ_2, respectively). The diagonal matrix elements are

$$\boldsymbol{G}(\boldsymbol{l},\boldsymbol{l};z) = \frac{1}{2\pi}\int_{-\pi}^{\pi} \mathrm{d}\varphi_1 \frac{1}{2\pi}\int_{-\pi}^{\pi} \mathrm{d}\varphi_2 \frac{1}{z-\varepsilon_0 - B\cos\varphi_1\cos\varphi_2} =$$

$$\frac{1}{2\pi}\int_{-\pi}^{\pi} \mathrm{d}\varphi_1 \frac{1}{\sqrt{(z-\varepsilon_0)^2 - B^2\cos^2\varphi_1}} =$$

$$\frac{1}{\pi(z-\varepsilon_0)}\int_0^{\pi} \frac{\mathrm{d}\varphi}{\sqrt{1-\lambda^2\cos^2\varphi}} = \frac{2}{\pi(z-\varepsilon_0)}K(\lambda) \tag{2.2.33}$$

where

$$\lambda = \frac{B}{z-\varepsilon_0}, B = 4\boldsymbol{V}_t \tag{2.2.34}$$

and $K(\lambda)$ is the complete elliptic integral of the first kind. When $z = E$ is a real number, $\boldsymbol{G}(\boldsymbol{l}, \boldsymbol{l};E)$ can be expressed as follows:

$$\left.\begin{aligned}
&\boldsymbol{G}(\boldsymbol{l},\boldsymbol{l};E) = \frac{2}{\pi(E-\varepsilon_0)}K\frac{B}{E-\varepsilon_0}, |E-\varepsilon_0| > B = 4\boldsymbol{V}_t \\
&\mathrm{Re}\boldsymbol{G}^{\pm}(\boldsymbol{l},\boldsymbol{l};E) = -\frac{2}{\pi B}K\left(\frac{E-\varepsilon_0}{B}\right), -B < E-\varepsilon_0 < 0 \\
&\mathrm{Re}\boldsymbol{G}^{\pm}(\boldsymbol{l},\boldsymbol{l};E) = \frac{2}{\pi B}K\left(\frac{E-\varepsilon_0}{B}\right), 0 < E-\varepsilon_0 < B \\
&\mathrm{Im}\boldsymbol{G}^{\pm}(\boldsymbol{l},\boldsymbol{l};E) = \pm\frac{2}{\pi B}K\left(\sqrt{1-(E-\varepsilon_0)^2/B^2}\right), |E-\varepsilon_0| < B
\end{aligned}\right\} \tag{2.2.35}$$

The site DOS is

$$\rho(E) = \pm\,\mathrm{Im}\boldsymbol{G}^{\pm}(\boldsymbol{l},\boldsymbol{l};E) = \frac{2}{\pi^2 B}\theta(B-|E-\varepsilon_0|)K\left(\sqrt{1-(E-\varepsilon_0)^2/B^2}\right) \tag{2.2.36}$$

In Fig. 2.2.3 we have plotted the real and imaginary parts of the site Green's function. The imaginary part, i.e., DOS, is nontrivial only within the energy band, and curves go to band edges with their slopes approaching zero. The real part exhibits logarithmic singularities at the band edges. These are the characteristics of a 2-d continuous spectrum, also shown by a 2-d free particle. Nevertheless, there is a singularity at the center of the energy band in Fig. 2. 2. 3, where the real part of the site Green's function shows discontinuity, too. This behaviour is different from a free space.

Computation results show that when z is at the exterior of the band, $\boldsymbol{G}(\boldsymbol{l},m;z)$ decreases exponentially with the distance $|\boldsymbol{l}-m|$ increasing. While as E in is the interior of the band, $\boldsymbol{G}(\boldsymbol{l},m;E)$ decreases in the form of $|\boldsymbol{l}-m|^{-1/2}$ accompanied by oscillation. This is different from 1-d case where the propagation probability amplitude does not decay if the energy is in the interior of the band. The difference also reflects the fact that the propagation probability amplitude embodied by the matrix element $\boldsymbol{G}(\boldsymbol{l},m;E)$ and the propagating wave Eq. (2. 2. 13) are not the same thing.

Numerical computation is inevitable to obtain the results of the Green's function. However, if one is interested in qualitative discussion about the Green's function, focusing on the behaviours around the energy band edges and neglecting other details, he can approximate the 2-d Green's function by a simple function that retains the correct analytic behavior near the band edge and gives one state per site. This simple approximation is

$$\boldsymbol{G}(\boldsymbol{l},\boldsymbol{l};z)=\frac{1}{2B}\ln\left(\frac{z-\varepsilon_0+B}{z-\varepsilon_0-B}\right) \tag{2.2.37}$$

which gives the following constant DOS:

$$\rho_0(E)=\frac{1}{2B}\theta(B-|E-\varepsilon_0|) \tag{2.2.38}$$

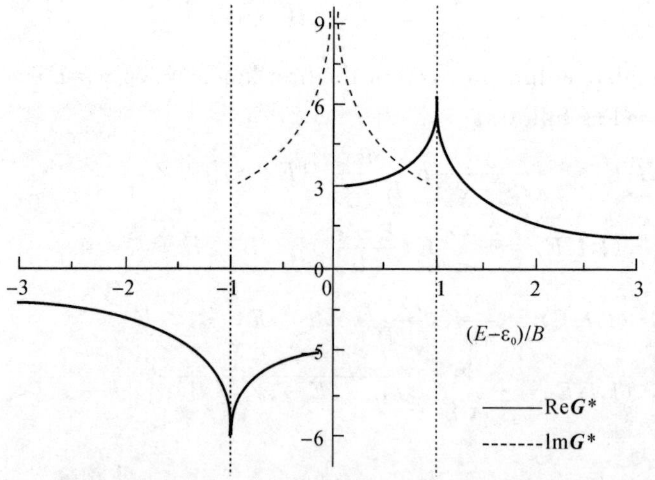

Fig. 2. 2. 3 The real and imaginary parts of the 2-d site Green's function Eq. (2. 2. 37) vs. real E

In Fig. 2. 2. 4 we have plotted the real and imaginary parts of the function Eq. (2. 2. 20) vs. real E. The curves show correct behaviours near the band edges as in Fig. 2. 2. 3. But there is no singularity in the interior of the band. Function Eq. (2.2.20) is not a Green's function, but just a replacer of the 2-d Green's function. Because of its particularly simple form while leaving the correct behaviour near the band edges, it can be employed to do quantitative discussion sometimes.

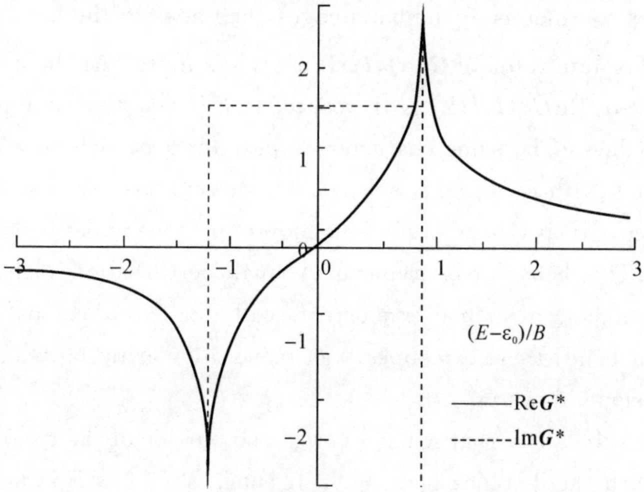

Fig. 2.2.4 The real and imaginary parts of the function Eq. (2.2.20) vs. real E

The 1BZ of a simple cubic lattice is the cube $-\pi/a \leqslant \boldsymbol{k}_1, \boldsymbol{k}_2, \boldsymbol{k}_3 \leqslant \pi/a$, where a is the lattice constant. Substituting Eq. (2.2.17) into Eq. (2.2.19) and introducing variables $\varphi_i = \boldsymbol{k}_i a$ $(i = 1,2,3)$, we obtain

$$\boldsymbol{G}(\boldsymbol{l},\boldsymbol{m};z) = \frac{1}{(2\pi)^3}\int_{-\pi}^{\pi} \mathrm{d}\varphi_1 \int_{-\pi}^{\pi} \mathrm{d}\varphi_2 \int_{-\pi}^{\pi} \mathrm{d}\varphi_3 \times \frac{\cos\left[(\boldsymbol{l}_1 - m_1)\varphi_1 + (\boldsymbol{l}_2 - m_2)\varphi_2 + (\boldsymbol{l}_3 - m_3)\varphi_3\right]}{z - \varepsilon_0 - 2\boldsymbol{V}_t(\cos\varphi_1 + \cos\varphi_2 + \cos\varphi_3)} \tag{2.2.39}$$

In particular, the diagonal element is simplified into

$$\boldsymbol{G}(\boldsymbol{l},\boldsymbol{l};z) = \frac{1}{(2\pi)^3}\int_{-\pi}^{\pi} \mathrm{d}\varphi_1 \int_{-\pi}^{\pi} \mathrm{d}\varphi_2 \int_{-\pi}^{\pi} \mathrm{d}\varphi_3 \times \frac{1}{z - \varepsilon_0 - 2\boldsymbol{V}_t(\cos\varphi_1 + \cos\varphi_2 + \cos\varphi_3)} \tag{2.2.40}$$

The integration over φ_1 and φ_2 can be done as in the case of 2-d square lattice and the result is

$$\boldsymbol{G}(\boldsymbol{l},\boldsymbol{l};z) = \frac{1}{2\pi^2 \boldsymbol{V}_t}\int_0^{\pi} \mathrm{d}\varphi t K(t) \tag{2.2.41}$$

where

$$t = \frac{4\boldsymbol{V}_t}{z - \varepsilon_0 - 2\boldsymbol{V}_t\cos\varphi} \tag{2.2.42}$$

$K(t)$ is again the complete elliptic integral of the first kind, so that Eq. (2.2.41) contains in fact double integrals. Numerical triple integrals of Eq. (2.2.41) can be done and the result of the diagonal matrix element $\boldsymbol{G}^{+}(\boldsymbol{l},\boldsymbol{l};E)$ vs. E is plotted in Fig. 2.2.5. $\boldsymbol{G}^{+}(\boldsymbol{l},\boldsymbol{l};E)$ near the band edges is continuous, a behaviour different from the low dimensional cases. As a result, the DOS is continuous at the band edges, and goes to the band edge in the way of $\sqrt{\delta}$ as $\delta \rightarrow 0$. The absolute value of $\boldsymbol{G}^{+}(\boldsymbol{l},\boldsymbol{l};E)$ is always finite. At the outside of the energy band, the derivative of Re $\boldsymbol{G}^{+}(\boldsymbol{l},\boldsymbol{l};E)$ with respect to E is negative, and goes to $-\infty$ as E approaches the band edges. The same characters appear for a particle in a free space and also for other 3-d systems with continuous spectra. The exceptions are that for face- and body-centered cubic lattices, their Green's functions blows up at the upper band edge and in the interior of the band. This behavior is atypical. A small perturbation such as inclusion of the second-nearest-neighbor transfer matrix elements will eliminate these pathological infinities. In the interior of the band, there can appear van Hove singularities at which the DOS curve is continuous but its slope blows up.

Computation results show that when z is real and outside of the band, $\boldsymbol{G}(\boldsymbol{l},\boldsymbol{m};z)$ decreases exponentially with the distance $|\boldsymbol{l}-\boldsymbol{m}|$ increasing. While as E is in the interior of the band, $\boldsymbol{G}(\boldsymbol{l},\boldsymbol{m};E)$ decays in the way of $|\boldsymbol{l}-\boldsymbol{m}|^{-1}$ accompanied by oscillation.

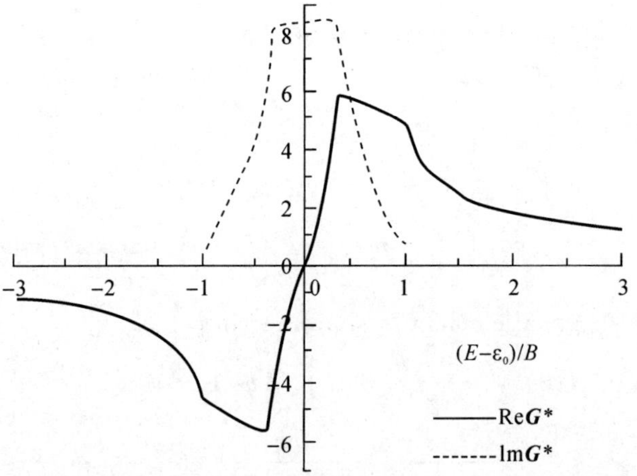

Fig. 2.2.5 Real and imaginary parts of the diagonal matrix element calculated from Eq. (2.2.41) vs. real E. The positions of the extrema of the real and imaginary parts are at the same E

In many cases where quantitative details are not important, it is very useful to have a simple approximate expression for diagonal matrix elements. This expression is

$$\boldsymbol{G}(\boldsymbol{l},\boldsymbol{l};z)=\frac{2}{z-\varepsilon_0+\operatorname{sgn}[\operatorname{Re}(z-\varepsilon_0)]\sqrt{(z-\varepsilon_0)^2-B^2}} \tag{2.2.43}$$

which is known as the Hubbard Green's function and taken as a replacer of the 3-d $\boldsymbol{G}(\boldsymbol{l},\boldsymbol{l};z)$ of the simple cubic lattice. In Eq. (2.2.43) the sign of Im $\sqrt{(z-\varepsilon_0)^2-B^2}$ is the same as the sign of Im z. We stress that Eq. (2.2.43) is not the Green's function of some lattice, but is used in discussion of some problems for convenience. In the next chapter we will see one example of using Eq. (2.2.43). The DOS corresponding to Eq. (2.2.43) is

$$\rho_0(E)=\frac{2\theta(B-|E-\varepsilon_0|)}{\pi B^2}\sqrt{B^2-(E-\varepsilon_0)^2} \tag{2.2.44}$$

We plot in Fig. 2.2.6 the real and imaginary parts of the function Eq. (2.2.44). The imaginary part, the DOS is of a semicircle form. Compared with the curves in Fig. 2.2.5, the following behaviours of the curves in Fig. 2.2.6 are correct: Re $\boldsymbol{G}^+(E)$ as E is outside the band and approaching the band edges, Im $\boldsymbol{G}^+(E)$ as E approaches the band edges. Note that the simple approximation Eq. (2.2.44) does not reproduce any van Hove singularity within the band.

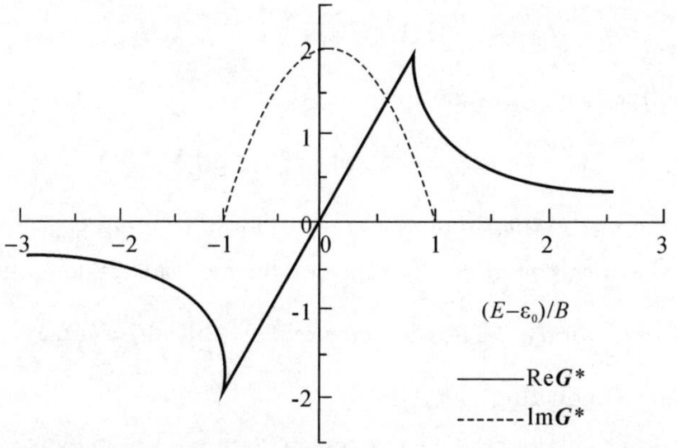

Fig. 2.2.6 The real and imaginary parts of the function Eq. (2.2.44) vs. real E

2.3 Nonequilibrium Green's functions

When an external field is applied, transport will occur in a system. In this case, the system is not in an equilibrium state, so that the nonequilibrium Green's functions have to be utilized to calculate the quantities in the problem of transport. If the field is not strong and the system is near the equilibrium state, a fundamental process to deal with electronic transport through a mesoscopic structure.

The structure under consideration is a central scattering region connected to several leads. Its Hamiltonian is

$$\boldsymbol{H}=\sum_{\beta}\boldsymbol{H}_{\beta}+\boldsymbol{H}_{c}+\boldsymbol{H}_{T} \tag{2.3.1}$$

It comprises three parts. The first part is the leads. The subscript β labels the βth lead.

Each lead has three contributions:

$$\boldsymbol{H}_\beta = \boldsymbol{H}_{\beta 1} + \boldsymbol{H}_{\beta 2} + \boldsymbol{H}_{\beta 3} \tag{2.3.2}$$

The first term

$$\boldsymbol{H}_{\beta 1} = \sum_{k\sigma} \varepsilon^0_{\beta k\sigma} a^+_{\beta k\sigma} a_{\beta k\sigma} \tag{2.3.3}$$

is the tight binding Hamiltonian of electrons. This form is in the momentum space after Fourier transformation and has been diagonalized. The subscript σ labels the spins of electrons. If the lead is ferromagnetic, the energies with spin up and down are different. The second term

$$\boldsymbol{H}_{\beta 2} = \sum_{k\sigma} \boldsymbol{V}_\beta(t) a^+_{\beta k\sigma} a_{\beta k\sigma} \tag{2.3.4}$$

is the energy level shift of electrons caused by the applied time-dependent external field $\boldsymbol{V}_\beta(t)$. $\boldsymbol{V}_\beta(t)$ may also involve a direct current bias. It is this field that brings the system to be nonequilibrium. The third term

$$\boldsymbol{H}_{\beta 3} = \sum_{k} [\Delta_\beta(t) a^+_{\beta,k\uparrow} a^+_{\beta,-k\downarrow} + \boldsymbol{H}_c] \tag{2.3.5}$$

indicates superconductive lead. The definition $\Delta_\beta(t)$ is

$$\Delta_\beta(t) = \Delta^0_\beta \mathrm{e}^{-\mathrm{i}\varphi_\beta} \exp\left[-2\mathrm{i}\int_0^t \mathrm{d}t_1 \boldsymbol{V}_\beta(t_1)\right] \tag{2.3.6}$$

where Δ^0_β and $\mathrm{e}^{-\mathrm{i}\varphi_\beta}$ are the energy gap function and its phase factor, respectively, of β lead. These two factors arise from the steady state of the superconductive lead. When a field $\boldsymbol{V}_\beta(t)$ is applied, it causes a phase shift of operator $a^+_{\beta k\sigma}$. Thus a factor $\exp\left[-2\mathrm{i}\int_0^t \mathrm{d}t_1 \boldsymbol{V}_\beta(t_1)\right]$ comes into Eq. (2.3.6). If the lead is not superconducting, let $\Delta^0_\beta = 0$.

$\boldsymbol{H}_c$ is the Hamiltonian of the central scattering region. The mesoscopic region consists of several sites with spins. It may comprise the transition between the sites, interaction between electrons, and so on. Thus it may be used to describe a variety of mesoscopic structures such as quantum dots, quantum lattices, carbon nanotubes, organic molecules, etc. The creation and annihilation operators in this region are denoted by c^+ and c. Here we merely generally put down this Hamiltonian as

$$\boldsymbol{H}_c = \boldsymbol{H}_c[\{c_{j\sigma}, c^+_{j\sigma}\}] \tag{2.3.7}$$

leaving the details undetermined. In Eq. (2.3.7) the suffix j may label the energy levels or sites in the central region.

Note that the central region and leads are independent of each other. This results in that the anticommutator between two fermion operators from different regions are always zero.

Lastly, $\boldsymbol{H}_T$ in Eq. (2.3.1) is the coupling Hamiltonian between the central region and leads:

$$\boldsymbol{H}_T = \sum_{\beta k\sigma} [\boldsymbol{V}^0_{\beta k\sigma} a^+_{\beta k\sigma} c_{i\sigma} + \boldsymbol{H}_c] \tag{2.3.8}$$

where $\boldsymbol{V}^0_{\beta k\sigma}$ is called tunnelling matrix element and may have a phase factor, i. e. , it may be a complex number. Eq. (2.3.8) is also named as tunnelling Hamiltonian.

In treating with the Hamiltonian, usually the central and lead parts are regarded as $\boldsymbol{H}_0$ and the coupling part is treated as perturbation. The Green's function associated with $\boldsymbol{H}_0$ can be easily obtained. However, in the present form, this is difficult to come true, because $\boldsymbol{H}_0$ depends on time, see Eq. (2.3.5). A proper transformation of the Hamiltonian can get rid of the difficulty.

The transformation transfers the time-dependent part in $\boldsymbol{H}_\beta$ to the coupling part $\boldsymbol{H}_T$. Suppose a unitary operator to be

$$U(t) = \exp\left\{-\mathrm{i}\sum_{\beta k\sigma}\left[\varphi_\beta/2 + \int_0^t \mathrm{d}t_1\, \boldsymbol{V}_\beta(t_1)\right]a^+_{\beta k\sigma} a_{\beta k\sigma}\right\} \tag{2.3.9}$$

where φ_β is just what was in Eq. (2.3.6). This operator has time as its argument. For the sake of convenience, we drop the argument temporarily, $U(t) = U$. The transformation of an annihilation operator is defined by

$$\bar{a}_{\beta k\sigma} = U a_{\beta k\sigma} U^+ \tag{2.3.10}$$

One ought to distinguish this definition with Heisenberg operator $a_{\beta k\sigma}(t) = \mathrm{e}^{iHt} a_{\beta k\sigma} \mathrm{e}^{-iHt}$. In the following the derivative with respect to t, $\frac{\partial}{\partial_t}$, is shortened by ∂_t, and that of U is further simplified by a dot on U: $\dot{U} = \partial_t U$. Now we take the time-derivatives of Eq. (2.3.9) and its hermitian conjugate:

$$\dot{U} = -\mathrm{i}\sum_{\beta k\sigma} \boldsymbol{V}_\beta(t) a^+_{\beta k\sigma} a_{\beta k\sigma} U \tag{2.3.11a}$$

$$\dot{U}^+ = -\mathrm{i}\sum_{\beta k\sigma} \boldsymbol{V}_\beta(t) a^+_{\beta k\sigma} a_{\beta k\sigma} U^+ \tag{2.3.11b}$$

Apparently, the operator $a^+_{\beta k\sigma} a_{\beta k\sigma}$ commutes to U. The time derivative of Eq. (2.3.10) is

$$\begin{aligned}
\partial_t a_{\alpha p\lambda} &= \dot{U} a_{\alpha p\lambda} U^+ + U a_{\alpha p\lambda} \dot{U}^+ = -\mathrm{i}U\sum_{\beta k\sigma} \boldsymbol{V}_\beta(t) a^+_{\beta k\sigma} a_{\beta k\sigma} a_{\alpha p\lambda} U^+ + \\
&\mathrm{i}U\sum_{\beta k\sigma} \boldsymbol{V}_\beta(t) a_{\alpha p\lambda} a^+_{\beta k\sigma} a_{\beta k\sigma} U^+ = \\
&-\mathrm{i}U\sum_{\beta k\sigma} \boldsymbol{V}_\beta(t)[a^+_{\beta k\sigma} a_{\beta k\sigma}, a_{\alpha p\lambda}]U^+ = \\
&\mathrm{i}U\sum_{\beta k\sigma} \boldsymbol{V}_\beta(t)\delta_{\alpha\beta}\delta_{kp}\delta_{\sigma\lambda} a_{\beta k\sigma} U^+ = \\
&\mathrm{i}U \boldsymbol{V}_\alpha(t) a_{\alpha p\lambda} U^+ = \mathrm{i}\boldsymbol{V}_\alpha(t)\bar{a}_{\alpha p\lambda}
\end{aligned} \tag{2.3.12}$$

It is solved from this equation that

$$\bar{a}_{\alpha p\lambda}(t) = \exp\left[-\mathrm{i}\int_0^t \mathrm{d}t_1\, \boldsymbol{V}_\alpha(t_1)\right]a_{\alpha p\lambda}(0) = \exp\left[-\mathrm{i}\int_0^t \mathrm{d}t_1\, \boldsymbol{V}_\alpha(t_1)\right]a_{\alpha p\lambda} \tag{2.3.13}$$

Thus, we have

$$\bar{a}^{+}_{\beta k\sigma}\,\bar{a}_{\beta k\sigma} = a^{+}_{\beta k\sigma}\,a_{\beta k\sigma} \tag{2.3.14a}$$

$$\bar{a}^{+}_{\beta k\uparrow}(t)\bar{a}^{+}_{\beta,-k\downarrow}(t) = \exp\left[2\mathrm{i}\int_0^t \mathrm{d}t_1\,\boldsymbol{V}_\beta(t_1)\right] a^{+}_{\beta,k\uparrow}\,a^{+}_{\beta,-k\downarrow} \tag{2.3.14b}$$

$$\Delta_\beta(t)\,\bar{a}^{+}_{\beta,k\uparrow}\,\bar{a}^{+}_{\beta,-k\downarrow} = \Delta^0_\beta\,\mathrm{e}^{-\mathrm{i}\varphi_\beta}\,\bar{a}^{+}_{\beta,k\uparrow}\,\bar{a}^{+}_{\beta,-k\downarrow} \tag{2.3.14c}$$

where Eq. (2.3.6) is used. $\boldsymbol{H}_c$ merely includes operators c and its conjugate, so it commutes with $a^{+}_{\beta k\sigma}\,a_{\beta k\sigma}$. That is to say, this transformation remains $\boldsymbol{H}_c$ unaltered. Let

$$\boldsymbol{V}^0_{\beta k\sigma}(t) = \boldsymbol{V}^0_{\beta k\sigma}\exp\left[i\varphi_\beta/2 + \mathrm{i}\int_0^t \mathrm{d}t_1\,\boldsymbol{V}_\alpha(t_1)\right] \tag{2.3.15}$$

In this way the Hamiltonian is transformed to be

$$\begin{aligned} U\boldsymbol{H}U^{+} = {} & \sum_{\beta k\sigma}[\varepsilon^0_{\beta k\sigma} + \boldsymbol{V}_\beta(t)]a^{+}_{\beta k\sigma}\,a_{\beta k\sigma} + \sum_k[\Delta^0_\beta\,\mathrm{e}^{-\mathrm{i}\varphi_\beta}\,a^{+}_{\beta,k\uparrow}\,a^{+}_{\beta,-k\downarrow} + \boldsymbol{H}_c] + \\ & \boldsymbol{H}_c + \sum_{\beta k\sigma}[\boldsymbol{V}^0_{\beta k\sigma}(t)a^{+}_{\beta k\sigma}\,c_{i\sigma} + \boldsymbol{H}_c] \end{aligned} \tag{2.3.16}$$

It is seen that the superconducting part becomes time-independent. However, this transformation does not meet our requirement in two aspects. One is that the lead Hamiltonian still contains time. The other is that this transformation is not the most proper one in respect of the Heisenberg equation that an operator should satisfies.

The equation satisfied by a Heisenberg operator $O(t)$ is

$$\mathrm{i}\partial_t O(t) = [O(t), \boldsymbol{H}] \tag{2.3.17}$$

If the operator $O(t)$ and Hamiltonian undergo a transformation, the transformed operator, denoted by $\overline{O}(t)$ and $\overline{\boldsymbol{H}}$, should obey the same equation:

$$\mathrm{i}\partial_t\,\overline{O}(t) = [\overline{O}(t), \overline{\boldsymbol{H}}] \tag{2.3.18}$$

In the present case, the transformation is Eq. (2.3.10):

$$\overline{O}(t) = UO(t)\,U^{+} \tag{2.3.19}$$

The time-derivative results in

$$\begin{aligned} \mathrm{i}\partial_t\,\overline{O}(t) = {} & \mathrm{i}\,\partial_t(UO(t)\,U^{+}) = \mathrm{i}\,\dot{U}O(t)U^{+} + U[\mathrm{i}\partial_t\,\overline{O}(t)]U^{+} + \mathrm{i}UO(t)\,\dot{U}^{+} = \\ & \mathrm{i}\,\dot{U}O(t)U^{+} + \mathrm{i}UO(t)\dot{U}^{+} + U[O(t), \boldsymbol{H}]U^{+} \end{aligned} \tag{2.3.20}$$

However, Eq. (2.3.18) gives a different result:

$$\begin{aligned} \mathrm{i}\partial_t\,\overline{O}(t) = {} & [\overline{O}(t), \overline{\boldsymbol{H}}] = \overline{O}(t)\overline{\boldsymbol{H}} - \overline{\boldsymbol{H}O}(t) = \\ & UO(t)U^{+}U\boldsymbol{H}\,U^{+} - U\boldsymbol{H}\,U^{+}UO(t)U^{+} = \\ & U[O(t), \boldsymbol{H} - \boldsymbol{H}O(t)]U^{+} \end{aligned} \tag{2.3.21}$$

The discrepancy between Eq. (2.3.21) and Eq. (2.3.20) reflects the fact that the transformation Eq. (2.3.16) is not the most appropriate one. It is found that the following transformation of the Hamiltonian will produce the result of Eq. (2.3.20):

$$\overline{\boldsymbol{H}} = U\boldsymbol{H}U^{+} - \mathrm{i}\dot{U}U^{+} \tag{2.3.22}$$

It follows from Eq. (2.3.11) that

$$\mathrm{i}\dot{U}U^{+} = -\mathrm{i}U\dot{U}^{+} = \sum_{\beta k\sigma} \boldsymbol{V}_{\beta}(t) a_{\beta k\sigma}^{+} a_{\beta k\sigma} \tag{2.3.23}$$

Thus,

$$\overline{\boldsymbol{H}}^{+} = U\boldsymbol{H}U^{+} + \mathrm{i}U\partial_t U^{+} = U\boldsymbol{H}U^{+} - \mathrm{i}(\partial_t U)U^{+} = \overline{\boldsymbol{H}} \tag{2.3.24}$$

which shows that after the transformation Eq. (2.3.22), the Hamiltonian is hermitian. Substitution of Eq. (2.3.22) into Eq. (2.3.17) gives

$$\begin{aligned}\mathrm{i}\partial_t \overline{O}(t) = [\overline{O}(t), \overline{\boldsymbol{H}}] = \overline{O}(t)\overline{\boldsymbol{H}} - \overline{\boldsymbol{H}}\overline{O}(t) = \\ UO(t)\boldsymbol{H}U^{+} - U\boldsymbol{H}O(t)U^{+} + \mathrm{i}UO(t)\dot{U}^{+} + \mathrm{i}\dot{U}O(t)U^{+} = \\ U[\overline{O}(t), \overline{\boldsymbol{H}}]U^{+} + \mathrm{i}UO(t)\dot{U}^{+} + \mathrm{i}\dot{U}O(t)U^{+}\end{aligned} \tag{2.3.25}$$

This result is exactly the same as Eq. (2.3.20). Therefore we are sure that the transformation of the Hamiltonian Eq. (2.3.22) is a right one. The transformed Hamiltonian is

$$\begin{aligned}\overline{\boldsymbol{H}} = \sum_{\beta}\left\{\sum_{\beta} \varepsilon_{\beta k\sigma}^{0} a_{\beta k\sigma}^{+} a_{\beta k\sigma} + \sum_{k}[\Delta_{\beta}^{0} \mathrm{e}^{-\mathrm{i}\varphi_{\beta}} a_{\beta,k\uparrow}^{+} a_{\beta,-k\downarrow}^{+} + \boldsymbol{H}_{c}]\right\} + \\ \boldsymbol{H}_{c} + \sum_{\beta k\sigma}[\boldsymbol{V}_{\beta k\sigma}^{0}(t) a_{\beta k\sigma}^{+} c_{i\sigma} + \boldsymbol{H}_{c}]\end{aligned} \tag{2.3.26}$$

In this form, the lead part does no longer depend on time, which meets the required condition. In the following we shall treat systems starting from this Hamiltonian. The superscript 0 may be omitted for convenience.

To be explicit, the coupling Hamiltonian is written in a matrix form:

$$\begin{aligned}\boldsymbol{H}_{T} &= \sum_{\beta k\sigma}[\boldsymbol{V}_{\beta k\sigma}^{*}(t) c_{i\sigma}^{+} - \boldsymbol{V}_{\beta k\sigma}(t) c_{i\sigma} a_{\beta k\sigma}^{+}] = \\ &\sum_{\beta k}\left[(c_{i\uparrow}^{+}, c_{i\downarrow})\begin{pmatrix}\boldsymbol{V}_{\beta k\uparrow}^{*}(t) & 0 \\ 0 & -\boldsymbol{V}_{\beta k\downarrow}(t)\end{pmatrix}\begin{pmatrix}a_{\beta k\uparrow} \\ a_{\beta,-k\downarrow}^{+}\end{pmatrix} + \boldsymbol{H}_{c}\right] = \\ &\sum_{\beta k}\left[(c_{i\uparrow}^{+}, c_{i\downarrow})\boldsymbol{V}_{\beta k}(t)\begin{pmatrix}a_{\beta k\uparrow} \\ a_{\beta,-k\downarrow}^{+}\end{pmatrix} + \boldsymbol{H}_{c}\right]\end{aligned} \tag{2.3.27}$$

Since the leads involve superconductivity, the wave functions in the leads should be written in the two-component form. Accordingly, the wave functions in the central region should also be in the two-component form. The coupling $\boldsymbol{V}_{\beta k}$ is a matrix of second order, which is diagonalized:

$$\boldsymbol{V}_{\beta k}(t, t') = \delta(t - t')\begin{bmatrix}\boldsymbol{V}_{\beta k\uparrow}^{*}(t) & 0 \\ 0 & -\boldsymbol{V}_{\beta k\downarrow}(t)\end{bmatrix} \tag{2.3.28}$$

because we assume that when electrons tunnel between the central region and leads, their spins are conserved.

The number of electrons with spin σ in β lead is

$$N_{\beta\sigma} = \sum_k a^+_{\beta k\sigma} a_{\beta k\sigma} \tag{2.3.29}$$

Its time-derivation is the electric current density with spin σ in this lead:

$$I_{\beta\sigma}(t) = -\mathrm{e}\langle \frac{\mathrm{d}}{\mathrm{d}t} N_{\beta\sigma}(t) \rangle = \mathrm{ie}\langle [N_{\beta\sigma}(t), \boldsymbol{H}] \rangle \tag{2.3.30}$$

The commutator $[N_{\beta\sigma}, \boldsymbol{H}]$ is easily evaluated. In the total Hamiltonian

$$\boldsymbol{H} = \sum_\beta \left\{ \sum_{k\sigma} \varepsilon_{\beta k\sigma} a^+_{\beta k\sigma} a_{\beta k\sigma} + \sum_k [\Delta_\beta \mathrm{e}^{-\mathrm{i}\varphi_\beta} a^+_{\beta k\uparrow} a^+_{\beta,-k\downarrow} + \boldsymbol{H}_c] \right\} +$$
$$\boldsymbol{H}_c + \sum_{\beta k\sigma} [\boldsymbol{V}_{\beta k\sigma}(t) a^+_{\beta k\sigma} c_{i\sigma} + \boldsymbol{H}_c] \tag{2.3.31}$$

all the parts except the βth lead and its coupling to the central region do not contain operator $a_{\beta k\sigma}$, so they commute with $N_{\beta\sigma}$. The operator $\sum_{k\sigma} \varepsilon_{\beta k\sigma} a^+_{\beta k\sigma} a_{\beta k\sigma}$ commutes with $N_{\beta\sigma}$. As a result, the nonzero part left in $[N_{\beta\sigma}, \boldsymbol{H}]$ is

$$[N_{\beta\sigma}, \boldsymbol{H}] = \left[N_{\beta\sigma}, \sum_k (\Delta_\beta \mathrm{e}^{-\mathrm{i}\varphi_\beta} a^+_{\beta,k\uparrow} a^+_{\beta,-k\downarrow} + \boldsymbol{H}_c) + \sum_{ik\alpha} (\boldsymbol{V}_{\beta k\alpha}(t) a^+_{\beta k\sigma} c_{i\sigma} + \boldsymbol{H}_c) \right] \tag{2.3.32}$$

What we need to do is to calculate $[N_{\beta\sigma}, \sum_k \Delta_\beta \mathrm{e}^{-\mathrm{i}\varphi_\beta} a^+_{\beta,k\uparrow} a^+_{\beta,-k\downarrow} + \sum_{ik_a} \boldsymbol{V}_{\beta k_a}(t) a^+_{\beta k\sigma} c_{ia}]$ plus its hermitian conjugate One recalls that the anticommutators between two fermion operators from different regions are always zero. As a consequence,

$$\left[a^+_{\beta k\sigma} a_{\beta k\sigma}, \sum_{ip\alpha} \boldsymbol{V}_{\beta p\alpha}(t) a^+_{\beta k\sigma} c_{i\sigma} \right] = \sum_i \boldsymbol{V}_{\beta p\alpha}(t) a^+_{\beta k\sigma} c_{i\sigma} \tag{2.3.33}$$

Another commutator $[a^+_{\beta k\sigma} a_{\beta k\sigma}, \sum_p \Delta_\beta \mathrm{e}^{-\mathrm{i}\varphi_\beta} a^+_{\beta,k\uparrow} a^+_{\beta,-p\downarrow}]$ is not zero, but the sum over spins is zero, so that this part does not contribute to current. Finally, we obtain

$$I_{\beta\sigma}(t) = \mathrm{ie}\langle [N_{\beta\sigma}, \boldsymbol{H}] \rangle = \mathrm{ie} \sum_k \sum_i \boldsymbol{V}_{\beta k\sigma}(t) a^+_{\beta k\sigma} c_{i\sigma} + \boldsymbol{H}_c =$$
$$\mathrm{ie} \sum_k \left[\sum_i \boldsymbol{V}_{\beta k\sigma}(t) \langle a^+_{\beta k\sigma} c_{i\sigma} \rangle + \boldsymbol{H}_c \right] =$$
$$\mathrm{e} \sum_k \left[\sum_i \boldsymbol{V}_{\beta k\sigma}(t) \boldsymbol{G}^<_{i,\beta k,\sigma\sigma} - \boldsymbol{H}_c \right] \tag{2.3.34}$$

Here $\boldsymbol{G}^<_{i,\beta k,\sigma\sigma}$ is the lesser Green's function describing the propagation from the βth lead to the central region. Now we see that the current has been expressed by the Green's function as it should be since we have known that the Green's function involves all physical information of a thermodynamic system. Thus, one task is to calculate the Green's function.

We use the capital letter $\boldsymbol{G}$ to denote the Green's functions of the whole system. If the couplings between the central region and leads are removed, the Green's functions of the isolated parts, the central region and leads, are denoted by lowercase $\boldsymbol{g}$. Dyson's equation introduced allows one to express $\boldsymbol{G}$ in terms of the uncoupled Green's function $\boldsymbol{g}$ self-energy Σ. The lesser

Green's function does not have its own Dyson's equation. Therefore, we have to calculate the causal Green's function.

The Green's function describing the propagation from the βth lead to central region is defined as

$$G_{i,\beta k,\sigma\sigma}(t_1, t_2) = \ll c_{i\sigma}(t_1) | a^{+}_{\beta k\sigma}(t_2) \gg \tag{2.3.35}$$

Obviously, when the central region and leads are all isolated, $g_{i,\beta k} = \ll c_{i\sigma}(t_1) | a^{+}_{\beta k\sigma}(t_2) \gg_0 = 0$, where the subscript 0 denotes the uncoupled system.

We are now at the stage to apply Dyson's equation, $\boldsymbol{G} = \boldsymbol{g} + \boldsymbol{G}\Sigma\boldsymbol{g}$. One can regard the βth lead as a site, the central region as another site, and the coupling Hamiltonian. With this parallelism, one gets

$$\boldsymbol{G}_{i,\beta k} = \boldsymbol{g}_{i,\beta k} + \Sigma_j \boldsymbol{G}_{ij} \boldsymbol{V}_{j,\beta k} \boldsymbol{g}_{\beta,kk} = \Sigma_j \boldsymbol{G}_{ij} \boldsymbol{V}_{j,\beta k} \boldsymbol{g}_{\beta,kk} \tag{2.3.36}$$

The Green's function describing the propagation from one energy level to another of the central region is denoted by $\boldsymbol{G}_{ij}$, and its Dyson's equation is

$$\boldsymbol{G}_{ij} = \boldsymbol{g}_{ij} + \Sigma_{\beta km} \boldsymbol{G}_{i,\beta k} \boldsymbol{V}^{*}_{m,\beta k} \boldsymbol{g}_{mj} \tag{2.3.37}$$

The total Green's function bridging the central region and leads $\boldsymbol{G}_{i,\beta k}$ appear on the right hand side, which is difficult to find. Fortunately, it can be eliminated by substitution of Eq. (2.3.36);

$$\boldsymbol{G}_{ij} = \boldsymbol{g}_{ij} + \Sigma_{n\beta km} \boldsymbol{G}_{in} \boldsymbol{V}_{n,\beta k} \boldsymbol{g}_{\beta,kk} \boldsymbol{V}^{*}_{m,\beta k} \boldsymbol{g}_{mj} \tag{2.3.38a}$$

Let us interpret the system in another point of view, The central region itself is considered as an "isolated" system, i. e., a noninteracting system. Its Hamiltonian is $\boldsymbol{H}_c$ and Green's function is $\boldsymbol{g}_{ij}$. When the leads are attached and the couplings between the central region and leads are applied, it is turned into an "interacting system" with Hamiltonian Eq. (2.3.26) and the associated Green's function is $\boldsymbol{G}_{ij}$. In this view point, Dyson's equation should be of the form

$$\boldsymbol{G}_{ij} = \boldsymbol{g}_{ij} + \Sigma_{\beta nm} \boldsymbol{G}_{in} \Sigma_{nm} \boldsymbol{g}_{mj} \tag{2.3.38b}$$

where Σ is the "self-energy" arising from the "interaction with the central region". Compared to Eq. (2.3.38a), the self-energy is

$$\Sigma_{ij} = \Sigma_{\beta k} \boldsymbol{V}_{i,\beta k} \boldsymbol{g}_{\beta,kk} \boldsymbol{V}^{*}_{j,\beta k} = \Sigma_\beta \Sigma_{\beta j} \tag{2.3.39}$$

The discussion above reveals a way how to find self-energy of a system.

By making use of Eq. (2.3.8) and Eq. (2.3.39), the expression of the current turns to be

$$\begin{aligned} I_{\beta}(t) = {} & e\Sigma_k [\Sigma_i (\boldsymbol{G}_{i,\beta k} \boldsymbol{V}^{*}_{\beta k})^{<}_{\sigma\sigma} + \boldsymbol{H}_c] = \\ & e\Sigma_k [\Sigma_i (\Sigma_j (\boldsymbol{G}_{ij} \boldsymbol{V}_{j,\beta k} \boldsymbol{g}_{\beta,kk} \boldsymbol{V}^{*}_{\beta k})^{<}_{\sigma\sigma} + \boldsymbol{H}_c)] = \\ & e\Sigma_{ij} [(\boldsymbol{G}_{ij\,\beta j})^{<}_{\sigma\sigma} + \boldsymbol{H}_c] \end{aligned} \tag{2.3.40}$$

It is now expressed by the Green's functions $\boldsymbol{G}_{ij}$ and corresponding self-energy Σ. The latter can be evaluated by the coupling Hamiltonian and the Green's functions of the isolated leads $g_{\beta,kk}$. The Green's function g_{ij} of the isolated central region belonging to H_c is supposed to have already known. The total Green's function G ; has to be solved by Eq. (2.3.38b).

The expression of the self-energy $\Sigma_{\beta ji}$ is

$$\Sigma_{\beta ji}(t_1,t)=\Sigma_k\begin{bmatrix}V^*_{\beta jk\uparrow}(t_1) & 0\\ 0 & -V_{\beta jk\downarrow}(t_1)\end{bmatrix}\times$$

$$\begin{bmatrix}\ll a_{\beta k\uparrow}(t_1)\mid a^+_{\beta k\uparrow}(t)\gg_0 & \ll a_{\beta k\uparrow}(t_1)\mid a_{\beta,-k\downarrow}(t)\gg_0\\ \ll a^+_{\beta,-k\downarrow}(t_1)\mid a^+_{\beta k\uparrow}(t)\gg_0 & \ll a_{\beta,-k\downarrow}(t_1)\mid a^+_{\beta,-k\downarrow}(t)\gg_0\end{bmatrix}\begin{bmatrix}V_{\beta k\uparrow}(t) & 0\\ 0 & -V^*_{\beta,-k\downarrow}(t)\end{bmatrix} \tag{2.3.41}$$

It is proportional to the Green's function of isolated lead. This simplicity allows one to easily obtain, by Langreth theorem, $\Sigma^R_{\beta ji}(\omega)$ and its hermitian conjugate $\Sigma^A_{\beta ji}(\omega)$, and eventually $\Sigma^<_{\beta ji}(\omega)$.

In order to make the calculation executable, some approximations are inevitable. A most often used approximation is that $V_{i,\beta k,\sigma}$ is assumed to be independent of wave vector k and spin σ . The independence of k arises from the consideration that the tunneling occurs near Fermi energy, $k\approx k_F$, and that of σ from that the tunneling usually does not violate spin-conservation. Thus,

$$\begin{bmatrix}V^*_{\beta jk\uparrow}(t_1) & 0\\ 0 & -V_{\beta jk\downarrow}(t_1)\end{bmatrix}=$$

$$V_{\beta j}\begin{bmatrix}\exp\left[-\mathrm{i}\varphi_{\beta j}-\mathrm{i}\int_0^{t_1}\mathrm{d}t'\,V_\beta(t')\right] & 0\\ 0 & -\exp\left[\mathrm{i}\varphi_{\beta j}+\mathrm{i}\int_0^{t_1}\mathrm{d}t'\,V_\beta(t')\right]\end{bmatrix}=$$

$$V_{\beta j}A_{\beta j}(t_1) \tag{2.3.42}$$

Eq. (2.3.41) becomes

$$\Sigma_{\beta ji}(t_1,t_2)=A_{\beta j}(t_1)\int\frac{\mathrm{d}\omega}{2\pi}\mathrm{e}^{-\mathrm{i}\omega(t_1-t_2)}\Sigma_{\beta ji}(\omega)A^*_{\beta j}(t_2)$$

where

$$\Sigma^R_{\beta ji}(\omega)=$$

$$V_{\beta j}V_\beta\Sigma_k\begin{bmatrix}u_k^2g_1(k,\omega)+V_k^2g_2(k,\omega) & -\mathrm{e}^{-\mathrm{i}\varphi_\beta}u_kV_k[g_1(k,\omega)+g_2(k,\omega)]\\ -\mathrm{e}^{-\mathrm{i}\varphi_\beta}u_kV_k[g_1(k,\omega)-g_2(k,\omega)] & u_k^2g_1(k,\omega)+V_k^2g_2(k,\omega)\end{bmatrix} \tag{2.3.43}$$

where the Green's function of the isolated lead. With

$$g_1(k,\omega)=\frac{1}{\omega-E_k+i\,0^+},\quad g_2(k,\omega)=\frac{1}{\omega+E_k+i\,0^+} \tag{2.3.44}$$

the diagonal element is

$$u_k^2 \boldsymbol{g}_1(\boldsymbol{k},\omega)+\boldsymbol{V}_k^2 \boldsymbol{g}_2(\boldsymbol{k},\omega) =$$

$$\frac{1}{2}\left[\left(1+\frac{\varepsilon_k}{E_k}\right)\frac{1}{\omega-E_k+\mathrm{i}0^+}+\left(1-\frac{\varepsilon_k}{E_k}\right)\frac{1}{\omega+E_k+\mathrm{i}0^+}\right]=$$

$$\frac{\varepsilon_k+\omega+\mathrm{i}0^+}{(\omega-E_k+i0^+)(\omega+E_k+\mathrm{i}0^+)} \tag{2.3.45}$$

The summation over $\boldsymbol{k}$ turns to the integration with respect to energy, in which a factor of density of states is inserted. Usually, the density of states is supposed to be constant taking the value at Fermi energy $D_{\beta s}(\varepsilon_F)$. Thus, the integration is

$$\Sigma_k[u_k^2 \boldsymbol{g}_1(\boldsymbol{k},\omega)+\boldsymbol{V}_k^2 \boldsymbol{g}_2(\boldsymbol{k},\omega)] =$$

$$\int \mathrm{d}\varepsilon \frac{D_{\beta s}(\varepsilon)(\Delta+\omega+\mathrm{i}0^+)}{(\omega-\sqrt{\varepsilon^2+\Delta^2}+\mathrm{i}0^+)(\omega+\sqrt{\varepsilon^2+\Delta^2}+\mathrm{i}0^+)}=$$

$$-D_{\beta s}(\varepsilon_F)\int_{-\infty}^{\infty}\mathrm{d}\varepsilon\frac{\omega+\mathrm{i}0^+}{(\varepsilon-\sqrt{(\omega+\mathrm{i}0^+)^2-\Delta^2})(\varepsilon+\sqrt{(\omega+\mathrm{i}0^+)^2-\Delta^2})}=$$

$$-D_{\beta s}(\varepsilon_F)\frac{2\pi\mathrm{i}}{2}\frac{\omega+\mathrm{i}0^+}{\sqrt{(\omega+\mathrm{i}0^+)^2-\Delta^2}}=-D_{\beta s}(\varepsilon_F)\pi\mathrm{i}\rho(\omega+\mathrm{i}0^+) \tag{2.3.46}$$

In implementing the integration, the integral limits are extended to infinity. The integrand is separated into two terms, the first of which is an odd function so that its integral is zero. The second integral is done by residue theorem. At last, the result is expressed by the density of states of quasiparticle in a superconductor,

$$\rho(\omega)=\frac{\omega}{\sqrt{\omega^2-\Delta^2}}=\begin{cases}\dfrac{|\omega|}{\sqrt{\omega^2-\Delta^2}}, & |\omega|>\Delta\\[2ex] \dfrac{\omega}{\mathrm{i}\sqrt{\Delta^2-\omega^2}}, & |\omega|<\Delta\end{cases} \tag{2.3.47}$$

with the shorthand notation

$$\boldsymbol{\Gamma}_{\beta i}=2\pi\boldsymbol{V}_{\beta i}\boldsymbol{V}_{\beta i}D_{\beta s}(\varepsilon_F) \tag{2.3.48}$$

the retarded self-energy of a superconducting lead is

$$\Sigma_{\beta i}^R(\omega)=-\frac{\mathrm{i}}{2}\boldsymbol{\Gamma}_{\beta i}\rho(\omega)\begin{bmatrix}1 & \dfrac{-\Delta \mathrm{e}^{-\mathrm{i}\varphi_\beta}}{\omega+\mathrm{i}0^+}\\[2ex] \dfrac{-\Delta\mathrm{e}^{\mathrm{i}\varphi_\beta}}{\omega+\mathrm{i}0^+} & 1\end{bmatrix} \tag{2.3.49}$$

Subsequently, the advanced and lesser self-energies Σ^A and $\Sigma^<$ are easily put down.

In practical computation, further approximations may still be needed. For example, $\boldsymbol{V}_\beta(t)$ is assumed to be a steady bias, or a sinusoidal alternating bias. If the Hamiltonian is independent of time, the Green's functions simply rely on time difference and Fourier transformation can be take, while if it is not, one has to be cautious about Fourier transformation.

A simple case is that the quantum dot is sandwiched by two conductors made of normal metals, and the couplings of the dot with the two leads have a proportional relationship. In this case the current formula looks concise.

As an illustration, we consider a very simple system comprising left and right leads without central region There is tunneling probability between the two conductors. This can be called tunneling model. Its Hamiltonian is

$$\boldsymbol{H} = \boldsymbol{H}_{\mathrm{R}} + \boldsymbol{H}_{\mathrm{L}} + \boldsymbol{H}_{T} = \boldsymbol{H}_0 + \boldsymbol{H}_T \tag{2.3.50}$$

where

$$\boldsymbol{H}_T = \sum_{k,p,\sigma} (\boldsymbol{T}_{k,p,\sigma}\, c_{k\sigma}^{+} c_{p\sigma} + \boldsymbol{H}_c) \tag{2.3.51}$$

is the tunneling Hamiltonian, and $\boldsymbol{H}_{\mathrm{R}}$ and $\boldsymbol{H}_{\mathrm{L}}$ are the Hamiltonians of the left and right leads, respectively Subscripts L and R label left and right, and bold letters $\boldsymbol{p}$ and $\boldsymbol{k}$ mean the wave vectors in the left and right leads, respectively. The two fermion operators from the two sides anticommute to each other. Because of the continuity of the current, it suffices to consider particles in one side, say, the left side:

$$N_{\mathrm{L}} = \sum_{p\sigma} c_{k\sigma}^{+} c_{p\sigma}$$

Its variation with time is electric current. Similar to Eq. (2.3.30),

$$-\frac{\mathrm{d}}{\mathrm{d}t} N_{\mathrm{L}} = \mathrm{i}[\boldsymbol{H}, N_{\mathrm{L}}] = \mathrm{i}[\boldsymbol{H}_T, N_{\mathrm{L}}] = \mathrm{i}\sum_{kp\sigma} (T_{kp\sigma}\, c_{k\sigma}^{+} c_{p\sigma} + \boldsymbol{H}_c) \tag{2.3.52}$$

In the averaging taken in the ensemble, $\mathrm{i}\langle c_{k\sigma}^{+} c_{p\sigma}\rangle = \boldsymbol{G}_{pk\sigma\sigma}^{<}$ emerges, which is the lesser Green's function describing propagation from the left to right lead. The causal Green's function $\boldsymbol{G}_{pk\sigma\sigma}$ across the two conductors is to be solved by Dyson's equation

$$\boldsymbol{G}_{pk\sigma\sigma} = \sum_{p_1 k_1} \boldsymbol{G}_{pp_1\sigma\sigma}\, T_{p_1 k_1 \sigma}\, \boldsymbol{G}_{k_1 k\sigma\sigma} \tag{2.3.53}$$

The following process will not be difficult if at least one lead is not superconducting, since one chooses to compute the current of the normal conductor. However, if both leads are superconducting, it would not be easy to compute $\boldsymbol{G}_{pp_1\sigma\sigma}$ because of the existence of off-diagonals.

Suppose that the bias applied is direct and not strong. We evaluate current by virtue of the linear response theory.

According to the nonequilibrium theorem, the operator Eq. (2.3.53) should take average in the whole system with Hamiltonian H. While the linear response theory merely requires to take average in unperturbed system with Hamiltonian H, which greatly simplifies the procedure in treating with the problem.

The variation of a quantity obeys:

$$\overline{\Delta D}(t) = -\mathrm{i}\int_{-\infty}^{t} \mathrm{d}t_1 \langle [\boldsymbol{H}_1^{i}(t_1), D(t)] \rangle_0 \tag{2.3.54}$$

The time-dependent operator is in the interacting picture. In the present section, $\boldsymbol{H}$ is the tunnelling Hamiltonian Eq. (2.3.51). Since $\boldsymbol{H}_1$ is steady, a convergent factor $e^{0^+ t_1}$ must be embedded in Eq. (2.3.54). We shall write this factor explicitly where necessary. Now the quantity to be changed is the current $-\frac{d}{dt}N_L(t)$. When the external field is absent, the current is zero, $-\frac{d}{dt}N_L(t)=0$.

There the current is just

$$I(t)=-e\langle\frac{d}{dt}N_L(t)\rangle = ie\int_{-\infty}^{t}\langle\left[\frac{d}{dt}N_L(t),\boldsymbol{H}_T(t_1)\right]\rangle_0 \, dt_1 \tag{2.3.55}$$

The tunneling Hamiltonian is written as

$$\boldsymbol{H}_T(t_1)=\sum_{kp\sigma}(\boldsymbol{T}_{kp\sigma}c_{k\sigma}^{+}(t_1)c_{p\sigma}(t_1)+\boldsymbol{T}_{pk\sigma}^{*}c_{p\sigma}^{+}(t_1)c_{k\sigma}(t_1))=A(t_1)+A^{+}(t_1) \tag{2.3.56}$$

where we have let

$$A(t)=\sum_{kp\sigma}\boldsymbol{T}_{kp\sigma}c_{k\sigma}^{+}(t)c_{p\sigma}(t) \tag{2.3.57}$$

Then Eq. (2.3.52) can be recast in a brief form:

$$\frac{d}{dt}N_L(t)=i\sum_{kp\sigma}(\boldsymbol{T}_{kp\sigma}c_{k\sigma}^{+}(t)c_{p\sigma}(t)-\boldsymbol{T}_{pk\sigma}^{*}c_{p\sigma}^{+}(t)c_{k\sigma}(t))= i[A(t)-A^{+}(t)] \tag{2.3.58}$$

The integral in Eq. (2.3.56) becomes

$$\begin{aligned}\langle\left[\frac{d}{dt}N_L(t),\boldsymbol{H}_T(t')\right]\rangle_0 &= i\,[A(t)-A^{+}(t),A(t')+A^{+}(t')]_0 = \\ &i\{\langle[A(t),A^{+}(t')]\rangle_0-\langle[A^{+}(t),A(t')]\rangle_0+ \\ &\langle[A(t),A(t')]\rangle_0-\langle[A^{+}(t),A^{+}(t')]\rangle_0\}\end{aligned} \tag{2.3.59}$$

The Hamiltonian Eq. (2.3.50) has not included the chemical potential yet. Because the Femi energies of the two leads may be different, the chemical potential should be considered, i. e., one ought to use the Hamiltonian $\boldsymbol{K}$ of a grand-canonical assembly instead of $\boldsymbol{H}$. $\boldsymbol{K}$ includes

$$\boldsymbol{K}_R=\boldsymbol{H}_R-\mu_R N_R,\ \boldsymbol{K}_L=\boldsymbol{H}_L-\mu_L N_L,\ \boldsymbol{K}_0=\boldsymbol{K}_R+\boldsymbol{K}_L \tag{2.3.60}$$

For free-particle systems, we have

$$\boldsymbol{H}_R=\sum_{k\sigma}\varepsilon_{k\sigma}c_{k\sigma}^{+}c_{k\sigma},\ \boldsymbol{K}_R=\sum_{k\sigma}\xi_{k\sigma}c_{k\sigma}^{+}c_{k\sigma},\ \xi_{k\sigma}=\varepsilon_{k\sigma}-\mu_R \tag{2.3.61}$$

and the associated Hamiltonian

$$\boldsymbol{H}_0=\boldsymbol{K}_0+\mu_R N_R+\mu_L N_L=\boldsymbol{K}_0+\mu_R\sum_{k\sigma}c_{k\sigma}^{+}c_{k\sigma}+\mu_L\sum_{p\sigma}c_{p\sigma}^{+}c_{p\sigma} \tag{2.3.62}$$

To gain the tunneling Hamiltonian in the interaction picture explicitly, following commutator is demanded,

$$\left[\mu_{\mathrm{R}}\sum_{k\sigma}c_{k\sigma}^{+}c_{k\sigma}+\mu_{\mathrm{L}}\sum_{p\sigma}c_{p\sigma}^{+}c_{p\sigma},c_{k\sigma}^{+}c_{p\sigma}\right]=(\mu_{\mathrm{R}}-\mu_{\mathrm{L}})c_{k\sigma}^{+}c_{p\sigma} \tag{2.3.63}$$

The chemical potential difference between the two conductors is also their electric energy potential difference, see Fig. 2. 3. 1, i. e. ,

$$\mu_{\mathrm{L}}-\mu_{\mathrm{R}}=\mathrm{e}\boldsymbol{V} \tag{2.3.64}$$

By virtue of Eq. (2. 3. 63), we obtain

$$\begin{aligned}\boldsymbol{H}_T(t)&=\mathrm{e}^{\mathrm{i}\boldsymbol{H}_0 t}\boldsymbol{H}_T\mathrm{e}^{-\mathrm{i}\boldsymbol{H}_0 t}=\mathrm{e}^{\mathrm{i}\boldsymbol{K}_0 t}\sum_{kp\sigma}(\boldsymbol{T}_{kp\sigma}\mathrm{e}^{-\mathrm{ie}\boldsymbol{V}t}c_{k\sigma}^{+}c_{p\sigma}+\boldsymbol{H}_c)\mathrm{e}^{-\mathrm{i}\boldsymbol{K}_0 t}=\\&\sum_{kp\sigma}(\boldsymbol{T}_{kp\sigma}c_{k\sigma}^{+}(t)c_{p\sigma}(t)\,\mathrm{e}^{-\mathrm{ie}\boldsymbol{V}t}+\boldsymbol{H}_c)=\\&\mathrm{e}^{-\mathrm{ie}\boldsymbol{V}t}A(t)+\mathrm{e}^{\mathrm{ie}\boldsymbol{V}t}A^{+}(t)\end{aligned} \tag{2.3.65}$$

In this picture, Eq. (2. 3. 59) becomes

$$\begin{aligned}&\left\langle\left[\frac{\mathrm{d}}{\mathrm{d}t}N_{\mathrm{L}}(t),\boldsymbol{H}_T(t')\right]\right\rangle_0=\\&\mathrm{i}\langle[\mathrm{e}^{-\mathrm{ie}\boldsymbol{V}t}A(t)-\mathrm{e}^{\mathrm{ie}\boldsymbol{V}t}A^{+}(t),\mathrm{e}^{-\mathrm{ie}\boldsymbol{V}t'}A(t')+\mathrm{e}^{\mathrm{ie}\boldsymbol{V}t'}A^{+}(t')]\rangle_0=\\&\mathrm{i}\{\mathrm{e}^{-\mathrm{ie}\boldsymbol{V}(t-t')}\langle[A(t),A^{+}(t')]\rangle_0-\mathrm{e}^{\mathrm{ie}\boldsymbol{V}(t-t')}\langle[A^{+}(t),A^{+}(t')]\rangle_0+\\&\mathrm{e}^{-\mathrm{ie}\boldsymbol{V}(t+t')}\langle[A(t),A(t')]\rangle_0-\mathrm{e}^{\mathrm{ie}\boldsymbol{V}(t+t')}\langle[A^{+}(t),A^{+}(t)]\rangle_0\}\end{aligned} \tag{2.3.66}$$

Thus, the expression of the current Eq. (2. 3. 55) contains four terms:

$$\begin{aligned}I(t)=\mathrm{e}\int_{-\infty}^{t}\mathrm{d}t'\{&\mathrm{e}^{-\mathrm{ie}\boldsymbol{V}(t-t')}\langle[A(t),A^{+}(t')]\rangle_0-\mathrm{e}^{\mathrm{ie}\boldsymbol{V}(t-t')}\langle[A^{+}(t),A(t')]\rangle_0+\\&\mathrm{e}^{-\mathrm{ie}\boldsymbol{V}(t+t')}\langle[A(t),A(t')]\rangle_0-\mathrm{e}^{\mathrm{ie}\boldsymbol{V}(t+t')}\langle[A^{+}(t),A^{+}(t')]\rangle_0\}\end{aligned} \tag{2.3.67}$$

The physical significances of the four terms are obvious. The first two terms demonstrate the process of destructing one electron on one side and creating one on the other side, describing one-electron transport. The latter two terms demonstrate the process of destructing (creating) a pair of electrons on one side and then creating (destructing) them on the other side, describing the transport of a pair of electrons between two superconductors,[5] the so-called Josephson effect.

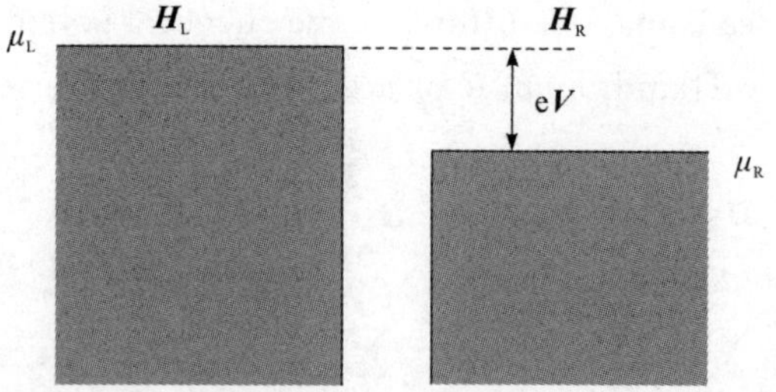

Fig. 2. 3. 1 The chemical potential difference between the two conductors is proportional to the applied voltage

In the following, we will only discuss the one-particle transport, and so ignore the other two terms:

$$I(t) = \mathrm{e}\int_{-\infty}^{t} \mathrm{d}t' \{ \mathrm{e}^{-\mathrm{ie}V(t-t')} \langle [A(t), A^{+}(t')] \rangle_0 - \mathrm{e}^{\mathrm{ie}V(t-t')} \langle [A^{+}(t), A(t')] \rangle_0 \}$$

(2.3.68)

The integrals are just the retarded Green's functions, denoted by X^R, constructed by $A(t)$ and $A^{+}(t')$:

$$I(t) = \mathrm{e}\int_{-\infty}^{\infty} \mathrm{d}t' \mathrm{e}^{-\mathrm{ie}V(t-t')} \theta(t-t') \langle [A(t), A^{+}(t')] \rangle_0 + \boldsymbol{H}_c =$$

$$i\mathrm{e}\int_{-\infty}^{\infty} \mathrm{d}t' \mathrm{e}^{-\mathrm{ie}V(t-t')} X^R(t-t') + \boldsymbol{H}_c =$$

$$i\mathrm{e}\int_{-\infty}^{\infty} \mathrm{d}t_1 \mathrm{e}^{-\mathrm{ie}Vt_1} X^R(t_1) + \boldsymbol{H}_c \qquad (2.3.69\mathrm{a})$$

Now that the operator A is the product of two fermion operators, having properties of a boson operator, so X^R is the bosonic retarded Green's function. After the variable substitution $t - t' = t_1$ in the last equal mark, it is seen that he current is actually independent of time, a natural result when a direct bias is applied. However, it should be addressed that the conclusion that a direct bias generates a direct current is valid for one-particle transport. In the case of two particle transport, a direct bias may produce an alternative current.

Eq. (2.3.68) is just the Fourier component of $X^R(t_1)$

$$I = \mathrm{ie}\, X^R(-\mathrm{e}V) + \boldsymbol{H}_c = i\mathrm{e}[X^R(-\mathrm{e}V) - X^A(-\mathrm{e}V)] =$$

$$-2\mathrm{eIm}\, X^R(-\mathrm{e}V) \qquad (2.3.69\mathrm{b})$$

It is twice the imaginary part of the retarded Green's function.

We exploit the relationship between the retarded and Matsubara Green's functions. The Matsubara Green's function constructed by A and A^{+} is

$$X(\mathrm{i}\omega) = -\int_0^{\beta} \mathrm{d}\tau\, \mathrm{e}^{\mathrm{i}\omega\tau} \langle \boldsymbol{T}_\tau [A(\tau) A^{+}(0)] \rangle_0 =$$

$$-\int_0^{\beta} \mathrm{d}\tau\, \mathrm{e}^{\mathrm{i}\omega\tau} \sum_{kp\sigma k_1 p_1 \sigma_1} \boldsymbol{T}_{kp\sigma} \boldsymbol{T}^{*}_{k_1 p_1 \sigma_1} \langle \boldsymbol{T}_\tau [c^{+}_{k\sigma}(\tau) c_{p\sigma}(\tau) c^{+}_{p_1\sigma_1} c_{k_1\sigma_1}] \rangle_0 =$$

$$\int_0^{\beta} \mathrm{d}\tau\, \mathrm{e}^{\mathrm{i}\omega\tau} \sum_{kp\sigma k_1 p_1 \sigma_1} \boldsymbol{T}_{kp\sigma} \boldsymbol{T}^{*}_{k_1 p_1 \sigma_1} \langle \boldsymbol{T}_\tau [c_{k_1\sigma_1} c^{+}_{k\sigma}(\tau)] \rangle_0 \langle \boldsymbol{T}_\tau [c_{p\sigma}(\tau) c^{+}_{p_1\sigma_1}] \rangle_0 \qquad (2.3.70)$$

The function $\langle T_\tau [A(\tau) A^{+}(0)] \rangle_0$ is constructed by the operators that are not fermion ones, so that the frequency ω should take boson values. The two leads are both normal metals. As a contraction in the second line leads to only one nonzero term, the operators in each side averaged in respective system as shown in the third line. Apparently, only when the pair of creation and annihilation operators possesses the same momenta and spins, can the averages be nonzero. Thus, the integral is just the product of the two Matsubara Green's functions of the two conductors:

$$X(\mathrm{i}\omega)=\sum_{kp\sigma}|\boldsymbol{T}_{kp\sigma}|^2\int_0^\beta \mathrm{d}\tau\, \mathrm{e}^{\mathrm{i}\omega\tau}\langle \boldsymbol{T}_\tau[c_{k\sigma}c^+_{k\sigma}(\tau)]\rangle_0\langle \boldsymbol{T}_\tau[c_{p\sigma}(\tau)c^+_{p\sigma}]\rangle_0 \tag{2.3.71}$$

The two functions are Fourier transformed:

$$X(\mathrm{i}\omega)=\sum_{kp\sigma}|\boldsymbol{T}_{kp\sigma}|^2\int_0^\beta \mathrm{d}\tau\, \mathrm{e}^{\mathrm{i}\omega\tau}\frac{1}{\beta^2}\sum_{m,n}\mathrm{e}^{\mathrm{i}\omega_m\tau-\mathrm{i}\omega_n\tau}\boldsymbol{G}_{\mathrm{R}\sigma}(\boldsymbol{k},\mathrm{i}\,\omega_m)\,\boldsymbol{G}_{\mathrm{L}\sigma}(\boldsymbol{p},\mathrm{i}\,\omega_n) \tag{2.3.72}$$

The integration over imaginary time eliminates all the terms of $m\neq n$ in the summation. The two Matsubara factors are constructed by fermion operators, so that frequencies ω_m and ω_n should take fermion values. As long as the Matsubara Green's functions are known, we are able to evaluate $X(\mathrm{i}\omega)$.

The summation in Eq. (2.3.69) concerns the product of two factors. We explore a general skill to evaluate the summation. Suppose a following summation

$$S=\frac{1}{\beta}\sum_n \boldsymbol{G}_{\mathrm{R}\sigma}(\boldsymbol{k},\mathrm{i}\,\omega_n-\mathrm{i}\omega)\boldsymbol{G}_{\mathrm{L}\sigma}(\boldsymbol{p},\mathrm{i}\,\omega_n) \tag{2.3.73}$$

A Matsubara Green's function has generally a form of

$$\boldsymbol{G}(\boldsymbol{p},\mathrm{i}\,\omega_n)=\frac{1}{\mathrm{i}\,\omega_n-\xi_p-\sum(\boldsymbol{p},\mathrm{i}\,\omega_n)} \tag{2.3.74}$$

The self-energy in the denominator arises from interaction. By means of the identity

$$\frac{1}{\beta\hbar}\sum_n\frac{\mathrm{e}^{\mathrm{i}\omega_n0^+}}{\mathrm{i}\,\omega_m+\mathrm{i}\omega-x}=\frac{1}{\beta\hbar}\sum_n\frac{\mathrm{e}^{\mathrm{i}\omega_n0^+}}{\mathrm{i}\,\omega_m-(x-\mathrm{i}\omega)}=\frac{1}{2\pi\mathrm{i}}\int_c \mathrm{d}z\,\frac{\mathrm{e}^{z0^+}f_+(z)}{z-(x-\mathrm{i}\omega)} \tag{2.3.75}$$

we have

$$\frac{1}{\beta}\sum_n\frac{1}{\mathrm{i}\,\omega_m-\mathrm{i}\omega-\xi_k-\sum(\boldsymbol{k},\mathrm{i}\,\omega_n-\mathrm{i}\,\omega_m)}\frac{1}{\mathrm{i}\,\omega_n-\xi_p-\sum(\boldsymbol{p},\mathrm{i}\,\omega_n)}=$$

$$\frac{1}{2\pi\mathrm{i}}\int_c \mathrm{d}z\,\frac{f_+(z)\mathrm{e}^{z0^+}}{z-\mathrm{i}\omega-\xi_k-\sum(\boldsymbol{k},z-\mathrm{i}\omega)}\frac{1}{z-\xi_p-\sum(\boldsymbol{p},z)}=$$

$$\frac{1}{2\pi\mathrm{i}}\int_c \mathrm{d}z\,f_+(z)\boldsymbol{G}(\boldsymbol{k},z-\mathrm{i}\omega)\boldsymbol{G}(\boldsymbol{p},z)\mathrm{e}^{z0^+} \tag{2.3.76}$$

The integral path C is deformed to be C' encircling the two poles in Eq. (2.3.76) as depicted in Fig. 2.3.2. Thus, the integral becomes those along four lines parallel to the real ω axis:

$$S=\frac{1}{\beta}\sum_n \boldsymbol{G}_{\mathrm{R}}(\boldsymbol{k},\mathrm{i}\,\omega_n-\mathrm{i}\omega)\boldsymbol{G}_{\mathrm{L}}(\boldsymbol{p},\mathrm{i}\,\omega_n)=$$

$$\frac{1}{2\pi\mathrm{i}}\left(\int_{C_1}+\int_{C_2}+\int_{C_3}+\int_{C_4}\right)\boldsymbol{G}_{\mathrm{R}}(\boldsymbol{k},z-\mathrm{i}\omega)\boldsymbol{G}_{\mathrm{L}}(\boldsymbol{p},z)f_+(z)\mathrm{d}z=$$

$$\frac{1}{2\pi}\int_{-\infty}^{\infty} d\varepsilon\, f_+(\varepsilon)[\boldsymbol{G}_R(\boldsymbol{k},\varepsilon-i\omega)A_L(\boldsymbol{p},\varepsilon)+\boldsymbol{G}_L(\boldsymbol{p},\varepsilon+i\omega)A_R(\boldsymbol{k},\varepsilon)] \tag{2.3.77}$$

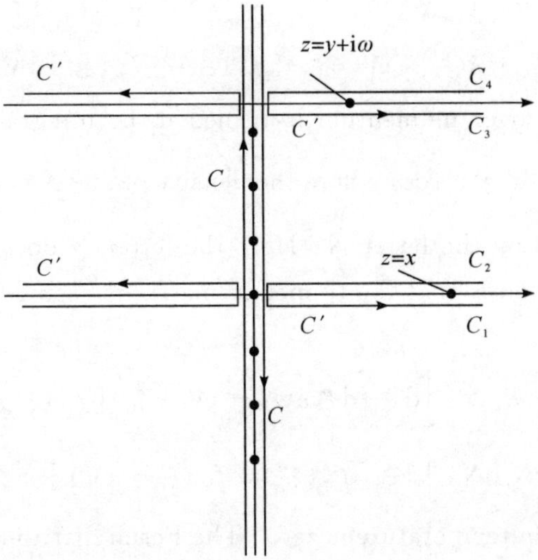

Fig. 2. 3. 2 Deformation of the integral path when the integrand is the product of two Matsubara functions

Let $i\omega \to eV + i0^+$. We obtain

$$I = -2e\mathrm{Im}\, X^R(-eV) =$$

$$-2e\mathrm{Im} \lim_{i\omega\to eV+i0^+} \frac{1}{\beta}\sum_{kp\sigma} |T_{kp\sigma}|^2 \sum_n \boldsymbol{G}_{R\sigma}(\boldsymbol{k}, i\omega_n - i\omega)\boldsymbol{G}_{L\sigma}(\boldsymbol{p}, i\omega_n) =$$

$$-\frac{e}{\pi}\sum_{kp\sigma} |T_{kp\sigma}|^2 \int_{-\infty}^{\infty} d\varepsilon[f_+(\varepsilon-eV)\mathrm{Im}\boldsymbol{G}_R(\boldsymbol{k},\varepsilon-i0^+)A_L(\boldsymbol{p},\varepsilon-eV)+$$

$$f_+(\varepsilon)\mathrm{Im}\,\boldsymbol{G}_L(\boldsymbol{p},\varepsilon-eV+i0^+)A_R(\boldsymbol{k},\varepsilon)] \tag{2.3.78}$$

In the first term, ε is replaced by $\varepsilon - eV$. Furthermore, the Green's functions are expressed by the spectral function:

$$I = -\frac{e}{2\pi}\sum_{kp\sigma} |T_{kp\sigma}|^2 \int_{-\infty}^{\infty} d\varepsilon[f_+(\varepsilon-eV)A_R(\boldsymbol{k},\varepsilon)A_L(\boldsymbol{p},\varepsilon-eV)-$$

$$f_+(\varepsilon)A_L(\boldsymbol{p},\varepsilon-eV)A_R(\boldsymbol{k},\varepsilon)] =$$

$$\frac{e}{2\pi}\sum_{kp\sigma} |T_{kp\sigma}|^2 \int_{-\infty}^{\infty} d\varepsilon\, A_L(\boldsymbol{p},\varepsilon-eV)A_R(\boldsymbol{k},\varepsilon)[f_+(\varepsilon)-f_+(\varepsilon-eV)]$$

$$\tag{2.3.79}$$

In the present case, both leads are normal metals. Suppose that there is no interaction between electrons. Then the retarded Green's function is $\boldsymbol{g}^R(\boldsymbol{k},\omega) = (\omega - \xi_k + i0^+)^{-1}$. Consequently, the spectral function is $A(\boldsymbol{k},\omega) = 2\pi\delta(\omega - \xi_k)$. Thus, Eq. (2. 3. 76) is simplified into

$$I = 2\pi e \sum_{kp\sigma} |\boldsymbol{T}_{kp\sigma}|^2 \int_{-\infty}^{\infty} d\varepsilon \delta(\varepsilon - e\boldsymbol{V} - \xi_p)\delta(\varepsilon - \xi_k)[f_+(\varepsilon) - f_+(\varepsilon - e\boldsymbol{V})] =$$

$$2\pi e \sum_{kp\sigma} |\boldsymbol{T}_{kp\sigma}|^2 \delta(\xi_k - e\boldsymbol{V} - \xi_p)[f_+(\xi_k) - f_+(\xi_k - e\boldsymbol{V})] \quad (2.3.80)$$

Now the summation over momentum is turned to be integral with respect to energy, $\sum_k \to \int \frac{d\boldsymbol{k}}{(2\pi)^3} \to \int N(\varepsilon) d\varepsilon \to N \int d\varepsilon$, where the density of states is approximated by its value at Fermi energy, denoted by the letter N. Here the letter N does not means particle number. Another approximation is that the tunneling matrix is assumed independent of momentum:

$$I = 2\pi e \sum_\sigma |\boldsymbol{T}_\sigma|^2 N_{L\sigma} N_{R\sigma} \int d\xi_L \int d\xi_R \delta(\xi_R - e\boldsymbol{V} - \xi_L)[f_+(\xi_R) - f_+(\xi_R - e\boldsymbol{V})] =$$

$$2\pi e \sum_\sigma |\boldsymbol{T}_\sigma|^2 N_{L\sigma} N_{R\sigma} \int d\xi_R [f_+(\xi_R) - f_+(\xi_R - e\boldsymbol{V})] \quad (2.3.81)$$

A simple case is when temperature is zero. The Fermi distribution function is simplified to be a step function, $\int d\xi_R [f_+(\xi_R) - f_+(\xi_R - e\boldsymbol{V})] = e\boldsymbol{V}$. The integration Eq. (2.3.81) is easily done:

$$I = 2\pi e^2 \boldsymbol{V} \sum_\sigma |\boldsymbol{T}_\sigma|^2 N_{L\sigma} N_{R\sigma} \quad (2.3.82)$$

The current is linearly proportional to applied voltage. The coefficient is the tunneling conductance:

$$\boldsymbol{G} = \frac{1}{\boldsymbol{V}} = 2\pi e^2 \sum_\sigma |\boldsymbol{T}_\sigma|^2 N_{L\sigma} N_{R\sigma} \quad (2.3.83)$$

This result demonstrates that in a tunneling model with no spin flip, the total tunneling conductance is the sum of the contributions from electrons with spins up and down. For each spin tunneling, the conductance is proportional to the product of densities of states at Fermi energies of the two conductors sandwiching the tunnel.

We consider the case that the two conductors are ferromagnetic, having magnetizations M_1 and M_2, respectively. The two conductors and a thin insulator in between compose a FM/I/FM tunneling conjunction, see Fig. 2. 3. 3. There are two simple configurations in respect to the orientations of the magnetizations, see Fig. 2. 3. 4. One is that the two magnetizations are parallel, referred to as P arrangement. and the other is antiparallel, referred to as AP arrangement. The electric resistance across the junction depends on the magnetization configuration, which is known as magnetoresistance (MR) effect of a FM/I/FM junction. The density of states at Fermi energy is denoted as N_+ (N_-) for the majority(minority) carriers no matter their spins are up or down.

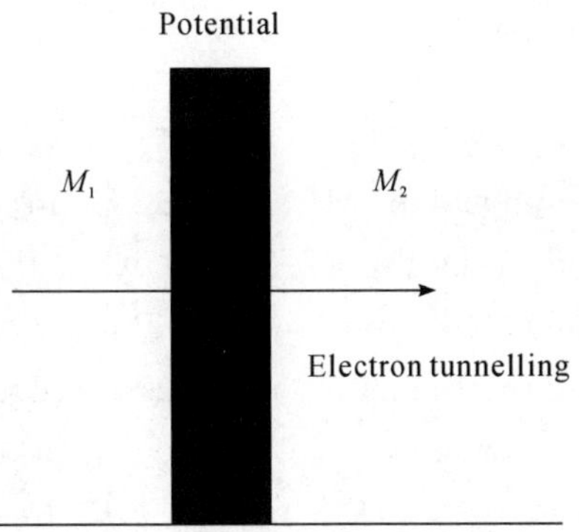

Fig. 2. 3. 3 Electron tunnelling through a FM/I/FM junction.
The magnetizations of the two ferromagnets are M_1 and M_2, respectively

In the case of P arrangement, the two magnetizations are parallel, both supposed up, see Fig. 2. 3. 4(a). The electrons with spin-up on the both sides are majority carriers and those with spin-down are minority carriers. The tunneling conductance is evaluated by Eq. (2. 3. 79):

$$\boldsymbol{G}_{\mathrm{P}} = 2\pi e^2 (|\boldsymbol{T}_+|^2 N_{\mathrm{L},+} N_{\mathrm{R},+} + |\boldsymbol{T}_-|^2 N_{\mathrm{L},-} N_{\mathrm{R},-}) \tag{2.3.84}$$

In the case of AP arrangement, the two magnetizations are antiparallel, see Fig. 2. 3. 4 (b). The electrons with spin-up are majority carriers on the left side, but are minority carriers on the right hand side. Thus, the calculated conductance is

$$\boldsymbol{G}_{\mathrm{AP}} = 2\pi e^2 (|\boldsymbol{T}|^2 N_{\mathrm{L},+} N_{\mathrm{R},-} + |\boldsymbol{T}|^2 N_{\mathrm{L},-} N_{\mathrm{R},+}) \tag{2.3.85}$$

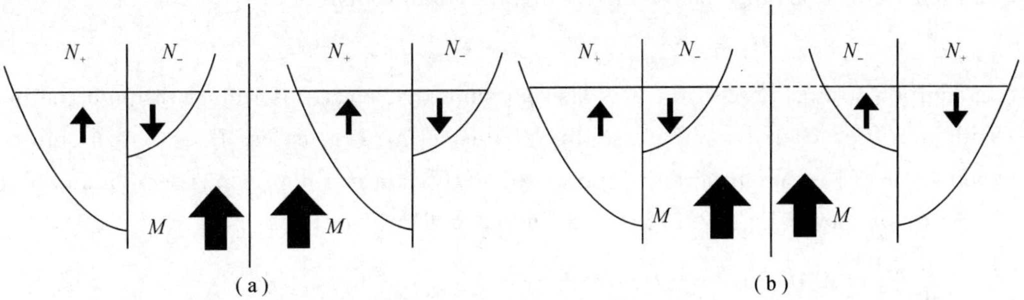

Fig. 2. 3. 4 Two arrangements of the magnetizations on the two sides of the junction
(a) P arrangement; (b) AP arrangement

The definition of the magnetoresistance effect is that the electric resistance of a material varies with applied external magnetic field. The resistance without and with a magnetic field $\boldsymbol{H}$ are denoted as $R(0)$ and $R(\boldsymbol{H})$. Then the magnetoresistance is defined by

$$M_R = \frac{R(0) - R(\boldsymbol{H})}{R(0)} \tag{2.3.86}$$

The conductance is the inverse of the resistance. The expression of the magnetoresistance in terms of the conductance is

$$M_R = \frac{G(H) - G(0)}{G(H)} \tag{2.3.87}$$

In the present case, suppose an AP arrangement in the absence of the field, which means $G(0) = G_{AP}$ already given by Eq. (2.3.85). After the field is applied, the magnetization on one side reverses, and the system turns to P arrangement, so that $G(H) = G_P$ given by Eq. (2.3.83). Assuming the tunnelling matrices T in Eq. (2.3.83) and Eq. (2.3.88) do not rely on spins, we obtain

$$M_R = \frac{G_P - G_{AP}}{G_P} = \frac{2 P_L P_R}{1 + P_L P_R} \tag{2.3.88}$$

where we have defined

$$P_{L,R} = \frac{N_{L,R,+} - N_{L,R,-}}{N_{L,R,+} + N_{L,R,-}} \tag{2.3.89}$$

Eq. (2.3.89) is called Julliere formula which is the result under the hypothesis of spin-conservation tunneling.

In some materials, electrons do retain their spins when tunnelling through the junction, as described by the tunnelling Hamiltonian Eq. (2.3.90). In other materials, electrons may flip their spins when tunnelling. This process should be included in the tunnelling Hamiltonian. The tunnelling Hamiltonians for P and AP arrangements have different forms, so they have to be discussed separately as follows.

In the case of P arrangement, the tunnelling Hamiltonian is

$$H_T = \sum_{kp\sigma} (T_{kp\sigma} c_{k\sigma}^+ c_{p\sigma} + T'_{pk\sigma} c_{k\sigma}^+ c_{p,-\sigma} + H_c) \tag{2.3.90}$$

In addition to Eq. (2.3.51), it is also possible that electrons tunnel through the junction with spin-flip. Usually, the possibility of this term $|T'|^2$ is less than that of spin-conservation term $|T|^2$. We proceed to repeat the derivation in the previous section and obtain the formulas similar to Eq. (2.3.59) and Eq. (2.3.61) as follows:

$$\begin{aligned}
\frac{d}{dt} N_L &= i[H, N_L] = i[H_T, N_L] = \\
&i\left[\sum_{k,p\sigma} (T_{kp\sigma} c_{k\sigma}^+ c_{p\sigma} + T'_{kp\sigma} c_{k\sigma}^+ c_{p,-\sigma} + H_c), \sum_{g\rho} c_{g\rho}^+ c_{g\rho}\right] = \\
&i\sum_{k,p\sigma} (T_{kp\sigma} c_{k\sigma}^+ c_{p\sigma} + T'_{kp\sigma} c_{k\sigma}^+ c_{p,-\sigma}) + H_c = i\{A(t) + A'(t)\} + H_c
\end{aligned} \tag{2.3.91}$$

$$\begin{aligned}
&\langle \left[\frac{d}{dt} N_L(t), H_T(t')\right] \rangle_0 = \\
&i\langle [A(t) + A'(t) - A^+(t) + A'^+(t), A(t') + A'(t') + A^+(t') + A'^+(t')] \rangle_0 = \\
&i(\{\langle [A(t), A^+(t)] \rangle_0 - \langle [A^+(t), A(t')] \rangle_0\} + \{\langle [A'(t), A'^+(t')] \rangle_0 - \\
&\langle [A'^+(t), A'(t')] \rangle_0\} + \{\langle [A(t), A'^+(t')] \rangle_0 - \langle [A^+(t), A'(t')] \rangle_0 +
\end{aligned}$$

$$\langle [A'(t), A^{+}(t')]\rangle_0 - \langle [A'^{+}(t), A(t')]\rangle_0 \}) \tag{2.3.92}$$

In the last equal mark, the terms describing the process of two-particle transport between two superconductors have been dropped. The time-dependent operators are in the interaction picture in the canonical ensemble. It is necessary to turn them to be in the grand-canonical ensemble.

$$\begin{aligned} \boldsymbol{H}_T(t) = \mathrm{e}^{\mathrm{i}\boldsymbol{H}_0 t}\boldsymbol{H}_T\mathrm{e}^{-\mathrm{i}\boldsymbol{H}_0 t} = & \\ & \sum_{k,p\sigma}(\boldsymbol{T}_{kp\sigma}\mathrm{e}^{\mathrm{i}(\mu_{\mathrm{R}k}-\mu_{\mathrm{L}p})t}c_{k\sigma}^{+}(t)c_{p\sigma}(t) + \\ & \boldsymbol{T}'_{kp\sigma}\mathrm{e}^{\mathrm{i}(\mu_{\mathrm{R}k}-\mu_{\mathrm{L}p})t}c_{k\sigma}^{+}(t)c_{p,-\sigma}(t) + \boldsymbol{H}_c) = \\ & A(t) + A'(t) + \boldsymbol{H}_c \end{aligned} \tag{2.3.93}$$

The operators $A(t)$ in Eq. (2.3.92) then are replaced by those defined in Eq. (2.3.93). The terms in Eq. (2.3.92) are partitioned into three groups, marked by curly brackets. In the first curly bracket, the first two terms in Eq. (2.3.93) are retrieved, indicating the spin-conservation tunnelling. Suppose the tunnelling matrices are independent of wave vector, $\boldsymbol{T}_{kp\sigma} = \boldsymbol{T}_\sigma$. Then their contributions are just Eq. (2.3.78):

$$I_1 = 2\pi\mathrm{e}\sum_\sigma |\boldsymbol{T}_\sigma|^2 N_{\mathrm{L}\sigma} N_{\mathrm{R}\sigma}\int \mathrm{d}\,\xi_{\mathrm{R}}[f_+(\xi_{\mathrm{R}}) - f_+(\xi_{\mathrm{R}} - \mathrm{e}\boldsymbol{V})] \tag{2.3.94}$$

In the second curly bracket of Eq. (2.3.92), the operators are the same as those in the first curly bracket, so that it has the same contribution as Eq. (2.3.94), except that the spins on the two sides are opposite. Again suppose the tunneling matrices are independent of wave vector, $\boldsymbol{T}'_{kp\sigma} = \boldsymbol{T}'_\sigma$. Thus, contribution to the current by the spin-flip tunneling is

$$I_2 = 2\pi\mathrm{e}\sum_\sigma |\boldsymbol{T}'_\sigma|^2 N_{\mathrm{L}\sigma} N_{\mathrm{R},-\sigma}\int \mathrm{d}\,\xi_{\mathrm{R}}[f_+(\xi_{\mathrm{R}}) - f_+(\xi_{\mathrm{R}} - \mathrm{e}\boldsymbol{V})] \tag{2.3.95}$$

These two parts of the current come from the processes that on each side one election is annihilated and another is created, both having parallel spins. In either conductor the spins are conserved. The terms in third curly bracket in Eq. (2.3.92), however, do not meet the spin conservation since they would always annihilate an electron and simultaneously create another with opposite spin. Such a possibility is necessarily zero.

In short, the total current is the sum of the contributions of Eq. (2.3.94) and Eq. (2.3.95):

$$I = 2\pi\,\mathrm{e}^2\boldsymbol{V}\sum_\sigma |\boldsymbol{T}_\sigma|^2 N_{\mathrm{L}\sigma} N_{\mathrm{R}\sigma} + 2\pi\,\mathrm{e}^2\boldsymbol{V}\sum_\sigma |\boldsymbol{T}'_\sigma|^2 N_{\mathrm{L}\sigma} N_{\mathrm{R},-\sigma} \tag{2.3.96}$$

We further assume $\boldsymbol{T}_\sigma = \boldsymbol{T}$ and $\boldsymbol{T}'_\sigma = \boldsymbol{T}'$. Let γ be the ratio

$$\gamma = \frac{\boldsymbol{T}'^2}{\boldsymbol{T}^2} \tag{2.3.97}$$

Subsequently, the conductance of P arrangement is

$$\boldsymbol{G}_{\mathrm{P}} = 2\pi\,\mathrm{e}^2\,\boldsymbol{T}^2\sum_{\sigma=\uparrow,\downarrow}(N_{\mathrm{L}\sigma}N_{\mathrm{R}\sigma} + \gamma N_{\mathrm{L}\sigma}N_{\mathrm{R},-\sigma}) =$$

$$2\pi e^2 T^2 (N_{L+} N_{R+} + \gamma N_{L+} N_{R-} + N_{L-} N_{R-} + \gamma N_{L-} N_{R+}) \quad (2.3.98)$$

In the case of AP arrangement, the tunnelling Hamiltonian is

$$H_T = \sum_{kp\sigma} (T_{kp\sigma} c_{k\sigma}^{+} c_{p,-\sigma} + T'_{kp\sigma} c_{k\sigma}^{+} c_{p,\sigma} + H_c) \quad (2.3.99)$$

In the arrangement, the possibility of spin-flip term $|T'|^2$ should be greater than that of spin-conservation term $|T|^2$. Once more the tunnelling matrices are supposed independent of wave vector. Another hypothesis is that the values of T and T' in this arrangement are the same as T' and T in P arrangement, respectively. Following the routine above we achieve the conductance of AP arrangement:

$$G_{AP} = 2\pi e^2 T^2 \sum_{\sigma=\uparrow,\downarrow} (N_{L\sigma} N_{R,-\sigma} + \gamma N_{L\sigma} N_{R\sigma}) =$$

$$2\pi e^2 T^2 (N_{L+} N_{R-} + \gamma N_{L+} N_{R+} + N_{L-} N_{R+} + \gamma N_{L-} N_{R-}) \quad (2.3.100)$$

The magnetoresistance is

$$\frac{G_P - G_{AP}}{G_P} = \frac{2(1-\gamma) P_L P_R}{1 + P_L P_R + \gamma(1 - P_L P_R)} = \frac{2 P_L P_R}{1 + P_L P_R + 2\gamma/(1-\gamma)} \quad (2.3.101)$$

This formula comprises the effect of spin-flip tunnelling. In the case that there is no such tunnelling, $T' = 0$, then $\gamma = 0$ and this formula degrades to that obtained by Julliere Eq. (2.3.88).

Finally, we would like to briefly discuss the general configuration. Up to now we have discussed the simplest two configurations, P and AP arrangement. Subject to a sufficiently strong magnetic field, the FM/I/FM conjunction will show P arrangement. Nevertheless, the initial configuration in the absence of the field is not necessarily AP arrangement. Rather, the relative orientation between the moments of the two ferromagnetic electrodes is an arbitrary angle θ. The P and AP arrangements are the special cases of $\theta=0$ and $\theta=\pi$, respectively. Suppose that the magnetizations of the left and right electrodes are along respective z and z' directions.

The Hamiltonian still consists of H_L, H_R and H_T three parts, where

$$H_L = \sum_{p\sigma} \varepsilon_{p\sigma} c_{p\sigma}^{+} c_{p\sigma} \quad (2.3.102a)$$

$$H_R = \sum_{k\sigma} \varepsilon_{k\sigma} c_{k\sigma}^{+} c_{k\sigma} \quad (2.3.102b)$$

and the tunneling Hamiltonian is Eq. (2.3.51). Due to the existence of the magnetizations, every electron feels a molecule field which shifts its energy by

$$\varepsilon_{p\sigma} = \varepsilon(\boldsymbol{p}) - \sigma\lambda_1 M_1 \quad (2.3.103a)$$

$$\varepsilon_{k\sigma} = \varepsilon(\boldsymbol{k}) - \sigma\lambda_2 M_2 \quad (2.3.103b)$$

where σ takes $+1(-1)$ when the electron spin is parallel (antiparallel) to the magnetization. This indicates that the electron energy will be lowered (raised) if its spin is parallel (antiparallel) to the magnetization.

Now the magnetizations of the left and right electrodes are $\boldsymbol{M}_1 = M_1(0,0,1)$ and $\boldsymbol{M}_2 = M_2(\sin\theta\cos\varphi, \sin\theta\sin\varphi, \cos\theta)$, *respectively. Since the spin quantization axis on the right side is along* (θ,φ) direction, the spin of an electron created or destructed by $c^+_{k\sigma}$ or $c_{k\sigma}$ is along this direction. Thus, the wave functions created from vacuum are

$$c^+_{k\uparrow}\left|0\right\rangle = \begin{bmatrix}1\\0\end{bmatrix}, c^+_{k\downarrow}\left|0\right\rangle = \begin{bmatrix}0\\1\end{bmatrix} \tag{2.3.104}$$

In order to consider the effect of spin-flip in the tunneling Hamiltonian, one has to make the spin quantization axes parallel. The following linear transformation 28 will work:

$$c^+_{k\sigma} = e^{\sigma i\varphi/2}\cos(\theta/2)b^+_{k\sigma} - \sigma\, e^{\sigma i\varphi/2}\sin(\theta/2)b^+_{k,-\sigma} \tag{2.3.105}$$

After the transformation, the wave functions created from vacuum are

$$b^+_{k\uparrow}\left|0\right\rangle = \begin{bmatrix}e^{i\varphi/2}\cos(\theta/2)\\ e^{i\varphi/2}\sin(\theta/2)\end{bmatrix}, b^+_{k\downarrow}\left|0\right\rangle = \begin{bmatrix}-e^{i\varphi/2}\sin(\theta/2)\\ e^{i\varphi/2}\cos(\theta/2)\end{bmatrix} \tag{2.3.106}$$

The Hamiltonian HR retains its form unchanged:

$$\boldsymbol{H}_R = \sum\nolimits_{k\sigma} \varepsilon_{k\sigma}\, b^+_{k\sigma}\, b_{k\sigma} \tag{2.3.107}$$

While the tunnelling Hamiltonian $\boldsymbol{H}_T$ is changed to the following form:

$$\begin{aligned}\boldsymbol{H}_T &= \sum\nolimits_{k,p,\sigma\sigma'} (\boldsymbol{T}_{kp\sigma\sigma'}\, c^+_{k\sigma}\, c_{p\sigma'} + \boldsymbol{H}_c) = \\ &\sum\nolimits_{k,p,\sigma\sigma'} (\boldsymbol{T}_{kp\sigma\sigma'}[e^{\sigma i\varphi/2}\cos(\theta/2)b^+_{k\sigma} - \sigma\, e^{\sigma i\varphi/2}\sin(\theta/2)b^+_{k,-\sigma}]c_{p\sigma'} + \boldsymbol{H}_c) = \\ &\sum\nolimits_{k,p,\sigma\sigma'} (\boldsymbol{T}_{kp\sigma\sigma'}[e^{\sigma i\varphi/2}\cos(\theta/2)b^+_{k\sigma} + \sigma\, e^{-\sigma i\varphi/2}\sin(\theta/2)b^+_{k,\sigma}]c_{p\sigma'} + \boldsymbol{H}_c) = \\ &\sum\nolimits_{k,p,\sigma\sigma'} (\boldsymbol{T}_{kp\sigma\sigma'}(\theta,\varphi)b^+_{k,\sigma}c_{p\sigma'} + \boldsymbol{H}_c)\end{aligned} \tag{2.3.108}$$

Here the spin-lip effect is also included. In the third equal mark, σ in the second term in the square bracket is replaced by $-\sigma$.

After the transformation of Eq. (2.3.105), the total Hamiltonian consists of Eq. (2.3.103a). The influence of the relative orientation (θ,φ) of spins is merged into the tunnelling matrix elements $\boldsymbol{T}_{kp\sigma\sigma'}(\theta,\varphi)$. The derivation process will not be presented here. If the field is not so strong that the linear response formula is valid, one can proceed to the calculation of the tunnelling conductance following the procedure. If the field strength is beyond this restriction, the linear response theory will not applicable and one has to implement the calculation by means of nonequilibrium Green's function technique. It should be reminded that because the spin-flip may happen when electrons tunnelling, the Green's function across the junction $\boldsymbol{G}_{pk\sigma\sigma'}(t_1, t_2) = \langle\langle c_{p\sigma'}(t_1) \mid b^+_{k\sigma}(t_2)\rangle\rangle$ ought to be written in a form of matrix of second order.

There are several reasons causing the magnetoresistance effect. One of them is the magnetoresistance of the FM/I/FM sandwich structure presented in this section. Since this kind of effect has a comparatively large magnetoresistance ($M_R > 20\%$), it is termed as giant magnetoresistance effect.

The tunnelling process discussed in this section is noncoherent, which means that in tunnelling energies of electrons are conserved but their momenta may not. For example, an electron from a quantum state $\boldsymbol{p}$ in the left electrode can occupy an arbitrary state $\boldsymbol{k}$ in the right electrode after tunnelling. The electron tunneling may also be coherent, which implies that not only the energies of electrons and spins of electrons, but also their momentum components parallel to the interface should be conserved. In this case, the momentum components perpendicular to the inter face should be one-to-one correspondence before and after tunneling due to the energy conservation. In other words, the quantum states of an electron are one-to-one correspondence before and after tunnelling.

2.4 Electronic transport through a mesoscopic structure

The Green's functions for a nonequilibrium system, or nonequilibrium Green's functions, are defined in:

$$\mathrm{i}\boldsymbol{G}_{12}^{--} = \mathrm{i}\,\boldsymbol{g}_{12}(x_1, x_2) = \langle \boldsymbol{T}_C (A_{H1} B_{H2}) \rangle \tag{2.4.1}$$

$$\mathrm{i}\,\boldsymbol{G}_{12}^{++} = \mathrm{i}\,\boldsymbol{g}_{12}^{F}(x_1, x_2) = \langle \widetilde{\boldsymbol{T}}_C (A_{H1} B_{H2}) \rangle \tag{2.4.2}$$

$$\mathrm{i}\boldsymbol{G}_{12}^{+-} = \mathrm{i}\,\boldsymbol{g}^{>}(x_1, x_2) = \langle A_{H1} B_{H2} \rangle \tag{2.4.3}$$

$$\mathrm{i}\boldsymbol{G}_{12}^{-+} = \mathrm{i}\,\boldsymbol{g}^{<}(x_1, x_2) = -\eta \langle A_{H1} B_{H2} \rangle \tag{2.4.4}$$

$$\mathrm{i}\,\boldsymbol{G}_{12}^{R} = \mathrm{i}\,\boldsymbol{g}^{R}(x_1, x_2) = \theta(t_1 - t_2)\langle [A_{H1}, B_{H2}]_{-\eta} \rangle \tag{2.4.5}$$

$$\mathrm{i}\,\boldsymbol{G}_{12}^{A} = \mathrm{i}\,\boldsymbol{g}^{A}(x_1, x_2) = -\theta(t_2 - t_1)\langle [A_{H1}, B_{H2}]_{-\eta} \rangle \tag{2.4.6}$$

For the sake of convenience and the consistency of constructing diagrams, we have employed the shorthand notations: the subscripts 1 and 2 of $\boldsymbol{G}$ represent the 4-d arguments $x_1 = (x_1, t_1)$ and $x_2 = (x_2, t_2)$, respectively, or, if the spins are considered, $x_1 = (x_1\ t_1\ \sigma_1)$ and $x_2 = (x_2\ t_2\ \sigma_2)$. The subscript H denotes that the operators are in the Heisenberg picture. The denotation $\boldsymbol{T}_C$ is called complex chronological operator, or contour-ordering operator. Its meaning is slightly different from the chronological operator $\boldsymbol{T}_t$, but their substances are the same: the operators with later time labels have to stand left to operators with earlier time labels. $\widetilde{\boldsymbol{T}}_C$ is a complex counter-chronological operator, or antitime-ordering operator. It arranges operators in a way inverse what $\boldsymbol{T}_C$ does, so they have contrary effects. The explicit meanings of the two operators will be presented below. These Green's functions except $\mathrm{i}\,\boldsymbol{G}_{12}^{++}$ have their correspondents in Eq. (2.4.1) to Eq. (2.4.5). The function $\mathrm{i}\,\boldsymbol{G}_{12}^{++}$ is a new one. $\mathrm{i}\,\boldsymbol{G}_{12}^{++}$ is named as anti-causal Green's function because the time order is governed by the anti-time-ordering operator $\widetilde{\boldsymbol{T}}_C$. We did not put it down there because it was useless in treating equilibrium systems. The function $\tilde{\boldsymbol{g}} = \boldsymbol{g}^{>} - \boldsymbol{g}^{<}$ defined in Eq. (2.4.6) is not copied here, though it can still be used. Other symbols in these definitions can refer to Eq. (2.4.1) to Eq. (2.4.5).

We have to stress here that the statistical average, denoted by $< \cdots >$, is now taken over all quantum states of the system, including possible nonequilibrium states, and not necessarily over the equilibrium states of a grand canonical ensemble. This difference between the equilibrium Green's functions defined above and makes the management of the former become extremely difficult. For example, for a noninteracting system,

$$\langle a_k^+ a_k \rangle_0 = n_k \tag{2.4.7}$$

should be the distribution function of particles of the nonequilibrium system, so one is unable to put down the expression of n_k. While the distribution function of an equilibrium system is available: it is either Fermi-Dirac or Bose-Einstein distribution function $f - \eta(\varepsilon_k) = 1/(e^{\beta\varepsilon_k} - \eta)$. Because of this difficulty, one often deals with the systems not far from equilibrium cases, referred to as near-equilibrium or quasi-equilibrium systems, where some quantities may be approximated by those of equilibrium systems. In Eq. (2.4.7) the suffix o denotes noninteracting systems.

In the case that A and B in Eq. (2.4.1) to Eq. (2.4.6) are fermion creation and annihilation operators, respectively. Let $t_1 = t_2 = t$ in Eq. (2.4.4), Then it becomes the one-particle density matrix:

$$\eta^{i\boldsymbol{G}^{-+}}(x_1 t, x_2 t) = N_\rho(t, x_1, x_2) \tag{2.4.8}$$

Here it does not matter from which side t_2 tends to the limit t_2, since $\boldsymbol{G}^{-+}$ is continuous when $t_1 = t_2$. The value of $i\boldsymbol{G}^{-+}$ with $t_1 = t_2$ is related to that of $i\boldsymbol{G}^{-+}$ by

$$i[\boldsymbol{G}^{-+}(x_1 t, x_2 t) - \boldsymbol{G}^{-+}(x_1 t, x_2 t)] = \delta(x_1 - x_2) \tag{2.4.9}$$

The identities of Eq. (2.4.14) to Eq. (2.4.19) are still valid here. We copy them with the present notations in the following:

$$\boldsymbol{G}_{12}^{--} = \theta(t_1 - t_2)\boldsymbol{G}_{12}^{+-} + \theta(t_2 - t_1)\boldsymbol{G}_{12}^{-+} \tag{2.4.10}$$

$$\boldsymbol{G}_{12}^{R} = \theta(t_1 - t_2)(\boldsymbol{G}_{12}^{+-} - \boldsymbol{G}_{12}^{-+}) = \boldsymbol{G}_{12}^{--} - \boldsymbol{G}_{12}^{-+} \tag{2.4.11a}$$

$$\boldsymbol{G}_{12}^{A} = -\theta(t_2 - t_1)(\boldsymbol{G}_{12}^{+-} - \boldsymbol{G}_{12}^{-+}) = \boldsymbol{G}_{12}^{--} - \boldsymbol{G}_{12}^{+-} \tag{2.4.11b}$$

$$\boldsymbol{G}_{12}^{R} - \boldsymbol{G}_{12}^{A} = \boldsymbol{G}_{12}^{+-} - \boldsymbol{G}_{12}^{-+} \tag{2.4.12}$$

Concerning $\boldsymbol{G}^{++}$, the four Green's functions defined in Eq. (2.4.1) to Eq. (2.4.4) are linearly related in the way of

$$\boldsymbol{G}^{--} + \boldsymbol{G}^{++} = \boldsymbol{G}^{-+} + \boldsymbol{G}^{+-} \tag{2.4.13}$$

which means that among them, three are independent. Combination of Eq. (2.4.13) with Eq. (2.4.10) to Eq. (2.4.12) yields three more identities involving $\boldsymbol{G}^{++}$:

$$\boldsymbol{G}_{12}^{++} = \theta(t_2 - t_1)\boldsymbol{G}_{12}^{+-} + \theta(t_1 - t_2)\boldsymbol{G}_{12}^{-+} \tag{2.4.14a}$$

$$G^{R}=G^{--}-G^{-+}=G^{+-}+G^{++} \quad (2.4.14b)$$

$$G^{A}=G^{--}-G^{+-}=G^{-+}-G^{++} \quad (2.4.14c)$$

Hereafter, we always assume A and B are a pair of annihilation and creation operators. From the definitions Eq. (2.4.1) to Eq. (2.4.6), one deduces the following hermitian conjugacy or anti-hermitian conjugacy relations. The retarded and advanced Green's functions are connected by hermitian conjugacy relation:

$$G_{12}^{A}=G_{21}^{R*} \quad (2.4.15)$$

The causal and anti-causal functions G^{--} and G^{++} are related by anti-hermitian conjugacy:

$$G_{12}^{--}=-G_{21}^{++*} \quad (2.4.16)$$

The lesser and greater functions G^{--} 和 G^{++} themselves are anti-hermitian conjugate:

$$G_{12}^{-+}=-G_{21}^{-+*}, \quad G_{12}^{+-}=-G_{21}^{+-*} \quad (2.4.17)$$

In taking the hermitian conjugacy in these equations, the arguments have to be interchanged.

In the steady state with spatial homogeneity, when all the functions depend only on differences $t=t_1-t_2$ and $x=x_1-x_2$, they can be Fourier transformed with respect to these variables. From Eq. (2.4.16) and Eq. (2.4.15) the Fourier components satisfy the equations

$$G^{--}(\boldsymbol{k},\omega)=-[G^{++}(\boldsymbol{k},\omega)]^{*}, \quad G^{A}(\boldsymbol{k},\omega)=[G^{R}(\boldsymbol{k},\omega)]^{*} \quad (2.4.18)$$

From Eq. (2.4.17) one has

$$G^{+-}(\boldsymbol{k},\omega)=-[G^{+-}(\boldsymbol{k},\omega)]^{*}, \quad G^{-+}(\boldsymbol{k},\omega)=-[G^{-+}(\boldsymbol{k},\omega)]^{*} \quad (2.4.19)$$

Thus, $G^{+-}(\boldsymbol{k},\omega)$ and $G^{-+}(\boldsymbol{k},\omega)$ are obviously imaginary.

The Green's functions defined in this sections apply to systems composed of fermions (bosons), phonons and photons. In the following, fermion (boson) systems are supposed, so that the Green's functions for the noninteracting systems obey Schrödinger equation. It is convenient to define an operator

$$G_0^{-1}=\mathrm{i}\hbar\frac{\partial}{\partial x}+\frac{\hbar^2}{2m}\nabla^2+\mu \quad (2.4.20)$$

to denote the time and coordinate derivatives, where the subscript 0 indicates that the functions acted pertains to a noninteracting system. Hence, its action on the noninteracting causal Green's function $G_{12}^{(0)--}$ leads to the equation

$$G_{01}^{-1}G_{12}^{(0)--}=\delta(x_1-x_2) \quad (2.4.21)$$

where the next subscript 1 in G_{01}^{-1} indicates that the differentiation is with respect to the variables x_1. Eq. (2.4.21) is just a slight extension of the differential Eq. (2.4.9), with the operator $L(\boldsymbol{r})$ replaced by the effective Hamiltonian of a grand-canonical ensemble and plus spin indices. The delta function in Eq. (2.4.21) is a compact form of three factors:

$$\delta(x_1 - x_2) = \delta_{\delta_1 \delta_2} \delta(t_1 - t_2) \delta(x_1 - x_2) \qquad (2.4.22)$$

It arises from the discontinuity of the function $G_{12}^{(0)--}$ at $t_1 = t_2$, implied by the step function. The function $G^{(0)++}$ has a discontinuity of the opposite sign at $t_1 = t_2$ and therefore

$$G_{01}^{-1} G_{12}^{(0)++} = -\delta(x_1 - x_2) \qquad (2.4.23)$$

The functions $G^{(0)R}$ and $G^{(0)A}$ have a similar discontinuity, so that they satisfy a similar equation. If the differentiation is with respect to the second, not the first, variables, G_{01}^{-1} ought to be replaced by G_{01}^{-1*}, e. g. ,

$$G_{02}^{-1*} G_{12}^{(0)--} = \delta(x_1 - x_2) \qquad (2.4.24)$$

The functions G^{+-} and G^{-+} are continuous at $t_1 = t_2$, so they satisfy the equations

$$G_{01}^{-1} G_{12}^{(0)+-} = 0 \,, \; G_{01}^{-1} G_{12}^{(0)-+} = 0 \qquad (2.4.25)$$

For phonon and photon Green's functions, the operator Eq. (2.4.20) has to be modified to the form of wave function equation, and the functions satisfy the equations of second-order time derivative.

We will now proceed to set up a diagram technique that is in principle suitable for calculating the Green's functions of a system in any nonequilibrium states. Before doing so, we have to explain the significance of the two complex chronological operators T_C and $\tilde{T}_C$. The total Hamiltonian is separated into two parts, the noninteracting H_0 and interacting part H^i, the latter being regarded the perturbation to the former. The time evolution operator $U\langle t_1, t_2 \rangle$ illustrates the change of a state with time. When the evolution time is from the remote past to the infinite future, it becomes the S matrix: $S = U\langle \infty, -\infty \rangle$.

$$U(\infty, 0) |\psi_H^0\rangle = U(\infty, -\infty) |\Phi_0\rangle = S |\Phi_0\rangle \qquad (2.4.26)$$

This equation concerns the ground state. It represents an ideal course as follows. At time $-\infty$ there is no interaction in the system, and the interaction is turned on sufficiently slowly. When time reaches $t = 0$ the system becomes the real one with interaction. Then the interaction is removed again sufficiently slowly until at time $+\infty$ when the system goes back to the noninteracting one. The whole course is reflected by the action of S matrix on the noninteracting ground state $|\Phi_0\rangle$, and the resultant is still $|\Phi_0\rangle$, at most with an additional phase factor. In the present chapter we are dealing with thermodynamic ensembles. For a thermodynamic state, one is unable to guarantee that it resumes the initial one after under going the action of the S matrix. A remediation is to apply an operator S^{-1} immediately after the action of the S matrix. Thus the change caused by the S matrix can be rigorously cancelled:

$$S^{-1} S = 1 \qquad (2.4.27)$$

It is obvious that for any state $|a\rangle$, one always has

$$S^{-1} S |a\rangle = |a\rangle \qquad (2.4.28)$$

Then, the average of a physical quantity $A_H(t)$ in a state is written as

$$\langle A_H(t)\rangle = \langle \boldsymbol{S}^{-1} U^+(0,\infty)U(0,t)A_I(t)U(t,0)U(0,\infty)\rangle_0 = \langle \boldsymbol{S}^{-1} \boldsymbol{T}_t[A_I(t)]\boldsymbol{S}\rangle_0 \tag{2.4.29}$$

where $\boldsymbol{T}_t$ is the chronological operator. The causal Green's function is

$$i\boldsymbol{G}_{12}^{--} = \langle \boldsymbol{T}_C(\psi_{H1}\psi_{H2}^+)\rangle = \langle \boldsymbol{S}^{-1} \boldsymbol{T}_t(\psi_{I1}\psi_{I2}^+\boldsymbol{S})\rangle_0 \tag{2.4.30}$$

The average of complex-chronological products of the Heisenberg operators in an interacting system can still be expressed by the average of chronological products ofthe operators in the interaction picture with respect to the corresponding noninteracting system, but with an additional operator $\boldsymbol{S}^{-1}$ as shown by Eq. (2.4.28) and Eq. (2.4.29). The $\boldsymbol{S}$ matrix turns the interaction on at zero temperature under the adiabatic assumption. Removing the interaction adiabatically will make the ground state go back exactly to the initial one $|\Phi_0\rangle$, at most with an additional phase factor e^{-iL}. This phase factor is a number, which can be removed out of the averaging and cancels with the same factor in the denominator, leaving the denominator to be $\langle\Phi_0 \mid \boldsymbol{S} \mid \Phi_0\rangle$. In this case, one can also apply the operator $\boldsymbol{S}^{-1}$ to eliminate the phase factors in the numerator and denominator, but it is needless to do so since $\boldsymbol{S}^{-1}$ merely yield a factor e^{iL}. In the present chapter, the averaging is taken over the whole ensemble including any possible state. The action of the $\boldsymbol{S}$ matrix does impose the interaction to a noninteracting excited state and then remove the interaction, whereas the final state cannot go back to the initial one, but become the superposition of other excited state, which may be intuitively regarded as the result of possible processes of mutual scattering of quasiparticle. This is a character that the ground state at zero temperature cannot possess. Therefore, the technique cannot be extended to general cases, even not the variable external field at $\boldsymbol{T}=0$ since such a field will impulse the system to enter excited states. The introduction of $\boldsymbol{S}^{-1}$ settles the problem satisfactorily, and Eq. (2.4.28) and Eq. (2.4.29) stand for any case.

Since the action of the time evolution operator in Eq. (2.4.29) evolves the state from the time at $-\infty$ to $+\infty$, and then back to $-\infty$, correspondingly, the integral with respect to time is from $-\infty$ to $+\infty$, and then back to $-\infty$, a closed path. Therefore, the Green's functions defined in this chapter are also termed as closed-time path Green's functions.

Using the unitary of $\boldsymbol{S}$ and the fact that the operator $\boldsymbol{H}^i$ is hermitian,

$$\boldsymbol{S}^{-1} = \boldsymbol{S}^+ = \tilde{\boldsymbol{T}}_t \exp\left[\frac{i}{\hbar}\int_{-\infty}^{\infty} \boldsymbol{H}_I^i(t)\,dt\right] \tag{2.4.31}$$

The effect of operator $\tilde{\boldsymbol{T}}_t$, is just contrary to that of $\boldsymbol{T}_t$. The expression of $\boldsymbol{S}$ and $\boldsymbol{S}^{-1}$ is to be expanded in powers of $\boldsymbol{H}^i$. Then Wick's theorem is used to make contractions of averaged product of operators. Nevertheless, the Wick's theorem at macroscopic limit $\boldsymbol{V}\to\infty$ is valid for any quantum state, even a nonequilibrium state.

In the case of two-body interaction, the perturbation expansion of the casual Green's function is

$$\mathrm{i}\boldsymbol{G}^{--}\langle x,y\rangle=$$

$$\sum_{m,n}\left\langle \begin{matrix}\left(-\dfrac{1}{\mathrm{i}\hbar}\right)^m \dfrac{1}{m!}\widetilde{\boldsymbol{T}}_t\left(\int \mathrm{d}^4 x_1\cdots \mathrm{d}^4 x_m \psi_{I1}^{+}\cdots\psi_{I2m}^{+}\boldsymbol{V}_{1,2}\boldsymbol{V}_{3,4}\cdots\boldsymbol{V}_{2m-1,2m}\psi_{I2m}\psi_{I1}\right)\times \\ \dfrac{1}{n!}\left(\dfrac{1}{\mathrm{i}\hbar}\right)^n \boldsymbol{T}_t\psi_{Ix}\psi_{Iy}^{+}\int \mathrm{d}^4 x_1\cdots \mathrm{d}^4 x_n \psi_{I1}^{+}\cdots\psi_{I2n}^{+}\boldsymbol{V}_{1,2}\boldsymbol{V}_{3,4}\cdots\boldsymbol{V}_{2n-1,2n}\psi_{I2n}\cdots\psi_{II}\end{matrix}\right\rangle_0 \tag{2.4.32}$$

where the shorthand notation $\boldsymbol{V}(x_1-x_2)=\boldsymbol{V}_{1,2}$ is employed. The expansions of other functions Eq. (2.4.2) to Eq. (2.4.30) are similar. The only discrepancy is that the two operators in $\boldsymbol{G}^{++}$ are arranged in a counter-chronological order. If in Eq. (2.4.31) one removes the operators $\psi_{Ix}\psi_{Iy}^{+}$ from the leftmost in the $\boldsymbol{T}_t$ product to the rightmost in the $\widetilde{\boldsymbol{T}}_t$ product, then he obtains the expansion of $\mathrm{i}\boldsymbol{G}^{++}$. The operators neither in the $\boldsymbol{T}_t$ product nor in the $\widetilde{\boldsymbol{T}}_t$ product do not take part in time-ordering. For example, the two operators in Eq. (2.4.3) do not subject to time-ordering. If in Eq. (2.4.32) one removes the operators $\psi_{Ix}\psi_{Iy}^{+}$ from the leftmost in the $\boldsymbol{T}_t$ product, and put them to the left of operator $\boldsymbol{T}_t$ but not into the $\widetilde{\boldsymbol{T}}_t$ product, then he obtains the expansion of $\mathrm{i}\boldsymbol{G}^{+-}$. Lastly, in Eq. (2.4.32) any two operators in the $\widetilde{\boldsymbol{T}}_t$ and $\boldsymbol{T}_t$ products, respectively, do not have necessary relation to each other in time, so there is no time ordering between them. Those operators that are in one $\boldsymbol{T}_t$ or in one $\widetilde{\boldsymbol{T}}_t$ products subject to time-ordering.

The followings are the zero-th, first- and second-order terms of $\mathrm{i}\boldsymbol{G}^{+-}$:

$$\mathrm{i}\boldsymbol{G}_{12}^{(0)--}=\langle \boldsymbol{T}_t(\psi_{I1}\psi_{I2}^{+})\rangle_0 \tag{2.4.33}$$

$$\mathrm{i}\boldsymbol{G}_{12}^{(1)--}=\frac{1}{\mathrm{i}\hbar}\left\langle \boldsymbol{T}_t\left[\psi_{I1}\psi_{I2}^{+}\int \mathrm{d}^4 x_3\,\mathrm{d}^4 x_4\,\frac{1}{2}\boldsymbol{V}_{34}\psi_{I3}^{+}\psi_{I4}^{+}\psi_{I4}\psi_{I3}\right]\right\rangle_0 -$$
$$\frac{1}{\mathrm{i}\hbar}\left\langle \widetilde{\boldsymbol{T}}_t\left(\int \mathrm{d}^4 x_3\,\mathrm{d}^4 x_4\,\frac{1}{2}\boldsymbol{V}_{34}\psi_{I3}^{+}\psi_{I4}^{+}\psi_{I4}\psi_{I3}\right)\boldsymbol{T}_t(\psi_{I1}\psi_{I2}^{+})\right\rangle_0 \tag{2.4.34}$$

$$\mathrm{i}\boldsymbol{G}_{12}^{(2)--}=\frac{1}{2}\left(\frac{1}{\mathrm{i}\hbar}\right)^2\left\langle \boldsymbol{T}_t\psi_{I1}\psi_{I2}^{+}\left[\int \mathrm{d}^4 x_3\,\mathrm{d}^4 x_4\,\mathrm{d}^4 x_5\,\mathrm{d}^4 x_6\,\frac{1}{4}\boldsymbol{V}_{34}\boldsymbol{V}_{56}\psi_{I3}^{+}\psi_{I4}^{+}\psi_{I4}\psi_{I3}\psi_{I5}^{+}\psi_{I6}^{+}\psi_{I6}\psi_{I5}\right]\right\rangle_0 +$$
$$\left(\frac{1}{\mathrm{i}\hbar}\right)\left(-\frac{1}{\mathrm{i}\hbar}\right)\left\langle \widetilde{\boldsymbol{T}}_t\left(\int \mathrm{d}^4 x_3\,\mathrm{d}^4 x_4\,\frac{1}{2}\boldsymbol{V}_{34}\psi_{I3}^{+}\psi_{I4}^{+}\psi_{I4}\psi_{I3}\right)\boldsymbol{T}_t\left(\psi_{I1}\psi_{I2}^{+}\int \mathrm{d}^4 x_5\,\mathrm{d}^4 x_6\,\frac{1}{2}\boldsymbol{V}_{56}\psi_{I5}^{+}\psi_{I6}^{+}\psi_{I6}\psi_{I5}\right)\right\rangle_0 +$$
$$\frac{1}{2}\left(-\frac{1}{\mathrm{i}\hbar}\right)^2\left\langle \widetilde{\boldsymbol{T}}_t\left(\int \mathrm{d}^4 x_3\,\mathrm{d}^4 x_4\,\mathrm{d}^4 x_5\,\mathrm{d}^4 x_6\,\frac{1}{4}\boldsymbol{V}_{34}\boldsymbol{V}_{56}\psi_{I3}^{+}\psi_{I4}^{+}\psi_{I4}\psi_{I3}\psi_{I5}^{+}\psi_{I6}^{+}\psi_{I6}\psi_{I5}\right)\boldsymbol{T}_t(\psi_{I1}\psi_{I2}^{+})\right\rangle_0 \tag{2.4.35}$$

Since the Wick's theorem at the macroscopic limit is valid, we are able to make pair contractions of operators in the averages.

Let us consider the first-order term Eq. (2.4.33). The second term is a characteristic of the situation in question. On averaging over the ground state at zero temperature, only the first term would have to be considered. In that case, the effect of operator $\boldsymbol{S}^{-1}$ is simply to separate a phase factor, as mentioned above, appearing as $\langle \Phi_0 \mid \boldsymbol{S} \mid \Phi_0 \rangle$ in the denominator which precisely cancels the disconnected parts of diagrams in the numerator, leaving the connected diagrams to be taken into account. In the present case, the effect of $\boldsymbol{S}^{-1}$ is not as simple as just a phase factor. Do we still partition the diagrams as connected and disconnected parts? The answer is yes. Fortunately, the combination of the two terms in Eq. (2.4.33) happens to cancel the disconnected diagrams exactly. Now we make contractions for the two terms by applying Wick's theorem and draw corresponding diagrams. We remind readers that the second term in Eq. (2.4.33) has a minus sign. The conclusion is that in Eq. (2.4.33) merely connected diagrams after contractions should be considered. Inspection of Eq. (2.4.34) leads to the same conclusion, since the sum of the disconnected diagrams of the first and third terms just cancels that of the second term. Obviously, this conclusion can be generalized to any-order term: all the disconnected diagrams can be discarded.

We look back the connected diagrams resulted from the first term in Eq. (2.4.33). They are distinguished to be two pairs. In each pair the diagrams are topologically equivalent, so that one of them is taken accompanied by the dropping of the factor 1/2 pertaining to $\boldsymbol{V}$. The conclusion is universal that only one of the topologically equivalent diagrams needs to be picked up. An n-th order connected diagram has n interaction lines. From this diagram 2^n topologically equivalent ones can be yielded by means of the interchange of the two vertices of every interaction line. Since the Integrations at the 2^n vertices are implemented, all the 2^n diagrams have exactly the same contribution. Therefore, only one topologically equivalent diagram is considered and the factor $1/2^n$ is dropped.

One point should be emphasized that distinguishes the diagrams of the present section. All the particles lines associated with the causal Green's function of free particle. While in the present case, because of the appearance of the $\tilde{\boldsymbol{T}}_t$ operator, one should discern the time directions of the particle and interaction lines. The modified line elements and their associated factors are depicted in Fig. 2.4.1. The Green's function $\mathrm{i}\boldsymbol{G}_{12}$ is associated with a line directed from 2 to 1, decorated by "+" or "−" symbol at the ends that is the same as the superscript of $\mathrm{i}\boldsymbol{G}_{12}$. An interaction line is decorated by "−"("+") symbols at its two ends if it is associated with factor $\boldsymbol{V}/\mathrm{i}\hbar$ ($\mathrm{i}\boldsymbol{V}/\hbar$).

The two topologically nonequivalent connected diagrams obtained after contraction are associated with the following expression:

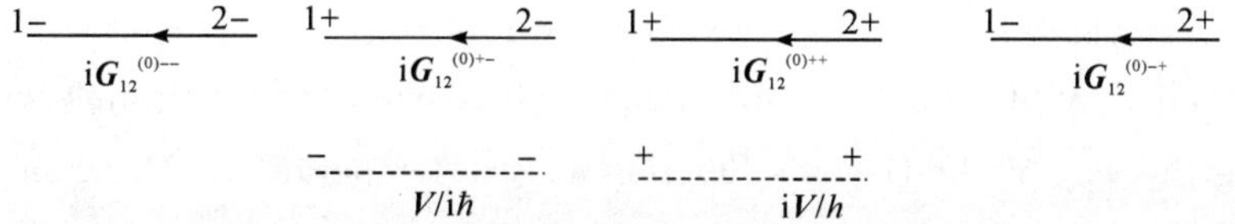

Fig. 2. 4. 1 Four single-particle lines and two interaction lines

$$(\pm\langle\psi_{I3}^{+}\psi_{I1}\rangle_0)\langle\tilde{\boldsymbol{T}}_t(\psi_{I3}\psi_{I4}^{+})\rangle_0\langle\psi_{I4}\psi_{I2}^{+}\rangle_0+(\pm\langle\psi_{I3}^{+}\psi_{I1}\rangle_0)\langle\psi_{I3}\psi_{I2}^{+}\rangle_0\langle\tilde{\boldsymbol{T}}_t(\psi_{I4}\psi_{I4}^{+})\rangle_0=$$
$$\mathrm{i}\,\boldsymbol{G}_{13}^{(0)-+}\mathrm{i}\,\boldsymbol{G}_{34}^{(0)++}\mathrm{i}\,\boldsymbol{G}_{42}^{(0)+-}+\mathrm{i}\,\boldsymbol{G}_{13}^{(0)-+}\mathrm{i}\,\boldsymbol{G}_{32}^{(0)+-}\mathrm{i}\,\boldsymbol{G}_{44}^{(0)++} \qquad (2.4.36)$$

If two operators do not coexist within $\boldsymbol{T}_t$ or within $\tilde{\boldsymbol{T}}_t$, then their times have no necessary relation. The contraction of such a pair of operator need not to be time-ordered, so that it contributes a factor $\mathrm{i}\boldsymbol{G}^{(0)-+}$ or $\mathrm{i}\boldsymbol{G}^{(0)+-}$. The contraction of a pair of operators in $\tilde{\boldsymbol{T}}_t$ operator contributes a factor $\mathrm{i}\boldsymbol{G}^{(0)++}$. With these conventions, one plots the four topologically nonequivalent diagrams in the first-order term as in Fig. 2. 4. 2. The associated expression is

$$\begin{aligned}\mathrm{i}\boldsymbol{G}_{12}^{(1)--}=\int\Big[&\frac{1}{\mathrm{i}\,\hbar}\mathrm{i}\,\boldsymbol{G}_{13}^{(0)--}\mathrm{i}\,\boldsymbol{G}_{34}^{(0)--}\mathrm{i}\,\boldsymbol{G}_{42}^{(0)--}\boldsymbol{V}_{34}+\\&\frac{1}{\mathrm{i}\,\hbar}\mathrm{i}\,\boldsymbol{G}_{13}^{(0)--}\mathrm{i}\,\boldsymbol{G}_{32}^{(0)--}\mathrm{i}\,\boldsymbol{G}_{44}^{(0)--}\boldsymbol{V}_{34}+\\&\frac{1}{\hbar}\mathrm{i}\,\boldsymbol{G}_{13}^{(0)-+}\mathrm{i}\,\boldsymbol{G}_{34}^{(0)++}\mathrm{i}\,\boldsymbol{G}_{42}^{(0)+-}\mathrm{i}\,\boldsymbol{V}_{34}+\\&\frac{1}{\hbar}\mathrm{i}\,\boldsymbol{G}_{13}^{(0)-+}\mathrm{i}\,\boldsymbol{G}_{32}^{(0)++}\mathrm{i}\,\boldsymbol{G}_{44}^{(0)++}\mathrm{i}\,\boldsymbol{V}_{34}\Big]\mathrm{d}^4x_3\,\mathrm{d}^4x_4\end{aligned} \qquad (2.4.37)$$

The analysis of the second-order term Eq. (2. 4. 34) arrives at following conclusions. There are 40 topologically nonequivalent diagrams, each having "−" symbols at its ends and two interact on lines. The 40 diagrams can be partitioned into four groups, each containing 10 diagrams. The difference between the groups is embodied in the symbols at the ends of interaction lines. In the first term of Eq. (2. 4. 34) the contractions lead to 10 diagrams, in which the two ends of every interaction line have "−" symbols. In the second term, the resulted 20 diagrams are partitioned into two groups each containing 10 diagrams. In every diagram of the first (second) group the first interaction line has"−"("+") symbols at its ends and the second interaction line has"+"("−") symbols at its ends. In the10 diagrams resulted from the third term, every interaction line has "+ "symbols at its ends.

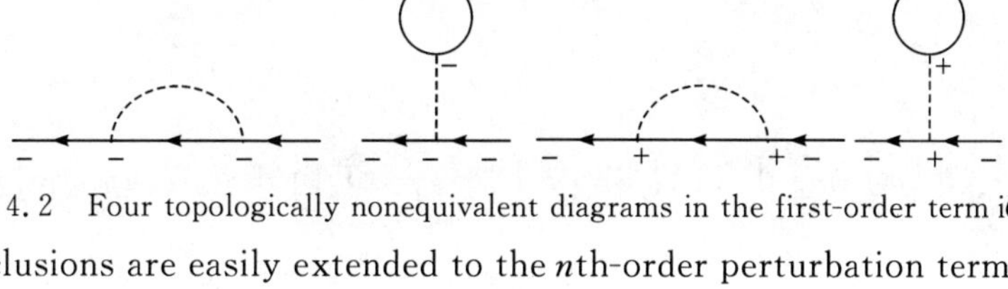

Fig. 2. 4. 2 Four topologically nonequivalent diagrams in the first-order term $\mathrm{i}\boldsymbol{G}_{12}^{(1)--}$

These conclusions are easily extended to the nth-order perturbation terms. The rules for the n th-order perturbation diagrams are written as follows:

(1) Draw all possible topologically nonequivalent connected diagrams with interaction lines following the way of the Green's function at zero temperature. Mark on the two outer ends with "−"symbols, too. This yields the first group of diagrams.

(2) In every diagram, select 1, 2, ⋯ , n interaction lines, and replace the marks on their two ends by"+" symbols. There are $C_n^1+C_n^2+\cdots+C_n^n=2^n-1$ possible ways of selection, which yields 2^n-1 new groups of diagrams. Together with the original group, there are totally 2^n groups of diagrams.

(3) Put down the expressions associated with these diagrams according to the denotations in Fig. 2. 4. 1. Integrate at each vertex with respect to 4-d coordinates: $\int d^4 x_i = \int dx_i \int dt_i$.

We have known by inspection of the first- and second-order diagrams that the former has $2^1=2$ groups, see Fig. 2. 4. 2, and the latter has $2^2=4$ groups.

The two outer ends of the diagrams of $i\boldsymbol{G}^{--}$ are marked by "−" symbols denoted as"−, −" for convenience. If they are replaced by"+, −""−, +"and "+, +", then one immediately obtains the diagrams of $i\boldsymbol{G}^{+-}$, $i\boldsymbol{G}^{-+}$ and $i\boldsymbol{G}^{++}$, respectively.

As for the rules for the diagrams in the momentum space, they can be easily put down by readers themselves.

Let us now turn to the case of external field. Like Eq. (2. 4. 31), the perturbation expansion of the casual Green's function is:

$$i\boldsymbol{G}^{--}(x,y)=$$

$$\sum_{m,n}\frac{1}{m!}\frac{1}{n!}\left(\frac{1}{i\hbar}\right)^m\left(\frac{i}{\hbar}\right)^n\left\langle\begin{matrix}\tilde{\boldsymbol{T}}_t\left(\int d^4x_1\cdots d^4x_m\,\boldsymbol{V}_1^e\boldsymbol{V}_2^e\cdots\boldsymbol{V}_m^e\,\psi_{I1}^+\psi_{I2}^+\cdots\psi_{Im}^+\psi_{Im}\cdots\psi_{I2}\psi_{I1}\right)\times\\ \boldsymbol{T}_t\left(\psi_{I_x}\psi_{I_y}^+\int d^4x_1\cdots d^4x_n\,\boldsymbol{V}_1^e\boldsymbol{V}_2^e\cdots\boldsymbol{V}_n^e\,\psi_{I1}^+\psi_{I2}^+\cdots\psi_{I_n}^+\psi_{I_n}\cdots\psi_{I2}\psi_{I1}\right)\end{matrix}\right\rangle_0 \tag{2.4.38}$$

where the shorthand notation $\boldsymbol{V}_1^e=\boldsymbol{V}^e(x_1)$ is employed. The expressions up to the second-order terms are

$$i\boldsymbol{G}_{12}^{(0)--}=\langle\boldsymbol{T}_t(\psi_{I1}\psi_{I2}^+)\rangle_0 \tag{2.4.39}$$

$$i\boldsymbol{G}_{12}^{(1)--}=\frac{1}{i\hbar}\left\langle\boldsymbol{T}_t\left(\psi_{I1}\psi_{I2}^+\int d^4x_3\,\boldsymbol{V}_3^e\psi_{I3}^+\psi_{I3}\right)\right\rangle_0+\frac{i}{\hbar}\left\langle\tilde{\boldsymbol{T}}_t\left(\int d^4x_3\,\boldsymbol{V}_3^e\psi_{I3}^+\psi_{I3}\right)\boldsymbol{T}_t(\psi_{I1}\psi_{I2}^+)\right\rangle_0 \tag{2.4.40}$$

$$
\begin{aligned}
\mathrm{i}\boldsymbol{G}_{12}^{(2)--} = & \frac{1}{2}\left(\frac{1}{\mathrm{i}\hbar}\right)^2 \left\langle \boldsymbol{T}_t\left(\psi_{I1}\psi_{I2}^{+}\int \mathrm{d}^4x_3\,\mathrm{d}^4x_4 \boldsymbol{V}_3^{\mathrm{e}}\boldsymbol{V}_4^{\mathrm{e}}\psi_{I3}^{+}\psi_{I3}\psi_{I4}^{+}\psi_{I4}\right)\right\rangle_0 + \\
& \frac{1}{\mathrm{i}\hbar}\left(-\frac{\mathrm{i}}{\hbar}\right)\left\langle \tilde{\boldsymbol{T}}_t\left(\int \mathrm{d}^4x_3 \boldsymbol{V}_3^{\mathrm{e}}\psi_{I3}^{+}\psi_{I3}\right)\boldsymbol{T}_t\left(\psi_{I1}\psi_{I2}^{+}\int \mathrm{d}^4x_4 \boldsymbol{V}_4^{\mathrm{e}}\psi_{I4}^{+}\psi_{I4}\right)\right\rangle_0 + \\
& \frac{1}{2}\left(-\frac{1}{\mathrm{i}\hbar}\right)^2 \left\langle \tilde{\boldsymbol{T}}_t\left(\int \mathrm{d}^4x_3\,\mathrm{d}^4x_4 \boldsymbol{V}_3^{\mathrm{e}}\boldsymbol{V}_4^{\mathrm{e}}\psi_{I3}^{+}\psi_{I3}\psi_{I4}^{+}\psi_{I4}\right)\boldsymbol{T}_t(\psi_{I1}\psi_{I2}^{+})\right\rangle_0
\end{aligned}
\tag{2.4.41}
$$

The denotations of the external field lines and their associated factors are shown in Fig. 2.4.3. An interaction line is marked by"−"("+") symbol at its one end if it is associated with factor $\boldsymbol{V}^{\mathrm{e}}/\mathrm{i}\hbar$ ($\mathrm{i}\boldsymbol{V}^{\mathrm{e}}/\hbar$). The two first-order diagrams obtained by the contractions of Eq. (2.4.35) are plotted in Fig. 2.4.4. The contractions of Eq. (2.4.40) give four second-order diagrams plotted in Fig. 2.4.5. The analysis of the diagrams is quite similar to, but simpler than, the case of two-body interaction.

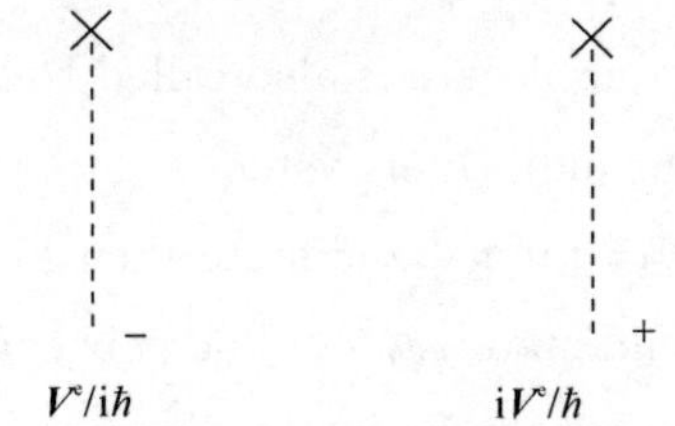

Fig. 2.4.3 The diagram elements of the external field lines and their associated factors

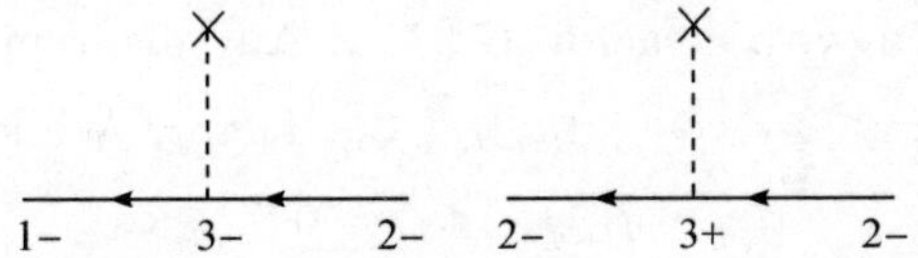

Fig. 2.4.4 Two first-order diagrams of $\mathrm{i}\boldsymbol{G}_{12}^{(1)--}$ in the case of external field

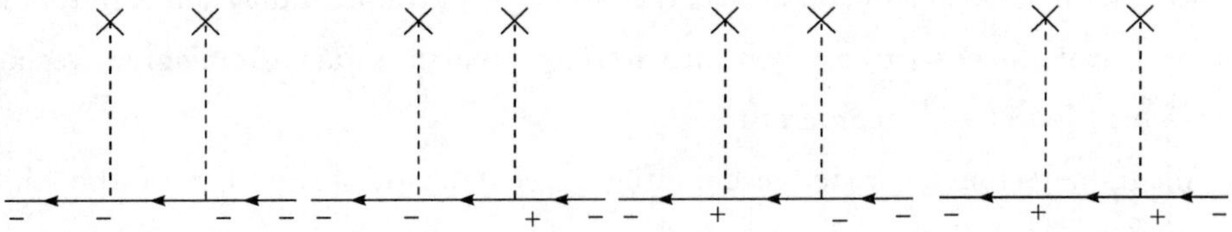

Fig. 2.4.5 Four second-order diagrams of $\mathrm{i}\boldsymbol{G}_{12}^{(1)--}$ in the case of external field

The rules for the n th-order perturbation diagrams are written as follows.

(1) Draw all possible topologically nonequivalent connected diagrams with n external field lines following the way of the Green's function at zero temperature in chapter 1. Mark on the two outer ends with "−" symbols. Mark each vertex with "−" symbol. There is only

are diagram.

(2)In the diagram, select 1, 2, ⋯ , n external field lines, and replace the marks on the corresponding vertices by "+" symbols. There are $C_n^1 + C_n^2 + \cdots + C_n^n = 2^n - 1$ possible ways of selection, which yields $2^n - 1$ new diagrams. Together with the original group, there are totally 2^n diagrams.

(3)Put down the expressions associated with these diagrams according to the denotations of the Green's functions in Fig. 2. 4. 1 and of the external field lines in Fig. 2. 4. 3. Integrate at each vertex with respect to 4-d coordinates: $\int d^4 x_i = \int d x_i \int dt_i$.

Again, the two outer ends of the diagrams of iG^{--} are marked by "−, −" symbols. If they are replaced by "+, −" "−, +" and "+, +", then one immediately obtains the diagrams of iG^{+-} , iG^{-+} and iG^{++} , respectively. The rules for the diagrams in the momentum space can be easily put down.

The diagram technique introduced here is also called Keldysh technique. This technique stands for any equilibrium or nonequilibrium system.

In the case that one deals with the ground state then Eq. (2. 4. 1) to Eq. (2. 4. 6) degrade to the averages in the ground state $|\psi_H^0\rangle$. The pair contractions result in the averages in the noninteracting ground state $|\Phi_0\rangle$. From the diagram rules above, any-order perturbation term has 2^n groups of diagrams. Take iG^{--} as an illustration. The particle lines in the first group are all associated with a factor $iG^{(0)--}$. Any diagram outside of the first group possess as at least one $iG_{12}^{(0)--} = \langle \Phi_0 \mid \psi_{I2}^+ \psi_{I1} \mid \Phi_0 \rangle$ factor. We know that

$$\psi_{I2}^+ \psi_{I1} |\Phi_0\rangle = 0 \tag{2. 4. 42}$$

This result remains unchanged for phonon field operators and for boson systems without Bose condensation. The conclusion is that the later $2^n - 1$ groups of diagrams are totally zero. Only the first group needs to be taken into account, which is just the diagram technique of the Green's function at zero temperature.

The diagram technique introduced in the present chapter and that of the Matsubara Green's functions have their own advantages. Keldysh technique can deal with any system (except boson systems with Bose condensation). It will automatically go back to the diagram technique of the Green's function at zero temperature when treating the ground state. The Matsubara Green's functions deal with equilibrium systems at finite temperature. One can also calculate the Physical quantities of the ground state by taking the zero temperature limit. This technique merely requires one group of diagrams, compared to the 2^n groups of Keldysh technique.

In each diagram, cutting off the two outer lines will leave the self-energy diagram. The diagrams in the preceding section reveal that there are four kinds of self-energy diagrams, denoted by $\boldsymbol{\Sigma}^{--}$, $\boldsymbol{\Sigma}^{+-}$, $\boldsymbol{\Sigma}^{-+}$ and $\boldsymbol{\Sigma}^{++}$, respectively. Dyson's equation for $\mathrm{i}\boldsymbol{G}^{--}$ is depicted in Fig. 2.4.6. If one self-energy diagram can be divided into two parts by cutting one particle line, then it is called the proper self-energy otherwise it will be named as improper self-energy. There are also four proper self-energy diagrams, denoted by $\boldsymbol{\Sigma}^{*--}$, $\boldsymbol{\Sigma}^{*+-}$, $\boldsymbol{\Sigma}^{*-+}$ and $\boldsymbol{\Sigma}^{*++}$, respectively. Fig. 2.4.7 illustrates how the self-energy $\boldsymbol{\Sigma}^{--}$ is constructed in terms of the proper self-energies. The other three self-energies are constructed in similar way. If the self-energy diagrams in Fig. 2.4.6 are expressed in terms of the proper self-energy diagrams, Dyson's equation will turn to be Fig. 2.4.8. We have to mention that the latter four terms represented by Figs. 2.4.8(a) (b) (c) and (d) are not one-to-one correspond to those in Fig. 2.4.6. Each term, say $\mathrm{i}\boldsymbol{G}^{--}$ in Fig. 2.4.8(a) in turn includes the contributions from all the four kinds of self-energy diagrams represented by Figs. 2.4.6(a) (b) (c) and (d). In some lowest-order diagrams of proper self-energy in the case of two-body interaction are plotted in Fig. 2.4.9. Dyson's equations for other three Green's functions $\mathrm{i}\boldsymbol{G}^{+-}$, $\mathrm{i}\boldsymbol{G}^{-+}$ and $\mathrm{i}\boldsymbol{G}^{++}$ can be achieved by change of symbols at the ends of the diagrams in Figs. 2.4.6 and 2.4.8. For example, when the upper and lower ends are r can be achieved by change of the symbols at the ends of the marked by "+" and "−" symbols, respectively, one gains Dyson's equation for $\mathrm{i}\boldsymbol{G}^{+-}$. Apparently, the four equations are coupled to each other and have to be solved simultaneously.

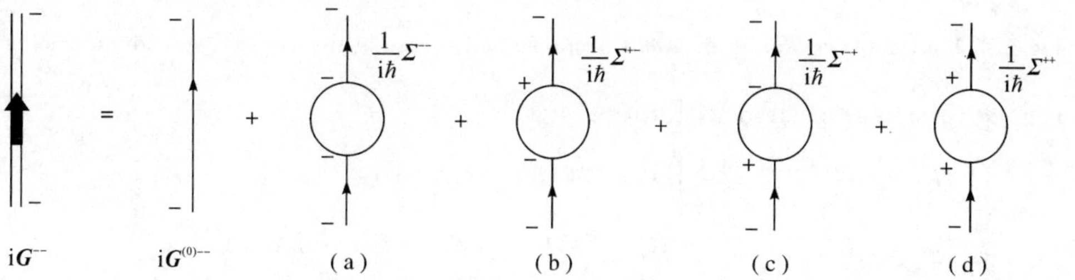

Fig. 2.4.6 Dyson's equation in which $\mathrm{i}\boldsymbol{G}^{--}$ is expressed in terms of the self-energies

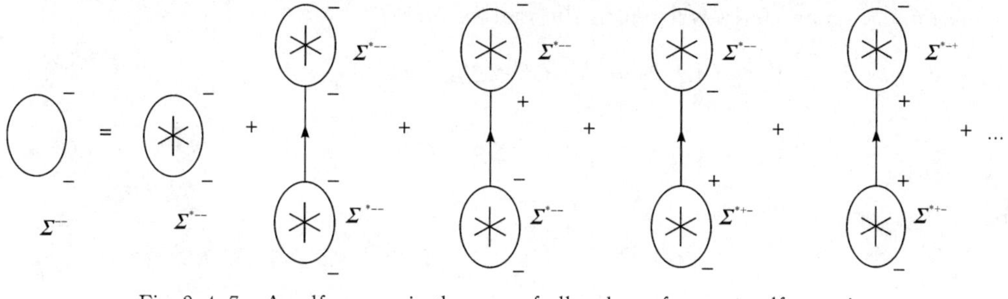

Fig. 2.4.7 A self-energy is the sum of all orders of proper self-energies

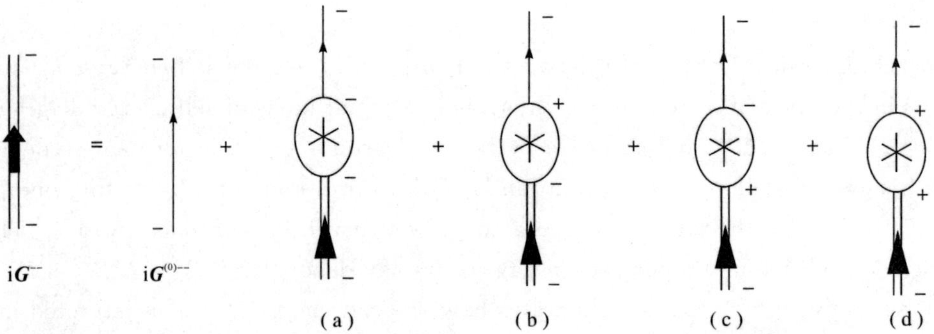

Fig. 2.4.8 Dyson's equation in which $\mathrm{i}\boldsymbol{G}^{--}$ is expressed in terms of the proper self-energies

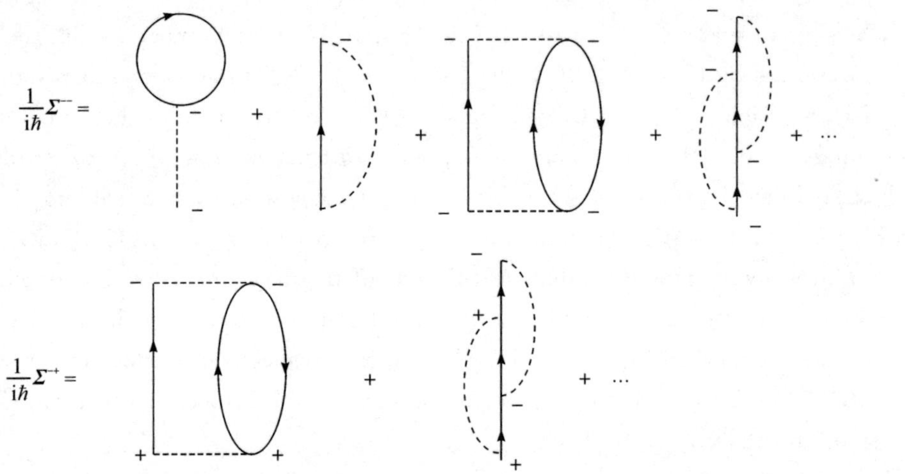

Fig. 2.4.9 Some lowest-order diagrams of proper self-energy in the case of two-body interaction

The expression associated with Fig. 2.4.8 is

$$\boldsymbol{G}_{12}^{--} = \boldsymbol{G}_{12}^{(0)--} + \frac{1}{\hbar}\int[\boldsymbol{G}_{14}^{(0)--}\boldsymbol{\Sigma}_{43}^{*--}\boldsymbol{G}_{32}^{--} + \boldsymbol{G}_{14}^{(0)-+}\boldsymbol{\Sigma}_{43}^{*+-}\boldsymbol{G}_{32}^{--} + \boldsymbol{G}_{14}^{(0)--}\boldsymbol{\Sigma}_{43}^{*-+}\boldsymbol{G}_{32}^{+-} + \boldsymbol{G}_{14}^{(0)-+}\boldsymbol{\Sigma}_{43}^{*++}\boldsymbol{G}_{32}^{+-}]\,\mathrm{d}^4x_3\,\mathrm{d}^4x_4 \tag{2.4.43}$$

Similarly, the other three expressions for $\mathrm{i}\boldsymbol{G}^{+-}$, $\mathrm{i}\boldsymbol{G}^{-+}$ and $\mathrm{i}\boldsymbol{G}^{++}$ can be put down. These four equations are compactly written in the matrix form:

$$\boldsymbol{G}_{12} = \boldsymbol{G}_{12}^{(0)} = \frac{1}{\hbar}\int \boldsymbol{G}_{14}^{(0)}\boldsymbol{\Sigma}_{43}^{*--}\boldsymbol{G}_{32}\,\mathrm{d}^4x_3\,\mathrm{d}^4x_4 \tag{2.4.44a}$$

$$\boldsymbol{G}_{12} = \boldsymbol{G}_{12}^{(0)} = \frac{1}{\hbar}\int \boldsymbol{G}_{14}\boldsymbol{\Sigma}_{43}^{*--}\boldsymbol{G}_{32}^{(0)}\,\mathrm{d}^4x_3\,\mathrm{d}^4x_4 \tag{2.4.44b}$$

where

$$\boldsymbol{G} = \begin{bmatrix}\boldsymbol{G}^{--} & \boldsymbol{G}^{-+}\\ \boldsymbol{G}^{+-} & \boldsymbol{G}^{++}\end{bmatrix}, \boldsymbol{\Sigma}^{*} = \begin{bmatrix}\boldsymbol{\Sigma}^{*--} & \boldsymbol{\Sigma}^{*-+}\\ \boldsymbol{\Sigma}^{*+-} & \boldsymbol{\Sigma}^{*++}\end{bmatrix} \tag{2.4.45}$$

Eq. (2.4.44) includes Eq. (2.4.43) and other three equations.

The differential Eq. (2.4.21) to Eq. (2.4.25) satisfied by the Green's functions of noninteracting systems are similarly written jointly as

$$\boldsymbol{G}_{01}^{-1}\,\boldsymbol{G}_{12}^{(0)} = \sigma_3\delta(x_1 - x_2) \tag{2.4.46}$$

where

$$\boldsymbol{\sigma}_3 = \begin{bmatrix} 1 & 0 \\ 0 & -1 \end{bmatrix} \tag{2.4.47}$$

Here the equation is written in virtue of Pauli matrix $\boldsymbol{\sigma}_3$ which is not of the physical significance of spin.

Let us now apply the operator $\boldsymbol{G}_{01}^{-1}$ defined by Eq. (2.4.20) on both sides of Eq. (2.4.44a). Using Eq. (2.4.46), we obtain a set of integro-differential equations

$$\boldsymbol{G}_{01}^{-1}\,\boldsymbol{G}_{12} = \boldsymbol{\sigma}_3\delta(x_1 - x_2) + \frac{1}{\hbar}\int \boldsymbol{\sigma}_3\,\boldsymbol{\Sigma}_{13}^{*}\,\boldsymbol{G}_{32}\,\mathrm{d}^4 x_3 \tag{2.4.48}$$

The application of $\boldsymbol{G}_{01}^{-1}$ on Eq. (2.4.44b) results in

$$\boldsymbol{G}_{02}^{-1}\,\boldsymbol{G}_{12} = \boldsymbol{\sigma}_3\delta(x_1 - x_2) + \frac{1}{\hbar}\int \boldsymbol{G}_{13}\,\boldsymbol{\Sigma}_{32}^{*}\,\boldsymbol{\sigma}_3\,\mathrm{d}^4 x_3 \tag{2.4.49}$$

where Eq. (2.4.24) is employed. The merit of Eq. (2.4.48) and Eq. (2.4.49) is that the Green's functions of noninteracting systems do not appear in these two equations, so they are formally simpler than Eq. (2.4.44). Their shortcoming is that both are integro-differential equations, which makes their solutions indefinite unless an integral constant is known. This integral constant is in fact the first term of Eq. (2.4.44), i. e., the Green's functions of a noninteracting system.

It should be addressed that because of the linear dependence of the four Green's functions disclosed by Eq. (2.4.13), two equations in Eq. (2.4.44) are independent. In order to show this characteristic explicitly in the matrix equation, a linear transformation is applied to the matrix $\boldsymbol{G}$, which will reduce one of its elements to be zero in virtue of Eq. (2.4.13). The linear transformation is

$$\boldsymbol{G}_g = \boldsymbol{R}^{-1}\boldsymbol{G}\boldsymbol{R} \tag{2.4.50}$$

$$\boldsymbol{R} = \frac{1}{\sqrt{2}}\begin{bmatrix} 1 & 1 \\ -1 & 1 \end{bmatrix},\ \boldsymbol{R}^{-1} = \frac{1}{\sqrt{2}}\begin{bmatrix} 1 & -1 \\ 1 & 1 \end{bmatrix} \tag{2.4.51}$$

are unitary matrices. It is easily seen that the transformed matrix is

$$\boldsymbol{G}_g = \frac{1}{2}\begin{bmatrix} \boldsymbol{G}^{--} - \boldsymbol{G}^{+-} - \boldsymbol{G}^{-+} + \boldsymbol{G}^{++} & \boldsymbol{G}^{--} - \boldsymbol{G}^{+-} + \boldsymbol{G}^{-+} - \boldsymbol{G}^{++} \\ \boldsymbol{G}^{--} + \boldsymbol{G}^{+-} - \boldsymbol{G}^{-+} - \boldsymbol{G}^{++} & \boldsymbol{G}^{--} + \boldsymbol{G}^{+-} + \boldsymbol{G}^{-+} + \boldsymbol{G}^{++} \end{bmatrix} = \begin{bmatrix} 0 & \boldsymbol{G}^{\mathrm{A}} \\ \boldsymbol{G}^{\mathrm{R}} & \boldsymbol{F} \end{bmatrix} \tag{2.4.52}$$

where Eq. (2.4.13) and Eq. (2.4.14) are used and an $\boldsymbol{F}$ function

$$F = G^{++} + G^{--} = G^{+-} + G^{-+} \tag{2.4.53}$$

is defined. When the matrices $\boldsymbol{G}^0$ and $\boldsymbol{\Sigma}^*$ are transformed in this way, the matrix Eq. (2.4.44) remain its form invariant. Furthermore, Eq. (2.4.13) also implies that the four proper self-energy functions cannot be independent of each other, so there must be a linear relation of them. To find this relation, we write Eq. (2.4.48) explicitly,

$$G_{01}^{-1}\begin{bmatrix} G^{--} & G^{-+} \\ G^{+-} & G^{++} \end{bmatrix} = \begin{bmatrix} 1 & 0 \\ 0 & -1 \end{bmatrix}\delta(x_1 - x_2) + \int d^4 x_3 \begin{bmatrix} 1 & 0 \\ 0 & -1 \end{bmatrix}\begin{bmatrix} \Sigma_{13}^{*--} & \Sigma_{13}^{*-+} \\ \Sigma_{13}^{*+-} & \Sigma_{13}^{*++} \end{bmatrix}\begin{bmatrix} G_{32}^{--} & G_{32}^{-+} \\ G_{32}^{+-} & G_{32}^{++} \end{bmatrix} \tag{2.4.54}$$

It follows from Eq. (2.4.13) that $G_{01}^{-1}(G^{--} + G^{++} + G^{-+} + G^{+-}) = 0$. Thus the sum of the four elements of the right hand side of Eq. (2.4.54) must be zero, too:

$$(\Sigma_{13}^{*--} + \Sigma_{13}^{*-+} + \Sigma_{13}^{*+-} + \Sigma_{13}^{*++})(G_{32}^{--} - G_{32}^{-+}) = 0$$

Hence, the linear relation of the four proper self-energy functions is

$$\Sigma^{*--} + \Sigma^{*-+} + \Sigma^{*+-} + \Sigma^{*++} = 0 \tag{2.4.55}$$

Please notice its difference to Eq. (2.4.13) in signs. After the linear transformation Eq. (2.4.51), the proper self-energy matrix becomes

$$\Sigma_g^* = R^{-1}\Sigma^* R = \begin{bmatrix} \Omega & \Sigma^{R} \\ \Sigma^{A} & 0 \end{bmatrix} \tag{2.4.56}$$

where we have defined

$$\left.\begin{aligned} \Omega &= \Sigma^{*--} + \Sigma^{*++} = -(\Sigma^{*-+} + \Sigma^{*+-}), \\ \Sigma^{R} &= \Sigma^{*--} + \Sigma^{*-+}, \ \Sigma^{A} = \Sigma^{*--} + \Sigma^{*+-} \end{aligned}\right\} \tag{2.4.57}$$

The equation transformed from Eq. (2.4.44) is

$$\begin{bmatrix} 0 & G_{12}^{A} \\ G_{12}^{R} & F_{12} \end{bmatrix} = \begin{bmatrix} 0 & G_{13}^{(0)A} \\ G_{12}^{(0)R} & F_{12}^{(0)} \end{bmatrix} + \int d^4 x_3\, d^4 x_4 \begin{bmatrix} 0 & G_{14}^{(0)A} \\ G_{14}^{(0)R} & F_{14}^{(0)} \end{bmatrix}\begin{bmatrix} \Omega_{43} & \Sigma_{32}^{A} \\ \Sigma_{43}^{A} & 0 \end{bmatrix}\begin{bmatrix} 0 & G_{32}^{A} \\ G_{32}^{R} & F_{32} \end{bmatrix} \tag{2.4.58}$$

Expanding this matrix equation, we obtain three equations. One of them is

$$G_{12}^{A} = G_{12}^{(0)A} + \int G_{14}^{(0)A}\Sigma_{43}^{A} G_{32}^{A}\, d^4 x_3\, d^4 x_4 \tag{2.4.59}$$

The element $\boldsymbol{G}^{R}$ satisfies a similar equation, but Eq. (2.4.15) reveals that it gives nothing new compared to Eq. (2.4.59). The functions $\boldsymbol{G}^{(0)R}$ and $\boldsymbol{G}^{(0)A}$ are independent of the distribution function of the noninteracting systems, and so is Eq. (2.4.59).

Lastly, the equation satisfied by $\boldsymbol{F}$ is

$$F_{12} = F_{12}^{(0)} + \int (G_{14}^{(0)R}\Omega_{43} G_{32}^{A} + G_{14}^{(0)A}\Sigma_{43}^{A} G_{32}^{A} + G_{14}^{(0)R}\Sigma_{43}^{R} F_{32})\, d^4 x_3\, d^4 x_4 \tag{2.4.60}$$

Since

$$\boldsymbol{F}_{12}^{(0)}\ \boldsymbol{G}_{01}^{-1} = 0 \tag{2.4.61}$$

$\boldsymbol{F}_{12}$ satisfies following integro-differential equation

$$\boldsymbol{G}_{01}^{-1}\ \boldsymbol{F}_{12} = \int (\boldsymbol{\Omega}_{13}\ \boldsymbol{G}_{32}^{\mathrm{A}} + \boldsymbol{\Sigma}_{13}^{\mathrm{R}}\ \boldsymbol{F}_{32})\mathrm{d}^4\ x_3 \tag{2.4.62}$$

Eq. (2.4.59) and Eq. (2.4.62) constitute in principle a complete description of the behavior of a nonequilibrium system. Specifically, the well-known Boltzmann transport equation is implied in Eq. (2.4.62). Since Eq. (2.4.62) is an integro-differential equation, its solution is arbitrary to some extent contrast, Eq. (2.4.59) is purely an integral equation, so there is no arbitrariness in its solution. Because of Eq. (2.4.8) and Eq. (2.4.9), the functions $\boldsymbol{G}^{-+}$, $\boldsymbol{G}^{+-}$ and $\boldsymbol{F}$ are directly related to the distribution function of a system.

An essential property of these two equations is that they contain two time variables t_1 and t_2. If the quasiclassical case is considered, there should be only one time variable. The so-called quasiclassical case means that the time interval τ and distance l over which all quantities vary significantly satisfy the inequalities

$$\tau\varepsilon_F \gg \hbar\ , l\,p_F \gg \hbar \tag{2.4.63}$$

In this limiting case, Eq. (2.4.62) can give the ordinary quasiclassical transport equation in which the explicit expressions of the "gain" and "loss" terms are achieved from the proper self-energy diagrams up to the second order.

Let us now explain the meaning of the complex chronological operator $\boldsymbol{T}_{\mathrm{C}}$ in detail. We have mentioned that the whole time path is a closed one: from $-\infty$ forward to $+\infty$, and then from $+\infty$ backward to $-\infty$. In this way, the state rigorously goes back to its initial one. However, the times in the forward and backward ways may be confused. In order to distinguish the two ways unambiguously, an infinitesimal positive (negative) imaginary part is pertained to the time in the forward (backward) way. Thus the times in the forward and backward ways are denoted by t^+ and t^-, respectively. Now the two ways are just above and below the real time axis, see Fig. 2.4.10, so that they are also named as the upper and lower ways (banks), respectively. It is because the time is a complex number, though its imaginary part is infinitesimal, that $\boldsymbol{T}_{\mathrm{C}}$ is called the complex chronological operator. In Eq. (2.4.1) to Eq. (2.4.6), all the time variables should have an infinitesimal imaginary part, so they are complex chronological Green's functions. If t_2 is later than t_1 in the closed path, then the exchange of A and B should be accompanied by a minus sign for fermions. One should keep in mind that the time in the lower bank is always later than that in the upper bank, no matter what value it is.

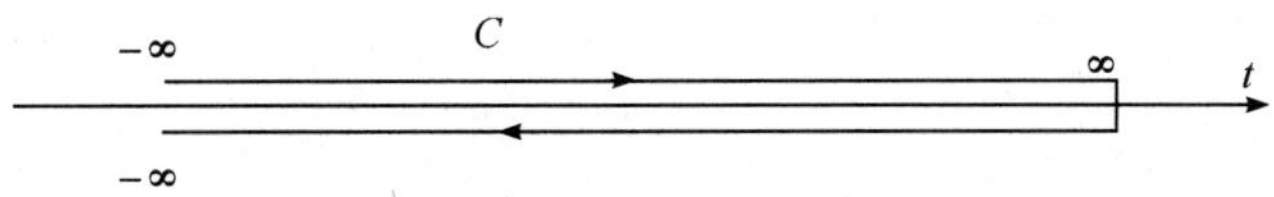

Fig. 2. 4. 10 Closed complex time path

Written the time more clearly, Eq. (2. 4. 1) to Eq. (2. 4. 4) are as follows:

$$\boldsymbol{G}_{12}^{--}=\boldsymbol{G}^{--}(t_1^+,t_2^+) \tag{2.4.64}$$

$$\boldsymbol{G}_{12}^{++}=\boldsymbol{G}^{++}(t_1^-,t_2^-) \tag{2.4.65}$$

$$\boldsymbol{G}^{+-}(t_1,t_2)=\boldsymbol{G}^{+-}(t_1^-,t_2^+) \tag{2.4.66}$$

$$\boldsymbol{G}^{-+}(t_1,t_2)=\boldsymbol{G}^{-+}(t_1^+,t_2^-) \tag{2.4.67}$$

In the causal Green's function Eq. (2. 4. 1), both time variables are in the forward way, meaning a chronological order. While in the anti-causal Green's function Eq. (2. 4. 2), both time variables are in the backward way, meaning an anti-chronological order. In the lesser and greater Green's functions $\boldsymbol{G}^<$ and $\boldsymbol{G}^>$, the two time variables are always on different banks. The retarded and advanced banks, because the step functions determine the time order, no matter which bank the times are on.

There is no conception of time for the Matsubara Green's functions, so that one is unable to define nonequilibrium Matsubara Green's functions.

With the prescribed complex time path, the relations Eq. (2. 4. 10) to Eq. (2. 4. 14) remain invariant.

The perturbation theorem for the nonequilibrium statistics is founded on the complex time Green's functions, while the measurable quantities are related to the real time Green's functions. The bridge connecting the two kinds of Green's functions is Langreth theorem. If there is an identity

$$C(t_1,t_2)=\int_C \mathrm{d}t A(t_1,t)B(t,t_2) \tag{2.4.68}$$

where times t_1 and t_2 are on the contour path shown in Fig. 2. 4. 11 and the integral is along this path, then one has

$$C^<(t_1,t_2)=\int_{-\infty}^{+\infty}\mathrm{d}t[A^{\mathrm{R}}(t_1,t)B^<(t,t_2)+A^<(t_1,t)B^A(t,t_2)] \tag{2.4.69}$$

$$C^>(t_1,t_2)=\int_{-\infty}^{+\infty}\mathrm{d}t[A^{\mathrm{R}}(t_1,t)B^>(t,t_2)+A^>(t_1,t)B^A(t,t_2)] \tag{2.4.70}$$

$$C^{\mathrm{R}}(t_1,t_2)=\int_{-\infty}^{+\infty}\mathrm{d}t\, A^{\mathrm{R}}(t_1,t)B^{\mathrm{R}}(t,t_2) \tag{2.4.71}$$

Note that the integrals on the right hand sides are not the close path, but just the forward way. The above four equations are usually written in the following shorthand forms:

$$(AB)^<=A^{\mathrm{R}}B^<+A^<B^A \tag{2.4.72}$$

$$(AB)^{>} = A^{R} B^{>} + A^{>} B^{A} \tag{2.4.73}$$

$$(AB)^{R} = A^{R} B^{R} \tag{2.4.74}$$

$$(AB)^{A} = A^{A} B^{A} \tag{2.4.75}$$

These equations are so-called Langreth theorem.

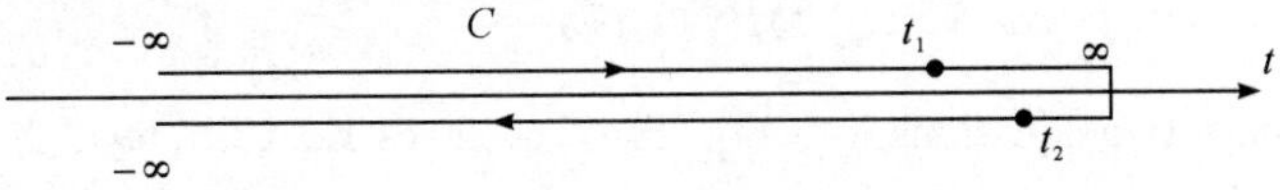

Fig. 2.4.11 The integral path of Eq. (2.4.68)

As an illustration, let us prove Eq. (2.4.69). Suppose that in Eq. (2.4.68) the times t_1 and t_2 are on upper and lower banks, respectively, which means that we are referring the lesser Green's functions $C^{<}(t_1, t_2)$. The integrals along the upper and lower banks when $t > \max(t_1, t_2)$ exactly cancel with each other. Assumed $t_1 < t_2$, the integral path in Fig. 2.4.11 becomes that in Fig. 2.4.12. The latter is then deformed to be the path in Fig. 2.4.13. Now the integral in Eq. (2.4.68) is partitioned to be the closed paths above and below the real axis:

$$C^{<}(t_1, t_2) = \int_C \mathrm{d}t A(t_1^+, t) B(t, t_2^-) = \int_{C_1} \mathrm{d}t A(t_1^+, t) B^{<}(t, t_2^-) + \int_{C_1} \mathrm{d}t A^{<}(t_1^+, t) B(t, t_2^-) \tag{2.4.76}$$

In the first term, $B(t, t_2^-)$ is relabelled as $B^{<}(t, t_2^-)$ for t is always before t_2^-. Similarly, $A(t_1^+, t)$ in the second term is relabelled to be $A^{<}(t_1^+, t)$. The first integral in Eq. (2.4.76) is

$$\begin{aligned}\int_{C_1} \mathrm{d}t A(t_1^+, t) B^{<}(t, t_2^-) &= \int_{-\infty}^{t_1} \mathrm{d}t A^{>}(t_1^+, t) B^{<}(t, t_2^-) + \int_{t_1}^{-\infty} \mathrm{d}t A^{<}(t_1^+, t) B^{<}(t, t_2^-) = \\ &\int_{-\infty}^{+\infty} \mathrm{d}t \theta(t_1^+ - t) A^{>}(t_1^+, t) B^{<}(t, t_2^-) - \int_{-\infty}^{t_1} \mathrm{d}t A^{<}(t_1^+, t) B^{<}(t, t_2^-) = \\ &\int_{-\infty}^{+\infty} \mathrm{d}t [\theta(t_1^+ - t) A^{>}(t_1^+, t) - \theta(t_1^+ - t) A^{<}(t_1^+, t)] B^{<}(t, t_2^-) = \\ &\int_{-\infty}^{+\infty} \mathrm{d}t A^{R}(t_1^+, t) B^{<}(t, t_2^-)\end{aligned} \tag{2.4.77a}$$

In the first equal mark, $A(t_1^+, t)$ in the first term is relabelled as $A^{>}(t_1^+, t)$ since t is before t_1^+. Similarly, $A(t_1^+, t)$ in the second term is relabelled as $A^{<}(t_1^+, t)$. In the third equal mark, the upper limit of the second integral is extended to infinity, so that a factor $\theta(t_1^+ - t)$ is inserted. In the last equal mark Eq. (2.4.77a) is employed. In a similar way, the second term of Eq. (2.4.76) is processed as following:

$$\begin{aligned}\int_{C_2} \mathrm{d}t A^{<}(t_1^+, t) B(t, t_2^-) &= \int_{-\infty}^{t_2} \mathrm{d}t A^{<}(t_1^+, t) B^{<}(t, t_2^-) + \int_{t_2}^{-\infty} \mathrm{d}t A^{<}(t_1^+, t) B^{>}(t, t_2^-) = \\ &\int_{-\infty}^{+\infty} \mathrm{d}t A^{<}(t_1^+, t) \theta(t_1^+ - t) B^{<}(t, t_2^-) - \int_{-\infty}^{t_2} \mathrm{d}t A^{<}(t_1^+, t) B^{<}(t, t_2^-) =\end{aligned}$$

$$\int_{-\infty}^{+\infty} \mathrm{d}t\, A^{<}(t_1^+, t)\theta(t_1^+ - t)B^{<}(t, t_2^-) -$$

$$\int_{-\infty}^{t_2} \mathrm{d}t\, A^{<}(t_1^+, t)\theta(t_1^+ - t)B^{<}(t, t_2^-) =$$

$$\int_{-\infty}^{+\infty} \mathrm{d}t\, A^{<}(t_1^+, t)B^{A}(t, t_2^-) \tag{2.4.77b}$$

Substitution of these two equations into Eq. (2.4.76) gives Eq. (2.4.69).

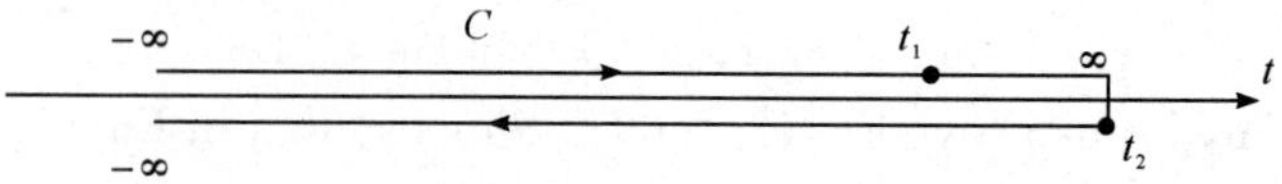

Fig. 2.4.12 The shortcut of the path in Fig. 2.4.11, where the integrals along the upper and lower banks when $t > \max(t_1, t_2)$ exactly cancel with each other

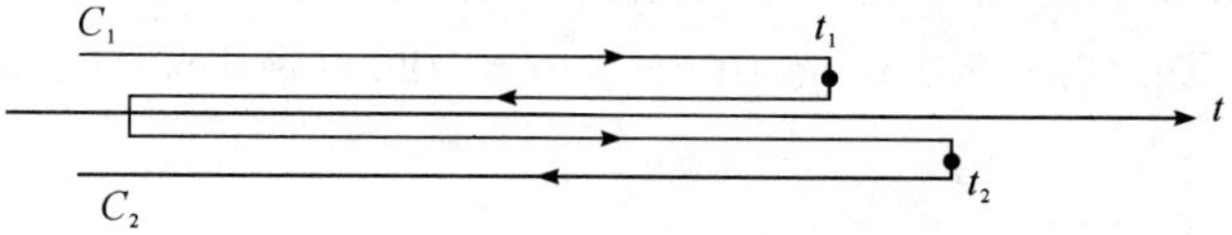

Fig. 2.4.13 The deformation of the path in Fig. 2.4.12

Dyson's equation is satisfied by the closed path Green's function

$$\boldsymbol{G}(t_1, t_2) = \boldsymbol{g}(t_1, t_2) + \int_C \mathrm{d}t_1\, \mathrm{d}t_2\, \boldsymbol{g}(t_1, t_3)\boldsymbol{\Sigma}(t_3, t_4)\boldsymbol{G}(t_4, t_2) \tag{2.4.78}$$

This equation is often shortened as

$$\boldsymbol{G} = \boldsymbol{g} + \boldsymbol{g\Sigma G} \tag{2.4.79}$$

Starting from this equation and using Langreth theorem, one can obtain following equations for real Green's functions:

$$\boldsymbol{G}^R = \boldsymbol{g}^R + \boldsymbol{g}^R \boldsymbol{\Sigma}^R \boldsymbol{G}^R \tag{2.4.80}$$

$$\boldsymbol{G}^A = \boldsymbol{g}^A + \boldsymbol{g}^A \boldsymbol{\Sigma}^A \boldsymbol{G}^A \tag{2.4.81}$$

$$\left.\begin{aligned} \boldsymbol{G}^{<} &= (1 + \boldsymbol{G}^R \boldsymbol{\Sigma}^R)\boldsymbol{g}^{<}(1 + \boldsymbol{\Sigma}^A \boldsymbol{G}^A) + \boldsymbol{G}^R \boldsymbol{\Sigma}^{<} \boldsymbol{G}^A = \\ &\quad \boldsymbol{G}^R(\boldsymbol{g}^R) - 1\, \boldsymbol{g}^{<}(\boldsymbol{g}^A) - 1\, \boldsymbol{G}^A + \boldsymbol{G}^R \boldsymbol{\Sigma}^{<} \boldsymbol{G}^A \\ \boldsymbol{G}^{>} &= (1 + \boldsymbol{G}^R \boldsymbol{\Sigma}^R)\boldsymbol{g}^{>}(1 + \boldsymbol{\Sigma}^A \boldsymbol{G}^A) + \boldsymbol{G}^R \boldsymbol{\Sigma}^{>} \boldsymbol{G}^A \end{aligned}\right\} \tag{2.4.82}$$

They form a complete set of equations describing the properties of nonequilibrium thermodynamic systems. Among them, three are independent due to Eq. (2.4.12). In obtaining these equations, Eq. (2.4.72) to Eq. (2.4.75) may be successively applied. Eq. (2.4.82) was called Keldysh equation. We now give the proof. It follows from Eq. (2.4.79) that

$$\boldsymbol{G}^{<} = \boldsymbol{g}^{<} + \boldsymbol{g}^R \boldsymbol{\Sigma}^R \boldsymbol{G}^{<} + \boldsymbol{g}^R \boldsymbol{\Sigma}^{<} \boldsymbol{G}^A + \boldsymbol{g}^{<} \boldsymbol{\Sigma}^A \boldsymbol{G}^A \tag{2.4.83}$$

and then

$$(1 - \boldsymbol{g}^R \boldsymbol{\Sigma}^R)\boldsymbol{G}^{<} = \boldsymbol{g}^{<}(1 + \boldsymbol{\Sigma}^A \boldsymbol{G}^A) + \boldsymbol{g}^R \boldsymbol{\Sigma}^{<} \boldsymbol{G}^A \tag{2.4.84}$$

Eq. (2.4.80) gives

$$(1-\boldsymbol{g}^{R}\boldsymbol{\Sigma}^{R})^{-1}\boldsymbol{g}^{R}=\boldsymbol{G}^{R} \tag{2.4.85}$$

From the identity

$$(1-\boldsymbol{g}^{R}\boldsymbol{\Sigma}^{R})(1+\boldsymbol{G}^{R}\boldsymbol{\Sigma}^{R})=1+(\boldsymbol{G}^{R}-\boldsymbol{g}^{R}-\boldsymbol{g}^{R}\boldsymbol{\Sigma}^{R}\boldsymbol{G}^{R})=1 \tag{2.4.86}$$

It is obtained that

$$(1-\boldsymbol{g}^{R}\boldsymbol{\Sigma}^{R})^{-1}=1+\boldsymbol{G}^{R}\boldsymbol{\Sigma}^{R} \tag{2.4.87}$$

Lastly, substitution of Eq. (2.4.86) and Eq. (2.4.87) leads to the first equal mark of Eq. (2.4.82), and the second equal mark employs Eq. (2.4.81) and Eq. (2.4.86).

The following two products do not concern integral:

$$C(t_1,t_2)=A(t_1,t_2)B(t_1,t_2) \tag{2.4.88}$$

$$D(t_1,t_2)=A(t_1,t_2)B(t_2,t_1) \tag{2.4.89}$$

Starting from the two equations, one easily verifies the following identities:

$$C^{<}(t_1,t_2)=A^{<}(t_1,t_2)B^{<}(t_1,t_2) \tag{2.4.90}$$

$$D^{<}(t_1,t_2)=A^{<}(t_1,t_2)B^{>}(t_1,t_2) \tag{2.4.91}$$

$$C^{R}(t_1,t_2)=A^{<}(t_1,t_2)B^{R}(t_1,t_2)+A^{R}(t_1,t_2)B^{<}(t_1,t_2)+A^{R}(t_1,t_2)B^{R}(t_1,t_2) \tag{2.4.92}$$

$$D^{R}(t_1,t_2)=A^{R}(t_1,t_2)B^{<}(t_2,t_1)+A^{<}(t_1,t_2)B^{A}(t_2,t_1)=$$
$$A^{<}(t_1,t_2)B^{A}(t_2,t_1)+A^{R}(t_1,t_2)B^{<}(t_2,t_1) \tag{2.4.93}$$

Finally, we prove a very useful identity:

$$\boldsymbol{G}^{R}-\boldsymbol{G}^{A}=\boldsymbol{G}^{R}(\boldsymbol{\Sigma}^{R}-\boldsymbol{\Sigma}^{A})\boldsymbol{G}^{A} \tag{2.4.94}$$

The inverse form of Dyson's Eq. (2.4.59) is $(\boldsymbol{G}^{R,A})^{-1}=(\boldsymbol{G}^{0})^{-1}-\boldsymbol{\Sigma}^{R,A}$, $(\boldsymbol{G}^{0})^{-1}=\boldsymbol{E}-\boldsymbol{H}_0$. Note that the difference of the retarded and advance Green's function is solely an infinitesimal imaginary part, which is not in denominator here so that can be ignored. Thus,

$$(\boldsymbol{G}^{A})^{-1}-(\boldsymbol{G}^{R})^{-1}=\boldsymbol{\Sigma}^{R}-\boldsymbol{\Sigma}^{A}$$

Multiplying $\boldsymbol{G}^{R}$ from the left and $\boldsymbol{G}^{A}$ from the fight, we obtain Eq. (2.4.94).

When a system is in an equilibrium state, the lesser Green's function can be expressed in terms of the retarded and advanced Green's functions,

$$\boldsymbol{g}^{<}(\boldsymbol{k},\omega)=-\eta[\boldsymbol{g}^{A}(\boldsymbol{k},\omega)-\boldsymbol{g}^{R}(\boldsymbol{k},\omega)]f_{-\eta}(\hbar\omega) \tag{2.4.95}$$

Chapter 3 Application in Molecular Devices

3.1 Salicylideneanilines-based optical molecular switch

Molecular electronic devices are currently attracting serious attention for their potential applications in the field of nanoscience and nanotechnology. With the advent of experimental techniques and theoretical methodology, many molecular devices with different functionalities like negative differential resistance (NDR), memory effects, molecular rectification, and electronic switching have been measured and designed. Among these devices, molecular switch has gained widespread interest from molecular electronics researchers because it is the important element for the design of molecular memory or logic devices. Molecular switch devices can be mainly divided into two categories. One is the electronic switch devices which control the conversion between on and off states by an external trigger such as the electric field, the tip of scanning tunneling microscopy (STM), and the redox process, etc. However, these triggering means are not ideal since they may interfere greatly with the function of a nanosize circuit and limit the real applications. Furthermore, all these means are relatively slow in response. The other is optical switch devices which exist in two thermally sufficiently stable states with a high conductance (on state) and a low conductance (off state) by light. On the contrary, light is a very attractive external stimulus for such switches including short response time, the ease of addressability and compatibility with a wide range of condensed phases.

In this regard, diarylethene – and azobenzene-based optical switch which are mainly based on light-induced conformational changes, namely ring-opening reactions and isomerization reactions of the molecular bridge, have been widely studied. Recently, the butadienimine-based optical switch which is based on photoinduced excited state hydrogen transfer in the molecular bridge has reported by Benesch et al.. This molecular switch comprises a butadienimine molecule with the enol and keto form, which can be reversed from one structure to another one upon photoexcitation. In contrast to most other mechanisms, hydrogen translocation within the molecular bridge has the advantage that the overall length and thus the molecule-electrode binding geometry of the junction is not changed significantly. Salicylideneanilines-based molecule which is characterized by having an OH – N hydrogen bond also exhibit photoinduced reversible enol-keto tautomerism. Generally, the enol-form (OH) of this molecule is the most stable form at room temperature. However, some references have reported that the keto-form (NH) also can be found in the solid-state and in solution. For example, Ogawa has undertaken low-temperature X-ray diffraction experiments of this molecule and observed that at 15K, the keto-form is present. Furthermore, Jonathan et al have reported that the highly stable NH salicylideneanilines have been prepared by reaction of 1,3,5-triformylphloroglucinol with aniline derivatives. And the NH form was confirmed by X-ray crystallographic data, as well as by NMR studies. As proton transfer in these systems causes a change in optical properties, these molecules can be considered to be the candidate for optical switch. However, little attention was paid to the electronic transport properties of these molecules. In the present work, by applying nonequilibrium Green's function (NEGF) formalism combined with first-principles density functional theory (DFT), we investigate the electronic transport properties of the salicylideneanilines-based optical molecular switch which consists of the enol and keto tautomers chemically bound via thiol groups to two Au electrodes. Furthermore, since the electron transport through the molecular device can be controlled by the substituent ligand, the molecule with donor/acceptor substituent is also analyzed in some detail.

Firstly, the structure of free thiol(SH) capped molecule with the above-mentioned two forms was first optimized. Based on experimental issues about self-assembled monolayers (SAMs), it is generally accepted that hydrogen atoms are dissociated upon adsorption to metal surfaces. So we construct a two-probe system, which the two terminal hydrogen atoms bonded to the sulfur atom are eliminated from the optimized structure, and the remained part is sandwiched between two parallel Au(111) surfaces that correspond to the surfaces of the Au electrodes. The optimized perpendicular distance between the Au surface and sulfur atom is set to 1.9 Å, which is a typical Au-S distance. The Au(111) surface is represented by a 4×4 supercell with the periodic boundary conditions so that it imitates bulk metal structures. The 4×4 supercell is large enough to avoid any interaction with molecules in the next supercell. In NEGF theory, the molecular wire junction is divided into three regions: left electrode (L), contact region (C), and right electrode (R). The contact region includes extended molecule and two layers of Au from each electrode (see Fig. 3.1.1). The contact region contains parts of the electrodes include the screening effects in the calculations. The semi-infinite electrodes are calculated separately to obtain the bulk self-energy.

The free molecular geometrical optimizations and the electronic transport properties of molecular junctions are calculated by ATK 2.0 package, which is based on the real-space, Keldysh NEGF formalism and the density functional theory (DFT). The molecular geometrical optimization used density functional theory (DFT) with the electronic wave functions expanded in a basis set of atom-based numerical orbitals. The geometry of the molecule is optimized by minimizing the atomic forces on atoms to be smaller than 0.05 eV/ Å. The main feature of the computational package is to model a nanostructure coupled to external electrodes with different electrochemical potentials and to realize the transport simulation of the whole two-probe system without inducing phenomenological parameters. In the molecule geometrical optimization and the electronic transport properties of the molecular switch, the double-zeta plus polarization (DZP) basis set for the organic and a single-zeta plus polarization (SZP) basis set for Au atoms are adopted. Earlier, Stokbro et al. have noticed that a change in the exchange-correlation (XC) functional from the local-density approximation (LDA) to the generalized-gradient approximation (GGA) has only minor effects on the transmission spectrum. Moreover, the use of the GGA functional in transport calculations takes enormous computational time. Based on these facts, in our calculations, the exchange-correlation potential is described by Ceperley-Alder local density approximation (LDA). Core electrons are modeled with Troullier-Martins nonlocal pseudopotential and valence electrons expanded in a SIESTA localized basis set. An energy cutoff of 150 Ry for the grid integration is set to present the accurate charge density.

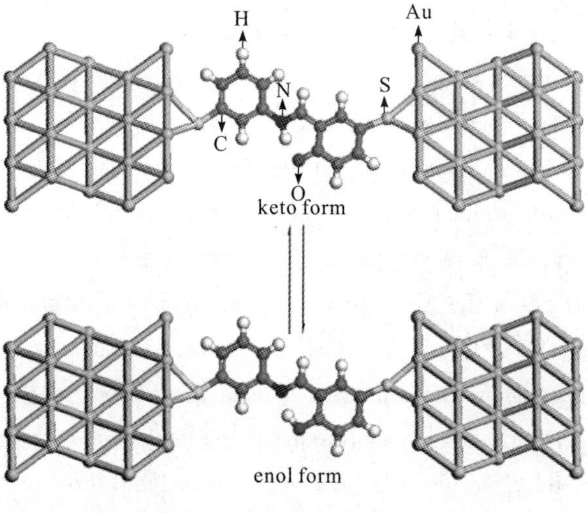

Fig. 3.1.1 Conversion between the keto and enol tautomers

In NEGF theory, the transmission function $T(E,V)$ of the system is the sum of transmission probabilities of all channels available at energy E under external bias V:

$$T(E,V) = \mathrm{Tr}[\Gamma_L(E,V)G^R(E,V)\Gamma_R(E,V)G^A(E,V)] \tag{3.1.1}$$

where $G^{R/A}$ are the retarded and advanced Green's functions, and coupling functions $\Gamma_{L/R}$ are the imaginary parts of the left and right self-energies, respectively. The self-energy depends on the surface Green's functions of the electrode regions and comes from the nearest-neighbor interaction between the extended molecule region and the electrodes.

For the system at equilibrium, the conductance G is evaluated by the transmission function $T(E)$

at the Fermi level E_F of the system:

$$G = G_0 T(E_F) \tag{3.1.2}$$

where $G_0 = 2e^2/h$ is the quantum unit of conductance, h is the Planck's constant, e is the electron charge.

The current through a molecular junction is calculated from the Landauer-Bütiker formula

$$I = \frac{2e}{h}\int [f(E-\mu_L) - f(E-\mu_R)]T(E,V)dE \tag{3.1.3}$$

where f is the Fermi function , $\mu_{L/R}$ is the electrochemical potential of the left/right electrode and the difference in the electrochemical potentials is given by eV with the applied bias voltage V , i. e. , $\mu_L = \mu(0) - eV/2$ and $\mu_R = \mu(0) + eV/2$. Furthermore, $\mu_{L/R}(0) = E_F$ is the Fermi level.

The calculated *I-V* characteristics of the enol and keto tautomers at a bias up to 1. 2 V in steps of 0. 2 V are plotted in Fig. 3. 1. 2. It should be pointed out that at each bias, the current is determined self-consistently under the non-equilibrium condition. The switching behavior can be clearly seen from Fig. 3. 1. 2. Although these two tautomers only differ by the position of a single hydrogen atom, their conductance properties are drastically different. The current of enol form is evidently larger than that of keto form over the entire bias range. Therefore, when the molecule in the switch changes from enol form to the keto form under photoexcitation, one can predict that there is a switch from on (low resistance) state to the off (high resistance) state.

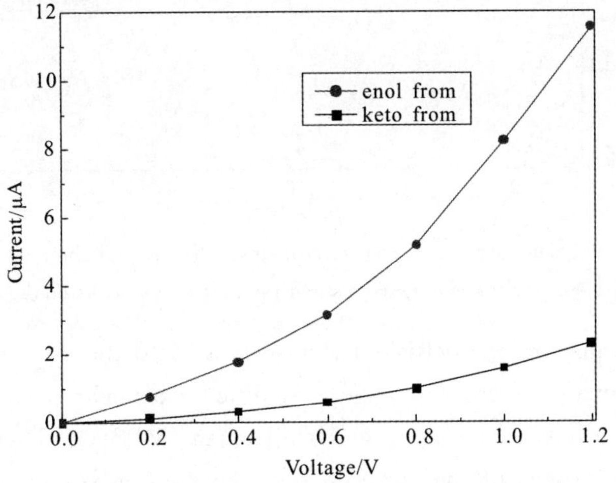

Fig. 3. 1. 2 The *I-V* characteristics of the molecular switch with two forms

The current in the system is calculated by theLandauer-Bütiker formula which is transmission spectra dependent. We calculate the transmission spectra $T(E,V)$ of the enol and keto forms under zero voltage to understand the dramatic difference in conductance appearing in I-V curve. As shown in Fig. 3.1.3, the average Fermi level is set as zero and the characters H and L represent the highest occupied molecular orbital (HOMO) and the lowest unoccupied molecular orbital (LUMO), respectively. The conductance enhancement is shown clearly in the transmission spectra. From Fig. 3.1.3, we can see that the system with enol form has a strong LUMO transmission peak at 0.56 eV above the Fermi level E_F ($\mu_{L/R}(0) = E_F$). However, for the system with keto form, there are just very weak LUMO transmission peak at 0.62 eV above the E_F. Meanwhile, the HOMO-LUMO gap of the enol form is smaller than that of the keto form. In such an open system, the Fermi level E_F aligns between HOMO and LUMO, and the barrier for the electron transport is intensively relevant to the HOMO-LUMO gap. The results show that the HOMO-LUMO gap increases from 1.73 to 2.07 eV when the enol form is switched to the keto form. As a result, the lack of any significant peak in the energy region and the lager HOMO-LUMO gap due to the molecular structure changes accounts for the low conductivity in the keto form.

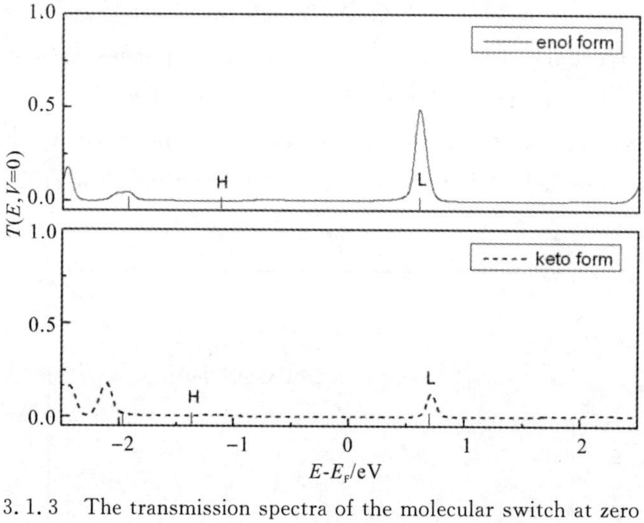

Fig. 3.1.3 The transmission spectra of the molecular switch at zero bias
(The short vertical bars near the energy axes stand for the energy levels of the extend molecule)

Generally, the transmission coefficient can be related to the molecular orbitals, which have been modified by the electrodes. These modified molecular orbitals can be obtained from the molecular projected self-consistent Hamiltonian (MPSH). Therefore, to get a further insight about the origins of the peaks in the transmission spectra and of the different transmission characteristics for two forms, we analyze the MPSH on the molecular junction.

Table 3.1.1 illustrates the MPSH orbitals near the Fermi level, i.e., the HOMO and LUMO orbitals. It can be seen that the LUMO is delocalized orbital for the enol form which leads to low barrier and provides good transport channels for electron transport. On the contrary, the HOMO and LUMO are both localized orbitals for the keto form leading to high barrier for electron transport which accounts for the low transmission strength in Fig. 3. 1. 3.

Table 3. 1. 1 The MPSH states corresponding to HOMO and LUMO

MPSH	HOMO	LUMO
enol form	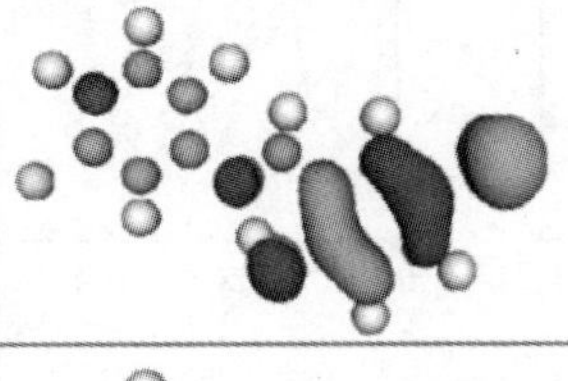	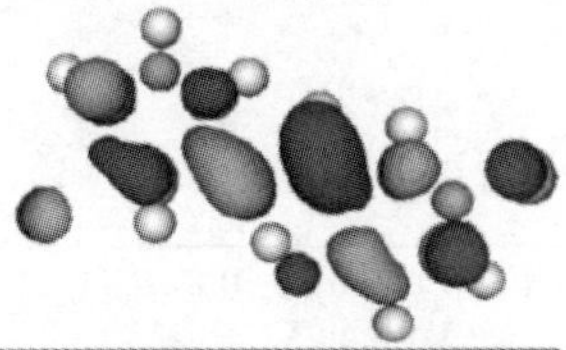
keto form	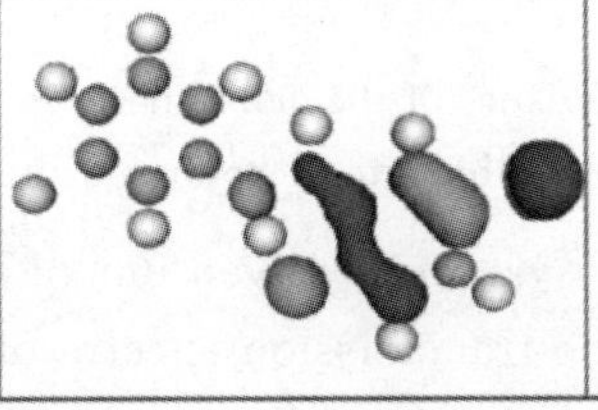	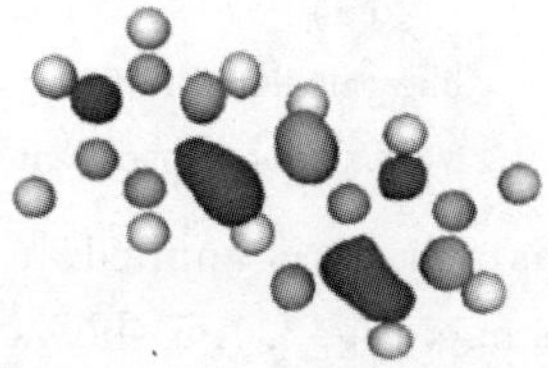

Furthermore, the electron transport through the molecular device can be controlled by the substituent ligand, in order to improve the performance of the switch, we compare the *I-V* characteristics for switches with different substituent, i. e. donor group NH_2, and acceptor group NO_2. Fig. 3. 1. 4 (a) and (b) show their *I-V* characteristic curves. Their geometry structures after optimization are displayed in the corresponding insets. The upper structure of the insets is the keto form, and the lower one is the enol form. As shown in Fig. 3. 1. 4 that the current through the switch is affected significantly by the substituent. The current of two forms with acceptor substituent increases evidently. On the contrary, the current of donor substituent decrease noticeably. To characterize the conduction change due to different substituents, we define the current ratio, Ratio= $I_{\text{enol-form}}/I_{\text{keto-form}}$, where $I_{\text{enol-form}}$ and $I_{\text{keto-form}}$ are the current of enol and keto form, respectively. The maximum current ratio is about 5. 1 at 1. 2 V for the switch without the substituent. When the molecule switch are substituted by donor group NH_2, and acceptor group NO_2, the maximum current ratio is changed to 3. 1 and 8. 2, respectively. It is clearly that the current ratio of the switch with the substituent NO_2 is enhanced evidently which suggests that the switching performance can be improved to some extent.

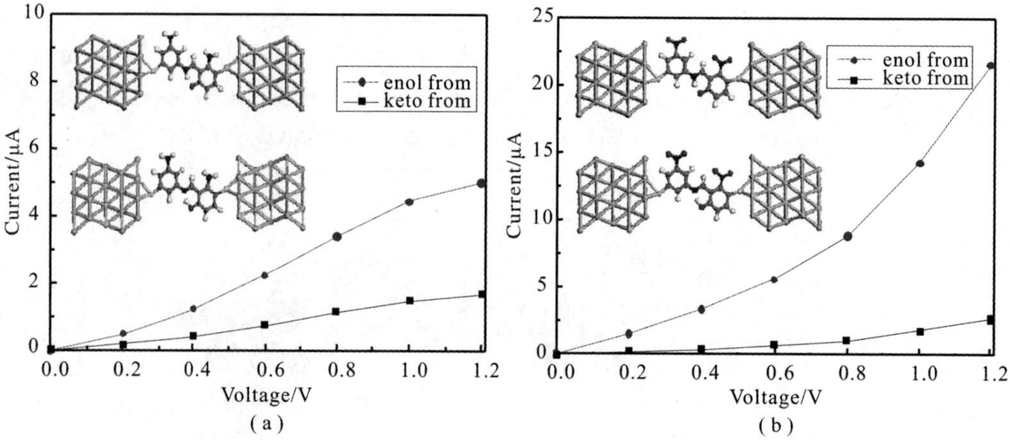

Fig. 3. 1. 4 The calculated *I*-*V* characteristics of the molecular switch with substituent

(a) with donor substituent NH_2; (b) with acceptor substituent NO_2

When the bias voltage is applied, the system is driven out of equilibrium and the electrode potential change. Fig. 3. 1. 5 shows the transmission spectra of these two models under voltages of 1. 2 V, which will give us a clear understanding of the different electron transport behavior. The regions between the two vertical dash lines indicate the bias window. When the molecular switch with substituent NO_2, the integral area entering the bias window increases compare to the system without substituent. However, in contrast to the acceptor substituent NO_2, the integral area entering the bias window decrease with the donor substituent NH_2, which leads the current to decrease. As a result, the current is changed evidently with different substituent.

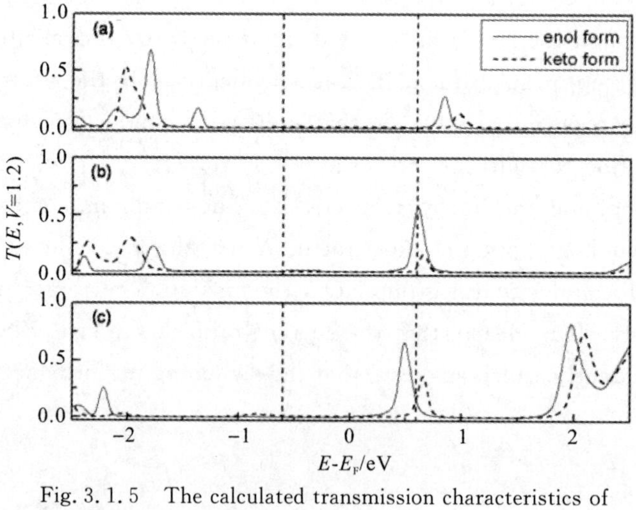

Fig. 3. 1. 5 The calculated transmission characteristics of

the molecular switch with and without substituents at 1. 2 V

(a) with donor substituent NH_2; (b) without substituent; (c) with acceptor substituent NO_2

To get a further insight about the different in transmission spectra, the MPSH of the molecular switch with different substituents is also analyzed. Table 3. 1. 2 illustrates the MPSH orbitals of HOMO and LUMO with substituents NH_2 and NO_2. It is found that the LUMO orbital which is the main transmission channel of the molecule with substituent NO_2 is more delocalized than other two systems . It is obvious that the π-overlapping and the π-electron conjugation of LUMO orbital are increased by the substituent NO_2, which is the main reason leads the current to increase.

Table 3. 1. 2 The spatial distribution of the MPSH states corresponding to HOMO and LUMO for enol and keto forms with substituents (a) with donor substituent NH_2; (b) with acceptor substituent NO_2

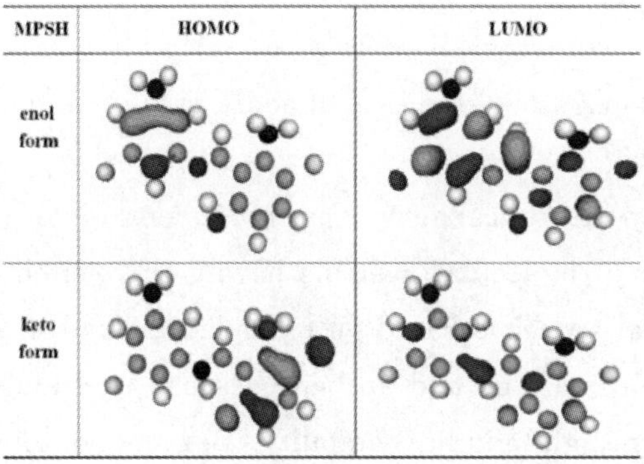
MPSH
HOMO
LUMO
enol form
keto form
(a)

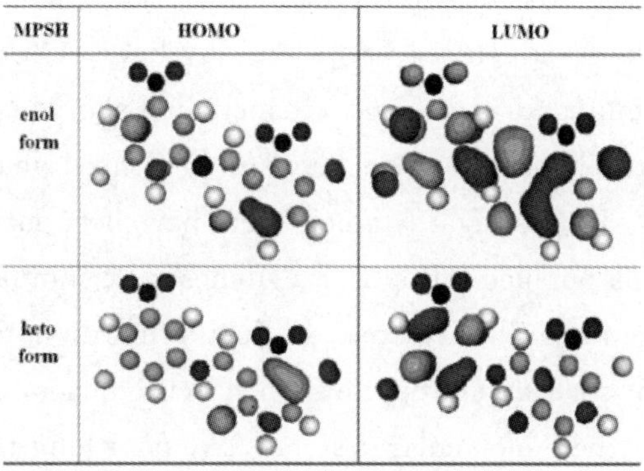
MPSH
HOMO
LUMO
enol form
keto form
(b)

In conclusion, we have investigated the electron transport properties of the salicylideneanilines-based optical molecular switch by using NEGF formalism combined with first-principles DFT. The dramatic difference in conductivity appearing in two tautomers can be observed. The physical origin of the switching behavior is interpreted based on the transmission spectra and spatial distribution of MPSH orbitals. Theoretical results show that the current through the enol form is significantly larger that that through the keto form. Furthermore, it can be found that the donor/acceptor substituent plays an important role in the electronic transport of molecular devices. The current ratio of the switch with the substituent NO_2 is enhanced evidently which suggests that the switching performance can be improved to some extent.

3. 2 Phenylazoimidazole optical molecular switch

Electronic devices based on single molecules have been considered as one of the most potentially promising alternatives to today's semiconductor-based electronics. Use molecular devices to realize the elementary functions in electronic circuits have been an active research area for a decade both experimentally and theoretically. Many interesting physical properties like negative differential resistance (NDR), memory effects, molecular rectification, and electronic switching in molecular devices have been reported. The most prominent among these is molecular switch, which is the important element for the design of molecular memory or logic devices. Molecular switch devices can be mainly divided into two categories. One is the electronic switch devices which control the conversion between on and off states by an external trigger such as the electric field, the tip of scanning tunneling microscopy (STM), and the redox process, etc. However, these triggering means are not ideal since they may interfere greatly with the function of a nanosize circuit and limit the real applications. Furthermore, all these means are relatively slow in response. The other is optical switch devices which exist in two thermally sufficiently stable states with a high conductance (on state) and a low conductance (off state) by light. On the contrary, light is a very attractive external stimulus for such switches including short response time, the ease of addressability and compatibility with a wide range of condensed phases.

In this regard, albeit there have been many theoretical as well as experimental investigations pursued on optical switches, the number of molecular systems having the switch feature are still very limited. Furthermore, optical switches based on extended π-electron system such as spiropyrans, benzochromenes, spiroxazines and azobenzenes are well-known, but these molecular systems have poor long-term light and heat stability. But phenylazoimidazole molecules which characterized by having a azoimine unit(-N=N-C=N) are known for their ability to undergo light-induced or thermal reversible cis-trans isomerism. These two forms can keep stable over a wider temperature range and reversibly switch from each other, which make it usable as one of the good candidates for light-driven molecular switches and may have some future applications in the molecular circuit . However, little attention was paid to the electronic transport properties of these molecules. In the present work, by applying nonequilibrium Green's function (NEGF) formalism combined with first-principles density functional theory (DFT), we investigate the electronic transport properties of the phenylazoimidazole optical molecular switch.

Fig. 3. 2. 1 illustrates the simulation setup. The pre-optimized phenylazoimidazole molecule with two forms are sandwiched via thiolate bonds between two parallel Au(111) surfaces that correspond to the surfaces of the Au electrodes. The optimized perpendicular distance between the Au surface and sulfur atom is set to 1. 9 Å, which is a typical Au-S distance. Each layer of Au electrodes is represented by a 4×4 supercell with the periodic boundary conditions so that it imitates bulk metal structures. The 4×4 supercell is large enough to avoid any interaction with molecules in the next supercell. In NEGF theory, the entire molecular switch is divided into three regions: left electrode (L), contact region (C), and right electrode (R). The contact region includes extended molecule and two layers of Au slab from each electrode to screen the perturbation effect from the central region and they are denoted as surface-atomic layers. The semi-infinite electrodes are calculated separately to obtain the bulk self-energy. The geometry of the central extended molecule is optimized by minimizing the atomic forces on the phenylazoimidazole molecule to be smaller than 0.05 eV/Å, while keeping all the Au atoms fixed. All the geometrical optimizations have been carried out by the SIESTA package.

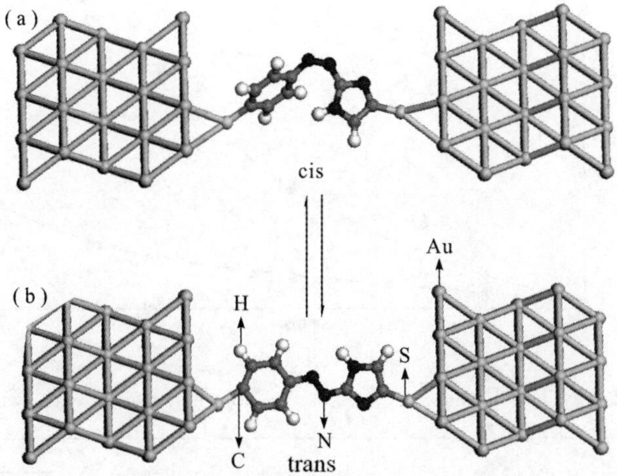

Fig. 3.2.1 Schematic illustration of the molecular switch with two forms
(a) cis form; (b) trans form

In Fig. 3.2.2, we show the self-consistently calculated *I*-*V* characteristics curves of the molecular switch with two forms in a bias range from 0 to 1.2 V in steps of 0.2 V. The switching behavior can be clearly seen from Fig. 3.2.2, the current through the trans form is evidently greater than that through the cis form over the entire bias range. Thus, one can predict that there is a switch from on (low resistance) state to the off (high resistance) state, when the phenylazoimidazole molecule reversed from trans form to cis form upon photoexitation.

The current in the system is calculated by the Landauer-Bütiker formula $I = G_0\int n(E)T(E, V)\mathrm{d}E$, which is transmission spectra dependent. Therefore, the switching characteristics of this molecule can be understood from the energy dependence of zero-bias transmission spectra, which are shown in Fig. 3.2.3. The short vertical bars near the energy axes stand for the energy levels of the extended molecule. The characters H and L represent the highest occupied molecular orbital (HOMO) and the lowest unoccupied molecular orbital (LUMO), respectively. In our calculation, the average Fermi level, which is the average value of the chemical potential of the left and right electrodes, is set as zero. Seen from the Eq. (3. 1. 3), we can expect that only electrons with energies within a range near the Fermi level E_F contribute to the total current. Therefore, a good approximation with the range of the bias window, i. e. , $[-V/2, +V/2]$ is enough to analyze a finite part of the transmission spectrum.

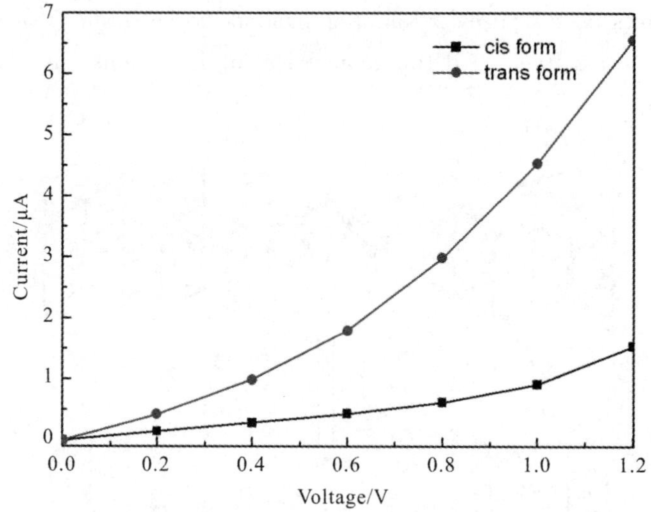

Fig. 3. 2. 2 The I-V characteristics of the molecular switch with two forms

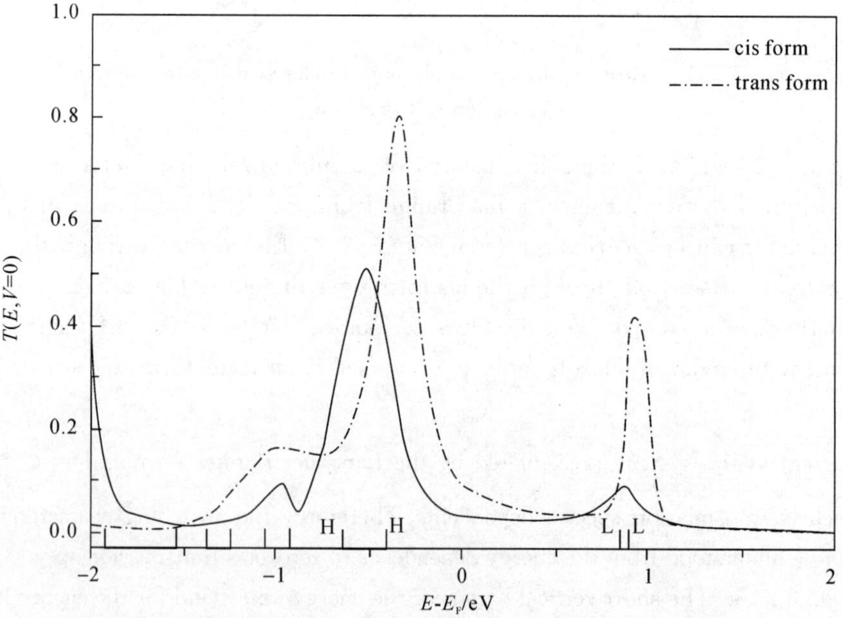

Fig. 3. 2. 3 The transmission spectra of the molecular switch at zero bias

(The short vertical bars near the energy axes stand for the energy levels of the extend molecule)

It is clear from Fig. 3. 2. 3 that the transmission spectra of two forms display extraordinarily different characteristics. The system with the trans form has a strong transmission peak at −0. 36 eV and 0. 9 eV below and above the E_F, respectively. However, for the system with the cis form, there are just very weak transmission peaks below and above the E_F. It is clear that the low transmission strength accounts for the low conductance in cis form. The sharp contrast between transmission spectra of these two forms reflects the vital significance of structure effects on transport properties in the phenylazoimidazole molecule. Meanwhile, the HOMO and LUMO levels are −0. 41 eV and 0. 85 eV for the trans form, −0. 64 eV and 0. 89 eV for the cis form, respectively. Thus, the HOMO-LUMO gap of the trans form is smaller than that of the cis form. In such an open system, the Fermi level E_F of the Au electrode aligns between HOMO and LUMO, and, therefore, the barrier for the electron transport is intensively relevant to the HOMO-LUMO gap. The results show that the HOMO-LUMO gap increases from 1. 26 to 1. 53 eV when the trans form is switched to the cis form. As a result, the lack of any significant peak in the region of [−2. 0 eV, 2. 0 eV] and the lager HOMO-LUMO gap due to the molecular structure changes accounts for the low conductivity in the cis form.

To further elucidate the origins of the peaks in the transmission spectra and of the different transmission characteristics for two forms, we analyze the molecular projected self-consistent Hamiltonian (MPSH), which can be obtained by projecting the self-consistent Hamiltonian of the molecular switch onto the Hilbert space spanned by the basis functions of the molecule. The MPSH is then the self-consistent Hamiltonian of the isolated molecule in the presence of the electrodes, namely the molecular part is extracted from the whole self-consistent Hamiltonian at the contact region. It is found from Table 3. 2. 1 that both the HOMO and the LUMO are delocalized orbitals which provide the main electronic transport channel for the trans form leading to low barrier for electron transport. However, the HOMO and LUMO are both localized orbitals for the cis form, which cannot provide good transport channels, because electrons that enter the molecule at the energy of these orbitals have low probability of reaching the other end. As a result, it leads to high barrier for electron transport which accounts for the low transmission strength in Fig. 3. 2. 3.

Table 3.2.1 Spatial distribution and the corresponding eigenvalues of HOMO and LUMO for cis and trans forms

MPSH	HOMO	LUMO
Cis Form	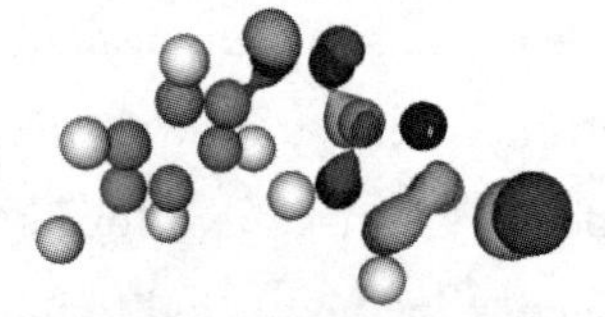	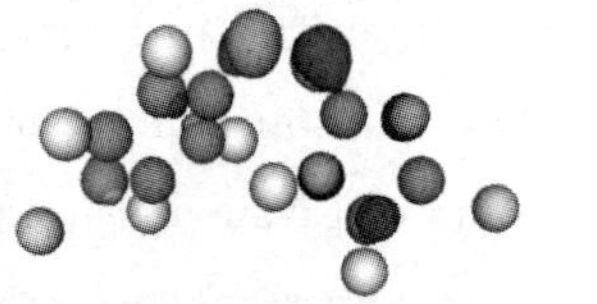
Trans Form	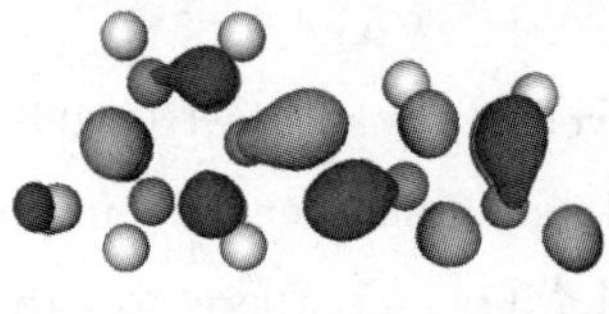	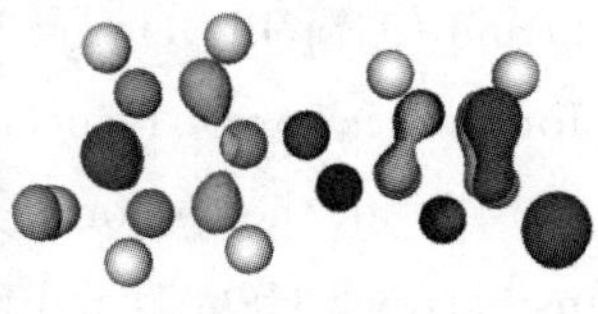

To conclude, the electronic transport properties of thephenylazoimidazole optical molecular switch have been investigated by using the DFT+NEGF first-principles method. The *I-V* characteristics, electronic transmission coefficients, and spatial distribution of MPSH orbitals corresponding to different forms are calculated and analyzed. The dramatic difference in conductivity appearing in two different forms can be observed. The physical origin of the switching behavior is interpreted based on the location of HOMO and LUMO, and the HOMO-LUMO gap. This suggests phenylazoimidazole molecule usable as one of good candidates for light-driven molecular switches and may have some future applications in the molecular circuit.

3.3 Naphthopyran-based optical molecular switch

In recent years, molecular devices based on single molecules have been attracting more and more attention due to their novel physical properties and potential for device application, such as negative differential resistance (NDR), memory effects, rectification, amplification and switching properties. Among these devices, most investigations are focused on molecular switch because it is the basic principle in any modern design of logic and memory circuits. Molecular switch devices can be mainly divided into two categories, including electronic switch devices which control the conversion between on and off states by an external trigger such as the electric field, the tip of scanning tunneling microscopy (STM), and the redox process, etc, and optical switch devices which exist in two thermally sufficiently stable states with a high conductance (on state) and a low conductance (off state) by light. Among them, the optical switches have attracted great attention, because light is a very attractive external stimulus for such switches including short response time, the ease of addressability and compatibility with a wide range of condensed phases.

In this regard, many differentoptical switches have been investigated and discussed, such as diarylethene-based and azobenzene-based optical switches. Albeit there have been many theoretical as well as experimental investigations pursued on optical switches, the number of molecular systems having the switch feature are still very limited. Especially, many optical switches based on extended π-electron system such as spiropyrans, benzochromenes, spiroxazines and azobenzenes have poor long-term light and heat stability. Recently, some experiments report that naphthopyran molecule which characterized by having a pyran ring to a naphthoquinone also exhibit photoinduced reversible open-closed form (see Fig. 3.3.1). Furthermore, the most significant features of this material are they present excellent photochromic properties including short response time, heat stability, and large changes of the absorption wavelengths between the two isomers, which make it usable as one of the candidates for light-driven molecular switches. However, little attention was paid to the electronic transport properties of these molecules.

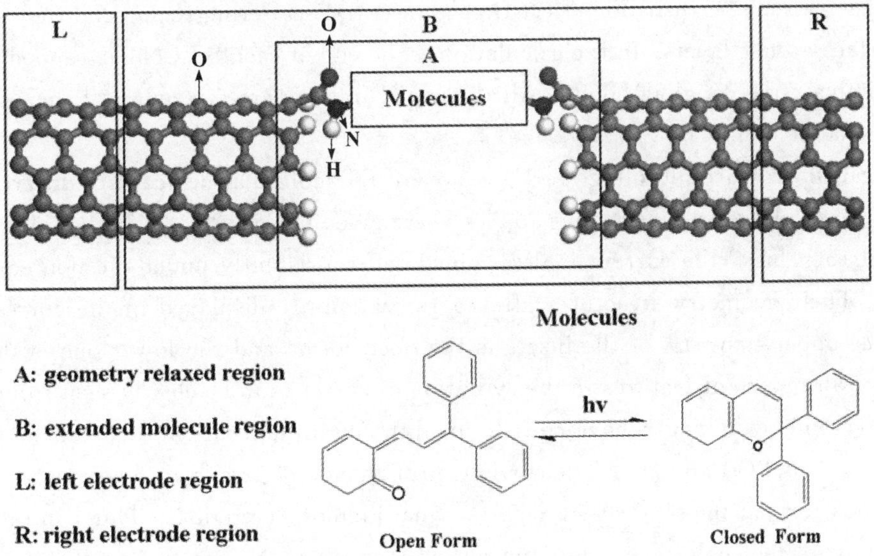

Fig. 3.3.1 The structure of the naphthopyran molecular switch with amide endgroup(CONH)

On the other hand, electrode materials as a part of the molecular device also play an important role in realizing the elementary functions in future molecular electronic circuits. However, common metallic electrodes materials like Au usually have many serious contact problems which is difficult to overcome. Recently, Cai et al. investigated the electronic transport through a single light-sensitive diarylethene molecule sandwiched between two graphene nanoribbons(GNRs) and observed strong rectification behavior of the closed diarylethen isomer with high rectification ratio. In addition to GNR electrodes, single-walled carbon nanotube (SWCNT) are among the most promising candidates for the construction of nanoscale devices due to their stable structures and unique electronic properties. In particularly, Guo et al. have reported that an individual molecule can covalently attach to SWCNT electrodes through the amide endgroup(CONH), which has proven to be very robust. Therefore, by applying nonequilibrium Green's function (NEGF) formalism combined with first-principles density functional theory (DFT), we investigate the switching characteristics of the naphthopyran optical molecular switch with two different SWCNT electrodes, i. e. (5,5) armchair/(9,0) zigzag SWCNT electrodes. The effects of the SWCNT's chirality on the quantum transport through the molecular device are discussed in detail.

The pre-optimizednaphthopyran molecule with two different forms are sandwiched between two (5,5) armchair SWCNT electrodes as shown in Fig. 3. 3. 1. Amide endgroups were used as linkages between the molecule and SWCNT electrodes. For each electrode, eight layers of carbon atoms are included into the extended molecule region(region B in Fig. 3. 3. 1) to screen the perturbation effect from the central scattering region and they are denoted as surface-atomic layers. In the calculations, the end of each SWCNT is capped by H atoms to eliminate the dangling bonds. All the configurations are relaxed until their force tolerance is smaller than 0. 05 eV/Å.

The calculated current-voltage (*I-V*) curves of the molecular device with different forms under the bias voltage varying from 0 to 2. 0 V are given in Fig. 3. 3. 2. It should be pointed out that at each bias, the current is determined self-consistently under the non-equilibrium condition. Their geometry structures after optimization are displayed in the corresponding insets. The upper structure of the insets is the open form, and the lower one is the closed form. Several important features in the evolution of current can be clearly seen from Fig. 3. 3. 2. Firstly, the switching behavior of the naphthopyran molecule system can be observed both in Figs. 3. 3. 2(a) and (b). The current through closed form is always greater than that through open form at the same bias no matter what kind of electrode is. Thus, in the process of naphthopyran molecule reversing from closed form to open form under photoexcitation, the current through the circuit can switch from on (high conductance) to off (low conductance), and vice versa. The on-off ratio of current, Ratio $= I_{\text{closed-form}} / I_{\text{open-form}}$, versus bias voltage V is also plotted in Fig. 3. 3. 2. Obviously, the zigzag junction shows better switching characteristics as compared with the armchair junction for the chosen system. The maximum on-off ratio reaches 292 at 1. 6 V for the zigzag junction. Such a significant on-off ratio is desirable for the real applications in future molecular switch technology. Furthermore, an apparent NDR effect appears in the zigzag-closed-zigzag molecular junction at the bias voltage of 1.2 V and no NDR behavior is observed in other three junctions. It is quite clear from Fig. 3.3.2(b) that initially the current of zigzag-closed-zigzag molecular junction increase with the increases in external voltage. However, this increase in current due to the variation in voltage is observed up to 1.0 V. Beyond 1.0 V, there is a rapid decrease in current with the increase in bias voltage. This decrease in current due to an increase in voltages is the manifestation of the NDR feature. NDR behavior is quite prominent up to a bias voltage of 1.2 V. But above 1.2 V the NDR feature is completely lost and the current trough the molecular junction increases again. In contrast, the current of other three junctions usually increase with the increase in bias voltage.

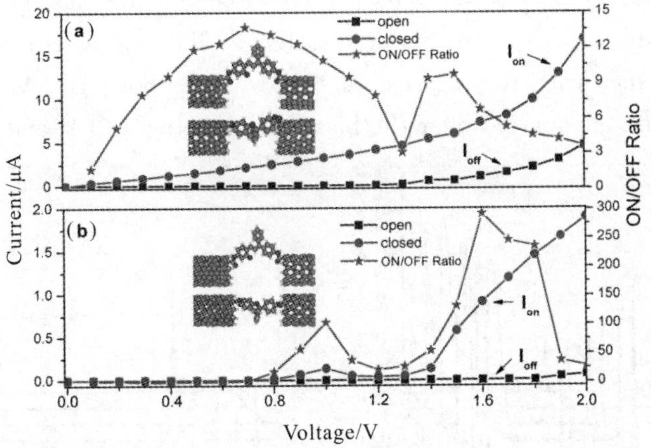

Fig. 3.3.2 The *I-V* characteristics and the on-off ratio of the molecular switch with (a) (5,5) armchair; (b) (9,0) zigzag SWCNT electrodes

In NEGF theory, the current of the system is calculated by the Landauer-Bütiker formula $I = \frac{2e}{h}\int[f(E-\mu_L) - f(E-\mu_R)]T(E,V)dE$, which is transmission spectra dependent. Thus, the particular mechanism can be further interpreted in terms of transmission spectra. Fig. 3.3.3 show the transmission spectra $T(E,V)$ of the closed and the open forms with different electrodes under various biases in a three-dimensional plot. In our calculation, the average Fermi level, which is the average value of the chemical potential of the left and right electrodes, is set as zero. Seen from the Eq. (3.1.3), we can expect that only electrons with energies within a range near the Fermi level E_F contribute to the total current. Therefore, a good approximation with the range of the bias window, i.e. $[-V/2, +V/2]$ is enough to analyze a finite part of the transmission spectrum.

The mechanism ofswitching behavior for naphthopyran molecule with different electrodes is shown clearly in the transmission spectrum from Fig. 3.3.3. The transmission coefficients of closed form are always bigger than those of open form in the bias window of [−1.0 eV, 1.0 eV]. Meanwhile, for the closed form, there is a narrow peak above the Fermi level, i.e., the lowest unoccupied molecular orbital (LUMO) transmission peak, which dominates the conductive behavior of the molecular junction under bias of 0 to 2.0 V. However, this narrow peak becomes very weak and shifts toward the higher energy when switch changes from the closed form to the open form under photoexcitation. As a result, the lack of any significant peak in the bias window of [−1.0 eV, 1.0 eV] and the low transmission coefficient due to the molecular structure changes accounts for the low conductivity in the open form. Furthermore, we also can see from Fig. 3.3.3(c) that the value of transmission peaks for the closed form with zigzag SWCNT electrodes in the bias window decrease with the increase in bias voltage at the range of 1.0 to 1.4 V. Therefore, current through the molecular device diminishes with the increase in bias voltage above 1.0 V and below 1.4 V. Thus the NDR feature is apparent over the bias voltage above 1.0 V and below 1.4 V.

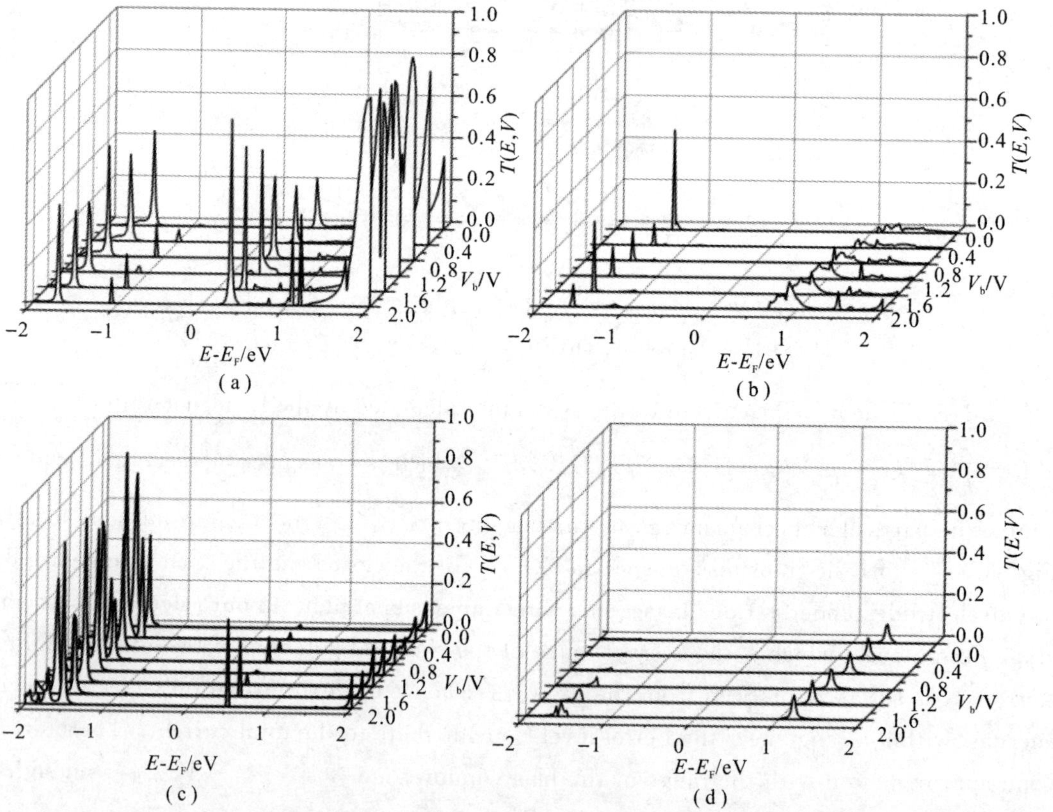

Fig. 3.3.3 The transmission spectrum under various bias voltages (0 to 2.0 V) for (a) closed form; (b) open form with (5,5) armchair SWCNT electrode; (c) closed form; (d) open form with (9,0) zigzag SWCNT electrodes

From Fig. 3. 3. 3, we can find that the main transmission peak of the closed form and open form are both contributed by orbital LUMO in the bias window of [−1. 0 eV, 1. 0 eV]. Therefore, to further elucidate the microscopic origin of the peaks in the transmission spectrum and of the different transmission characteristics for two forms, we analyze the local density of states (LDOS) for LUMO, which can reflect the spatial distribution of electronic states are plotted in Table 3. 3. 1. It is found obviously that the density of LUMO is more delocalized in the closed form both in armchair and zigzag junctions, which leads to low barrier and provides good transport channels for electron transport. On the contrary, this transmission channel is localized orbital for the open form leading to high barrier for electron transport which accounts for the low transmission strength in Fig. 3. 3. 3. As a result, the current is decreased when the molecule switch from closed to open form upon photoexcitation, which is in good agreement with the trends of *I-V* curves shown in Fig. 3. 3. 2.

Table 3. 3. 1 The LDOS of the molecular junction reflect the spatial distribution of electronic states (The profile shows the mount of the electrons on the atom in the molecule contributing to the energy level)

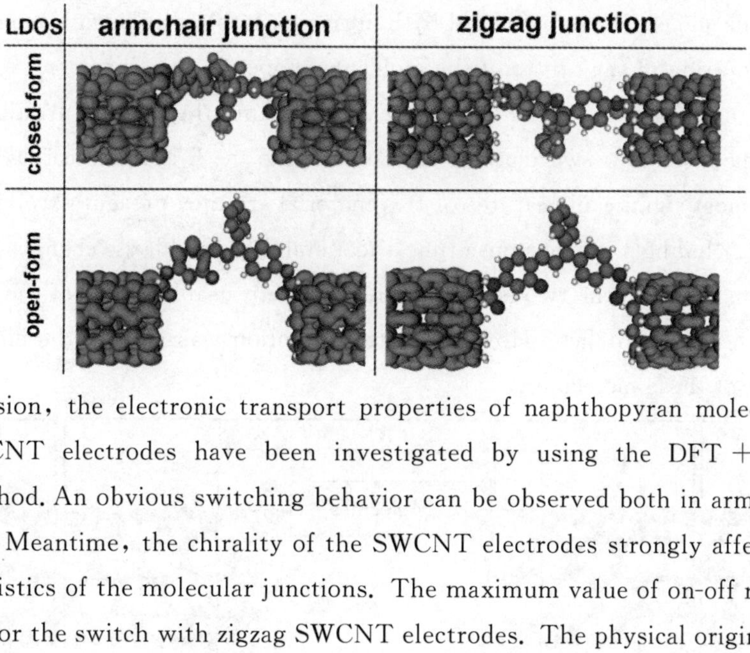
LDOS
armchair junction
zigzag junction
closed-form
open-form

In conclusion, the electronic transport properties of naphthopyran molecule with two different SWCNT electrodes have been investigated by using the DFT + NEGF first-principles method. An obvious switching behavior can be observed both in armchair and zigzag junctions. Meantime, the chirality of the SWCNT electrodes strongly affects the switching characteristics of the molecular junctions. The maximum value of on-off ratio can reach 292 at 1. 6 V for the switch with zigzag SWCNT electrodes. The physical origin of the switching behavior is interpreted based on the transmission spectra and the LDOS. A reversible change between the two forms would realize a molecular nanoswitch, which suggests that this system has attractive potential application in future molecular circuit.

3.4 Effect of carbon nanotubes chirality on the E-C photo-isomerization switching behavior in molecular device

With the miniaturization of traditional electronic devices, using molecules as components in atomic-scale circuits has become an attractive field. Molecular devices are capable of providing features like negative differential resistance (NDR), memory effects, rectification, amplification and switching properties, which open the way for a variety of electronic applications. Among all of these, molecular switches, especially optical switches, have drawn considerable attention in recent years due to their potential applications in future logic and memory. In this regard, many different optical switches using photochromic molecules which can change their isomer type upon light irradiation have been investigated and discussed. Albeit there have been many theoretical as well as experimental investigations pursued on optical switches, the number of molecular systems having the switch feature are still very limited. Especially, many optical switches such as spiropyrans, benzochromenes, spiroxazines and azobenzenes have poor long-term light and heat stability. Recently, some experiments report that fulgide molecule can be transformed by light in reversible way from E isomer into C isomer. The carbon ring in the center of the molecule is open in the E isomer. When irradiated with UV light the ring closes and the moelcule switches into the C state. Irradiation with visible light opens the ring again, switching the fulgide back into the E isomer (see Fig. 3.4.1). Furthermore, the most significant features of this material are they present excellent photochromic properties including short response time, heat stability, and large changes of the absorption wavalengths between the two isomers, which make it usable as one of the candidates for light-driven molecular switches. However, little attention was paid to the electronic transport properties of these molecules.

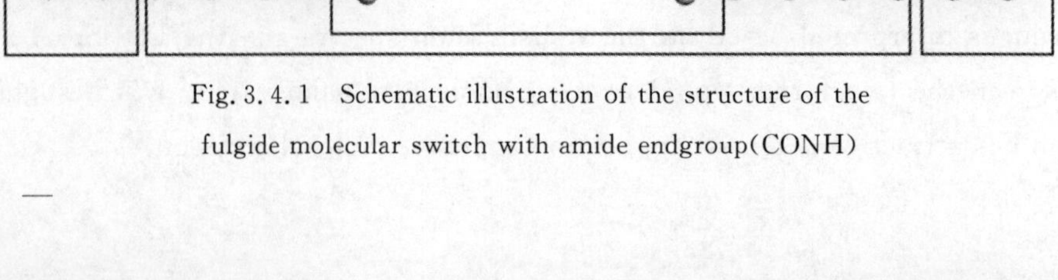

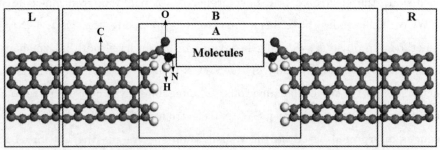

Fig. 3.4.1 Schematic illustration of the structure of the fulgide molecular switch with amide endgroup(CONH)

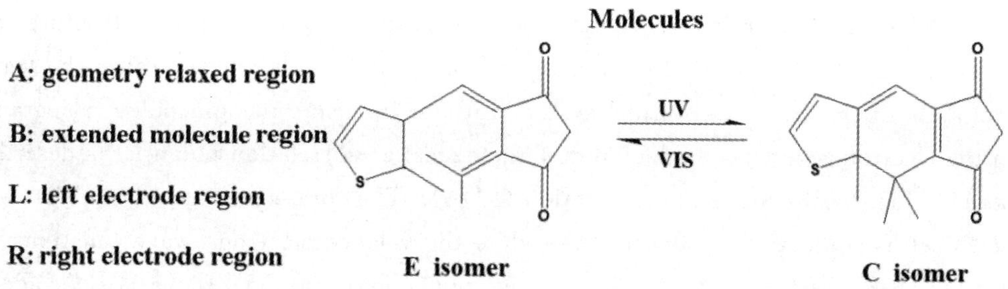

Continued Fig. 3. 4. 1 Schematic illustration of the structure of the fulgide molecular switch with amide endgroup(CONH)

On the other hand, electrode materials as a part of the molecular device also play an important role in realizing the elementary functions in future molecular electronic circuits. However, common metallic electrodes materials like Au usually have many serious contact problems which are difficult to overcome. Recent experiments show that carbon nanotubes (CNTs) have become a potential ideal material for functional devices and interconnect in nanoelectronics because of their stable structures and rich electronic properties. In particularly, Guo et al. have reported that an individual molecule can covalently attach to SWCNT electrodes through the amide endgroup(CONH), which has proven to be very robust. Therefore, by applying nonequilibrium Green's function (NEGF) formalism combined with first-principles density functional theory (DFT), we investigate the switching characteristics of the fulgide optical molecular switch with two different SWCNT electrodes, i. e. (5,5) armchair/(9,0) zigzag SWCNT electrodes. The effects of the SWCNT's chirality on the switching behavior in the molecular device are discussed in detail.

The pre-optimizedfulgide molecule with two isomers are sandwiched between two (5,5) armchair SWCNT electrodes as shown in Fig. 3. 4. 1. Amide endgroups were used as linkages between the molecule and SWCNT electrodes. For each electrode, eight layers of carbon atoms are included into the extended molecule region(region B in Fig. 3. 4. 1) to screen the perturbation effect from the central scattering region and they are denoted as surface-atomic layers. In the calculations, the end of each SWCNT is capped by H atoms to eliminate the dangling bonds. All the configurations are relaxed until their force tolerance is smaller than 0. 05 eV/Å.

All the geometrical optimizations and the electronic transport properties of the molecular junctions are calculated by a fully self-consistent NEGF formalism combined with first-principles DFT, which is implemented in Atomistix Tool Kit (ATK) software package (version

2012.8.2). The main feature of the computational package is to model a nanostructure coupled to external electrodes with different electrochemical potentials and to realize the transport simulation of the whole two-probe system without inducing phenomenological parameters. In the electronic transport calculations, the exchange-correlation potential is described by Ceperley-Alder local density approximation (LDA). The core electrons are modeled with Troullier-Martins nonlocal pseudopotential, while the valance electrons wave functions are expanded by a SIESTA basis set. The double-zeta plus polarization (DZP) basis set is adopted for all atoms. The Brillouin zone is set to be $5 \times 5 \times 100$ points following the Monkhorst-Pack k-point scheme. The cut-off energy and the iterated convergence criterion for total energy are set to 150 Rydberg and 10^{-5}, respectively.

The calculated current-voltage (I-V) curves of the E-C isomers with two different SWCNT electrodes under the bias voltage varying from 0 to 1.0 V are given in Fig. 3.4.2. It should be pointed out that at each bias, the current is determined self-consistently under the non-equilibrium condition. Their geometry structures after optimization are displayed in the corresponding insets. The upper structure of the insets is the E isomer, and the lower one is the C isomer. The switching behavior of the fulgide molecule system can be clearly seen from Fig. 3.4.2. From the figure, the current through C isomer is always greater than that through E isomer at the same bias. Thus, in the process of fulgide molecule reversing from C isomer to E isomer under photoexcitation, the current through the circuit can switch from on (high conductance) to off (low conductance), and vice versa. Furthermore, it also can be seen from Fig. 3.4.2 that the electronic transport properties through the molecular junction are affected significantly by the SWCNT's chirality. The current through the armchair-fulgide junction is 2-3 orders of magnitude larger than that through the zigzag-fulgide junction. Meanwhile, an apparent NDR effect appears in zigzag-E-zigzag junction with a maximum current at 0.8 V and a minimum current at 0.9 V. However, no NDR behavior is seen in other models. In order to characterize the conduction change due to different SWCNT electrodes, we define the on-off ratio, as shown in Fig. 3.4.3, Ratio$= I_{\text{C-isomer}} / I_{\text{E-isomer}}$, where $I_{\text{C-isomer}}$ and $I_{\text{E-isomer}}$ are the current of C and E isomer, respectively. Obviously, the zigzag junctions show better switching characteristics as compared with the armchair junctions for the chosen system.

In NEGF theory, the current of the system is calculated by the Landauer-Bütiker formula $I = \frac{2e}{h}\int [f(E-\mu_L) - f(E-\mu_R)] T(E,V) \mathrm{d}E$, which is transmission spectra dependent. Thus, we calculate the transmission spectra $T(E,V)$ of the C and E isomer with different SWCNT electrodes under zero voltage to understand the dramatic difference in conductance appearing in I-V curves. In our calculation, the average Fermi level, which is the average value of the chemical potential of the left and right electrodes, is set as zero. The mechanism of switching behavior for fulgide molecule with different electrodes is shown clearly in the transmission spectrum from Fig. 3. 4. 4. The transmission coefficients of C isomer are always bigger than those of E isomer. Meanwhile, for the C isomer, there is a narrow peak above the Fermi level, which dominates the conductive behavior of the molecular junction under the small bias. However, this narrow peak becomes very weak and shifts toward the higher energy when switch changes from the C isomer to the E isomer under photoexcitation. As a result, the lack of any significant peak and the low transmission coefficient due to the molecular structure changes accounts for the low conductivity in the E isomer.

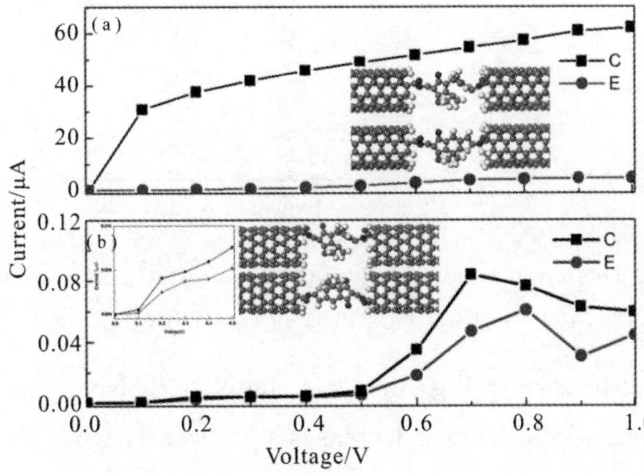

Fig. 3. 4. 2 The *I-V* characteristics of the molecular switch with (a) (5,5) armchair; (b) (9,0) zigzag SWCNT electrodes

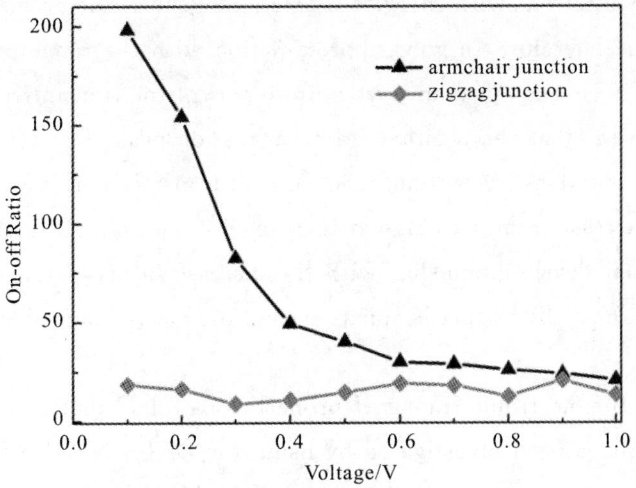

Fig. 3. 4. 3 The on-off ratio of the molecular switch with (5,5) armchair and (9,0) zigzag SWCNT electrodes

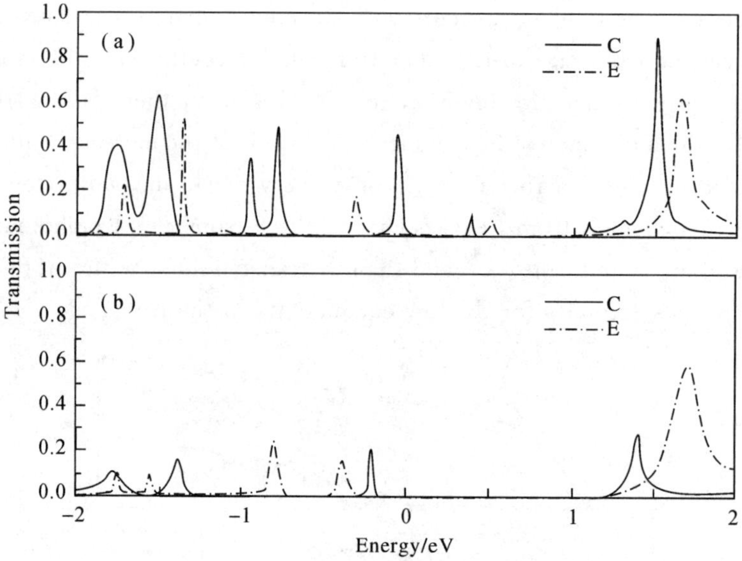

Fig. 3. 4. 4 The zero-bias transmission spectrum for the molecular switch with (a) (5,5) armchair; (b) (9,0) zigzag SWCNT electrodes

Another interesting feature in Fig. 3. 4. 2 is the NDR behavior in zigzag-E-zigzag junction. The current through the E isomer decreases significantly when the bias exceeds 0. 8 V. To investigate the mechanism responsible for NDR behavior, we calculated the transmission spectra under bias of 0. 7 to 1. 0 V as shown in Fig. 3. 4. 5. Seen from the Eq. (3. 4. 3), we can expect that only electrons with energies within a range near the Fermi level E_F contribute to the total current. Therefore, a good approximation with the range of the bias window, i. e. $[-V/2, +V/2]$ is enough to analyze a finite part of the transmission spectrum. From Fig. 3. 4. 5, we can see that the positions of transmission peak shift toward low-energy region with the increase in bias. Meantime, the value of transmission peaks in the bias window decrease with the increase in bias voltage at the range of 0. 8 to 1. 0 V. As a result, current through the molecular device diminishes with the increase in bias voltage above 0. 8 V and below1. 0 V. Thus the NDR feature is apparent over the bias voltage above 0. 8 V and below 1. 0 V.

In conclusion, the electronic transport properties of fulgid molecule with two different SWCNT electrodes have been investigated by using the DFT+NEGF first-principles method. An obvious switching behavior can be observed both in armchair and zigzag junctions. Meantime, the chirality of the SWCNT electrodes strongly affects the switching characteristics of the molecular junctions. The current through the armchair-fulgide junction is 2-3 orders of magnitude larger than that through the zigzag-fulgide junction and the armchair junctions show better switching characteristics as compared with the zigzag junctions for the chosen system. The physical origin of the switching behavior is interpreted based on the transmission spectra. A reversible change between the two isomers would realize a molecular nanoswitch, which suggests that this system has attractive potential application in future molecular circuit.

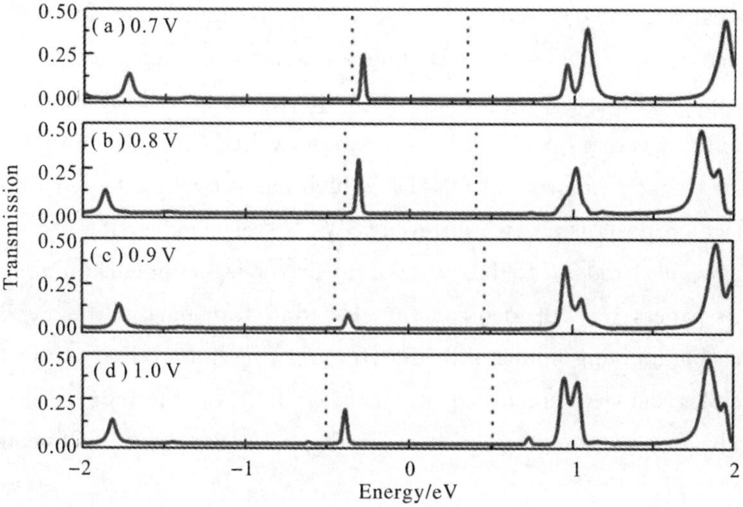

Fig. 3. 4. 5 The transmission spectra of the molecular switch under biases (a) 0. 7 V; (b) 0. 8 V; (c)0. 9 V; (d)1. 0 V

3. 5 A reversible hydrogen transfer in single organic molecular device

Recently, molecular devices have been considered as one of the most promising candidates for nano-electronics in the future. Many potentially functionalities, such as negative differential resistance (NDR), rectification, amplification and switching properties, have been performed both in experimental and theoretical studies. In this regard, the design of molecular switches using photochromic molecules which can change their isomer type upon light irradiation is of particular interest in molecular devices. Especially, diarylethene- and azobenzene-based optical switch which are mainly based on light-induced conformational changes, namely ring-opening reactions and isomerization reactions of the molecular bridge, have been widely studied. However, the mechanism for these optical switching has the disadvantage that the overall length and the molecule-electrode binding geometry of the junction are changed significantly. Recently, salicylideneaniline molecule which is characterized by having an OH - N hydrogen bond has been reported. This molecule can convert between enol and keto isomers upon photoinduced excited state hydrogen transfer in the molecular bridge (see Fig. 3. 5. 1). In contrast to most other mechanisms, hydrogen translocation within the molecular bridge has the advantage that the overall length is not changed significantly, which suggests that this molecule can be considered to be the candidate for optical switch.

On the other hand, electrode materials as a part of the molecular device also play an important role in realizing the elementary functions in future molecular electronic circuits. Currently, carbon nanotubes (CNTs) have become a potential ideal electrode material because of their rich electronic and mechanical properties. In particularly, recent experiment shows that an individual molecule can covalently attach to single-walled carbon nanotube(SWCNT) electrodes through the amide endgroup(CONH), which has proven to be very robust by Guo et al. Therefore, we propose using two different SWCNT electrodes, i. e. (5,5) armchair/(9,0) zigzag SWCNT electrodes, to fabricate light-driven salicylideneaniline-molecular based switches. In this paper, we will focus on the electronic transport of this new kind of optical molecular switch by applying nonequilibrium Green's function (NEGF) formalism combined with first-principles density functional theory (DFT). Furthermore, the effects of the SWCNT's chirality on the switching behavior in the molecular device will be also discussed in detail.

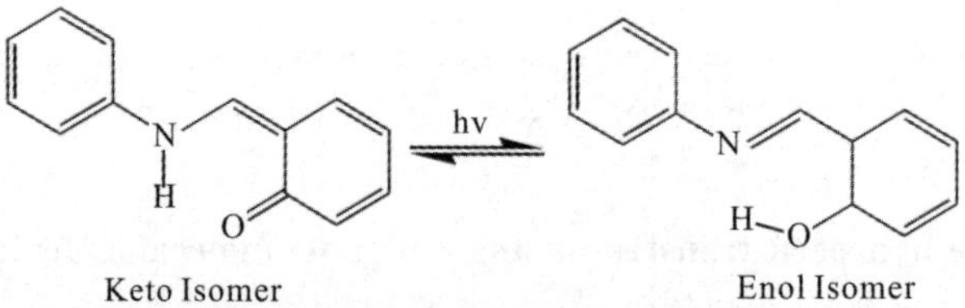

Fig. 3. 5. 1 Scheme of a photoinduced hydrogen transfer reaction between two tautomers corresponding to keto and enol isomers of molecule

The pre-optimizedsalicylideneaniline molecule with two isomers is sandwiched between two SWCNT electrodes as shown in Fig. 3. 5. 2. Amide endgroups were used as linkages between the molecule and SWCNT electrodes. For each electrode, eight layers of carbon atoms are included into the extended molecule region(region B in Fig. 3. 5. 2) to screen the perturbation effect from the central scattering region and they are denoted as surface-atomic layers. In the calculations, the end of each SWCNT is capped by H atoms to eliminate the dangling bonds. The geometry, including the H atoms, end C atoms and the central molecule (region A in Fig. 3. 5. 2), is fully optimized by minimizing the atomic forces on those atoms to be smaller than 0.05 eV/Å.

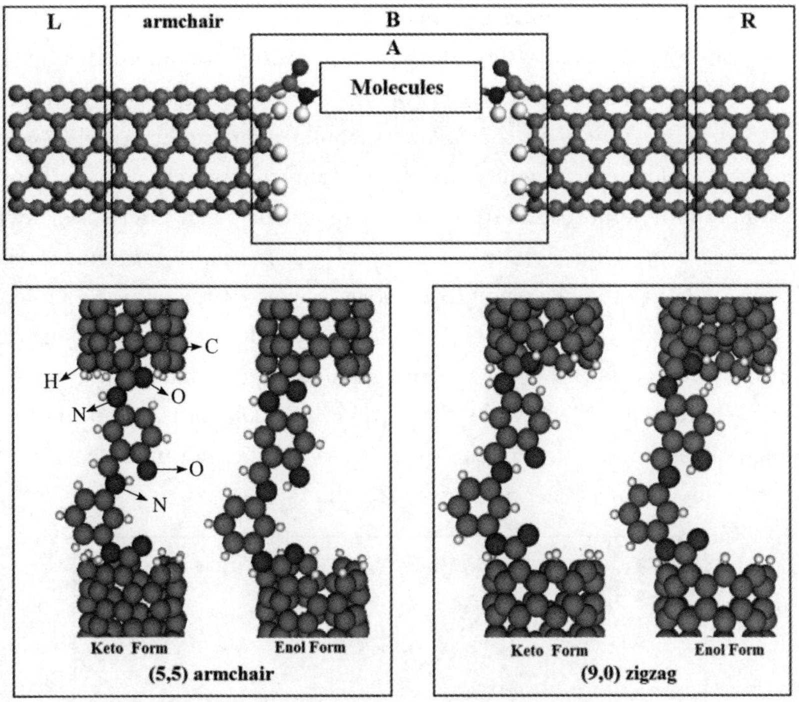

Fig. 3. 5. 2 Optimized structures of the salicylideneaniline molecular switch with two different kinds of SWCNT electrode

All the geometrical optimizations and the electronic transport properties of the molecular junctions are calculated by a fully self-consistent NEGF formalism combined with first-principles DFT, which is implemented in Atomistix Tool Kit (ATK) software package (version 2012. 8. 2) . The main feature of the computational package is to model a nanostructure coupled to external electrodes with different electrochemical potentials and to realize the transport simulation of the whole two-probe system without inducing phenomenological parameters. In the electronic transport calculations, the exchange-correlation potential is described by Ceperley-Alder local density approximation (LDA) . The core electrons are modeled with Troullier-Martins nonlocal pseudopotential , while the valance electrons wave functions are expanded by a SIESTA basis set . The double-zeta plus polarization (DZP) basis set is adopted for all atoms. The Brillouin zone is set to be $5 \times 5 \times 100$ points following the Monkhorst-Pack *k*-point scheme. The cut-off energy and the iterated convergence criterion for total energy are set to 150 Rydberg and 10^{-5}, respectively.

The calculated current-voltage (*I-V*) curves of the molecular device with different isomers under the bias voltage varying from 0 to 2. 0 V are given in Fig. 3. 5. 3. It should be pointed out that at each bias, the current is determined self-consistently under the nonequilibrium condition. The switching behavior of the salicylideneaniline molecule system can be clearly seen from Fig. 3.5.3. Although these two tautomers only differ by the position of a single hydrogen atom, their conductance properties are drastically different. The current of enol isomer is evidently larger than that of keto isomer over the entire bias range regardless of the electrode type. Therefore, when the molecule in the switch changes from enol isomer to the keto isomer under photoexcitation, one can predict that there is a switch from on (high conductance) to off (low conductance), and vice versa. Furthermore, it also can be seen from Fig. 3.5.3 that the electronic transport properties through the molecular junction are affected significantly by the SWCNT's chirality. The current through the armchair-salicylideneaniline junction is 2-3 orders of magnitude larger than that through the zigzag-salicylideneaniline junction. Meanwhile, an apparent NDR effect appears in zigzag-enol-zigzag junction with a maximum current at 1.6 V and a minimum current at 1.8 V. However, no NDR behavior is seen in other models.

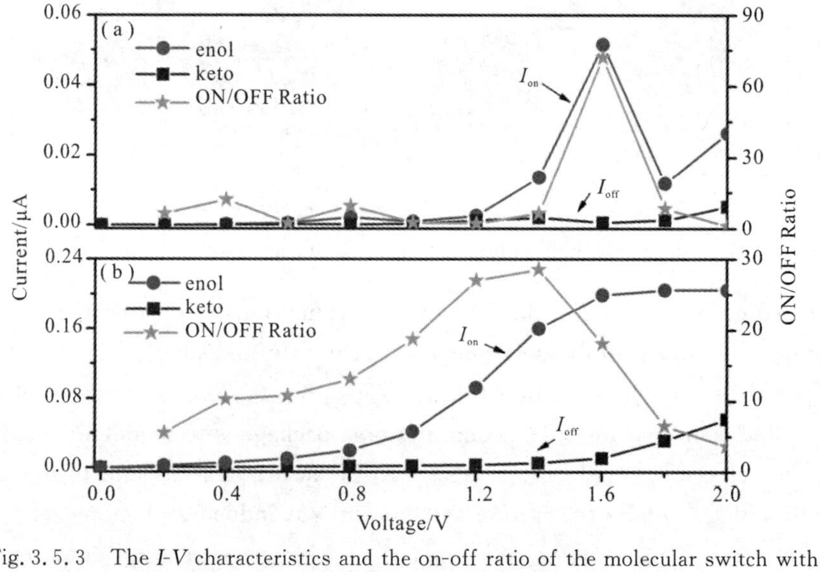

Fig. 3.5.3 The *I-V* characteristics and the on-off ratio of the molecular switch with (a) (9,0) zigzag; (b) (5,5) armchair SWCNT electrodes

In NEGF theory, the current of the system is calculated by the Landauer-Bütiker formula $I=\frac{2e}{h}\int[f(E-\mu_L)-f(E-\mu_R)]T(E,V)dE$, which is transmission spectra dependent. Thus, we calculate the transmission spectra of the enol and keto isomer with different SWCNT electrodes under zero voltage to understand the dramatic difference in conductance appearing in *I-V* curves. As shown in Fig. 3.5.4, the average Fermi level is set as zero and the positions of the highest occupied molecular orbital (HOMO) and the lowest unoccupied molecular orbital (LUMO) are marked with squares for the enol isomer and with circles for the keto isomer, respectively. The mechanism of switching behavior for salicylideneaniline molecule with different electrodes is shown clearly in the transmission spectrum from Fig. 3. 5. 4. The enol isomer presents an overall much bigger transmission coefficient than the keto isomer regardless of the electrode types. Furthermore, the barrier for the electron transport is intensively relevant to the HOMO-LUMO gap in such an open system, which the Fermi level E_F of the electrode aligns between HOMO and LUMO. From Fig. 3. 5. 4, we also can see that the HOMO-LUMO gap increases from 1. 91 to 2. 59 eV, from 1. 59 to 2. 20 eV for the zigzag and armchair junction, respectively, when the enol isomer is switched to the keto isomer. As a result, the low transmission coefficient and the lager HOMO-LUMO gap due to the molecular structure changes accounts for the low conductivity in the keto isomer.

Generally, the transmission coefficient can be related to the molecular orbitals, which have been modified by the electrodes. These modified molecular orbitals can be obtained from the molecular projected self-consistent Hamiltonian (MPSH). From Fig. 3. 5. 4, it is notable that the LUMO is the main transmission channel because it is close to the Fermi energy in both of these two tautomers. Therefore, to get a further insight about the origins of the peaks in the transmission spectra and of the different transmission characteristics for two tautomers, we analyze the LUMO-MPSH on the molecular junction. As shown in Table 3. 5. 1, it is found obviously that the density of LUMO-MPSH is more delocalized in the enol isomer both in armchair and zigzag junctions, which leads to low barrier and provides good transport channels for electron transport. On the contrary, this transmission channel is localized orbital for the keto isomer leading to high barrier for electron transport which accounts for the low transmission strength in Fig. 3. 5. 4. As a result, the current is decreased when the molecule switch from enol to keto isomer upon photoexcitation, which is in good agreement with the trends of *I-V* curves shown in Fig. 3. 5. 3. Another interesting feature in Fig. 3. 5. 3 is the NDR behavior in zigzag-enol-zigzag junction. The current through the enol isomer decreases significantly when the bias exceeds 1. 6 V. To investigate the mechanism responsible for NDR behavior, we calculated the transmission spectra under bias of 1. 4 to 2. 0 V as shown in Fig. 3. 5. 5. Seen from the Eq. (3. 5. 3), we can expect that only electrons with energies within a range near the Fermi level E_F contribute to the total current. Therefore, a good approximation with the range of the bias window, i. e. $[-V/2, +V/2]$ is enough to analyze a finite part of the transmission spectrum. From Fig. 3. 5. 5, we can see that the positions of transmission peak shift toward low-energy region with the increase in bias. Meantime, the value of transmission peaks in the bias window decrease with the increase in bias voltage at the range of 1. 6 to 1. 8 V. As a result, current through the molecular device diminishes with the increase in bias voltage above 1. 6 V and below 2. 0 V. Thus the NDR feature is apparent over the bias voltage above 1. 6 V and below 2. 0 V.

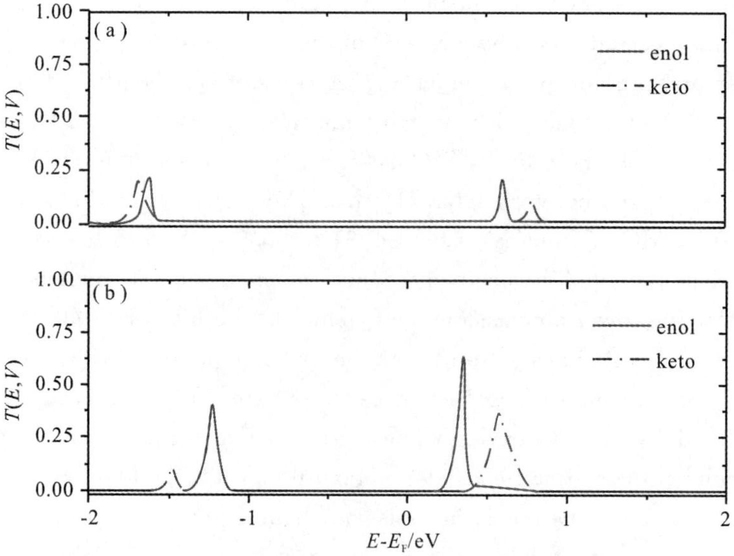

Fig. 3. 5. 4 Transmission spectra of the molecular switch with (a) (9,0) zigzag; (b) (5,5) armchair SWCNT electrodes at zero bias

Table 3. 5. 1 The MPSH of LUMO for enol and keto isomer at zero bias

MPSH	armchair junction	zigzag junction
Keto isomer	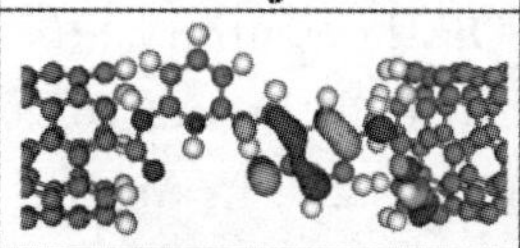	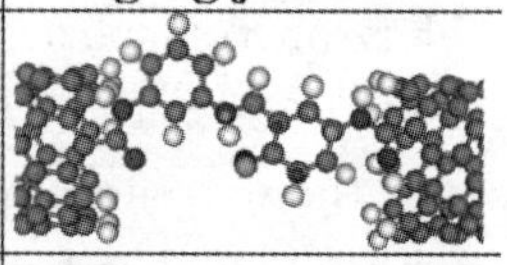
Enol isomer	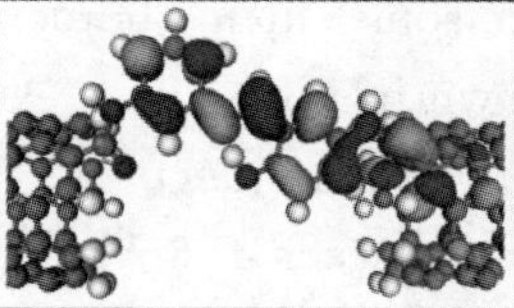	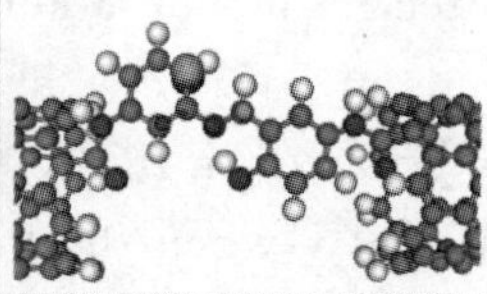

In conclusion, the electronic transport properties ofsalicylideneaniline molecule with two different SWCNT electrodes have been investigated by using the DFT+NEGF first-principles method. An obvious switching behavior can be observed both in armchair and zigzag junctions. Meantime, the chirality of the SWCNT electrodes strongly affects the switching characteristics of the molecular junctions. The maximum value of on-off ratio can reach 72 at 1. 6 V for the switch with zigzag SWCNT electrodes. The physical origin of the switching behavior is interpreted based on the transmission spectra and the MPSH. A reversible change between the two isomers would realize a molecular nanoswitch, which suggests that this system has attractive potential application in future molecular circuit.

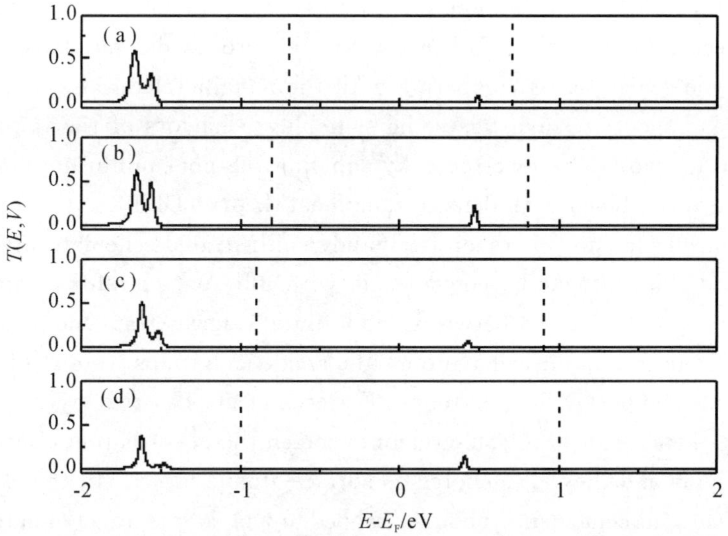

Fig. 3. 5. 5 The transmission spectra of the molecular switch under biases (a) 1. 4 V; (b) 1. 6 V; (c) 1. 8V; (d) 2. 0 V

3. 6 Single chiroptical molecular switch

Molecular devices which used molecules as electronic transport channels have attracted more and more attention due to theirnovel physical properties and potential application in nanoelectronic devices. Several molecular devices with different functionalities have been designed and measured theoretically and experimentally in the past years, such as negative differential resistance (NDR), rectification, memory effects, amplification, switch and others. As the basic element of logic in molecular devices, molecular switches, especially optical molecular switches, have been extensively investigated and discussed. However, most mechanisms for optical molecular switch considered so far are based on the ring-opening reactions of the molecular bridge, such as diarylethene, naphthopyrans, dihydroazulenes and benzochromenes. The mostly disadvantage for this mechanism is that the overall length of molecule are changed significantly when the molecule convert between two different isomers. Recently, some experiments report that the thioxanthene-based molecule can exhibit different chirality ,namely cis-isomer and trans-isomer, by ultraviolet or visible irradiation. These two isomers only differ by the position of upper half. The switching between the different chirality is very efficient and shows excellent reversibility, which make it usable as one of the candidates for light-driven molecular switches. However, to our knowledge, the electronic transport properties of the thioxanthene-based molecule is not reported so far.

Moreover, following the progress in fabrication of graphene nanoribbons (GNRs) at room temperature, GNRs have become one of the most promising candidates for the next generation of electrodes materials. Because GNRs electrodes did not have serious contact problems at the molecular scales comparing with the metallic electrodes materials like Au. Therefore, in this paper, we investigate the switching behaviors of the thioxanthene-based molecule coupled to two GNRs electrodes by applying the nonequilibrium Green's function (NEGF) formalism combined with density functional theory (DFT).

The molecular device for our theoretical study is illustrated schematically in Fig. 3. 6. 1. The pre-optimized thioxanthene-based molecule with different chirality, namely cis-isomer and trans-isomer, are sandwiched between two infinite 6-zigzag-graphene nanoribbon(named as 6-ZGNRs) electrodes. In our calculations, the amide endgroups were used as linkages between the molecule and 6-ZGNRs electrodes. For each electrode, two layers of carbon atoms are included into the extended molecule regionto screen the perturbation effect from the central scattering region and they are denoted as surface-atomic layers. In the calculations, the end of each zigzag-graphene nanoribbon is capped by H atoms to eliminate the dangling bonds.

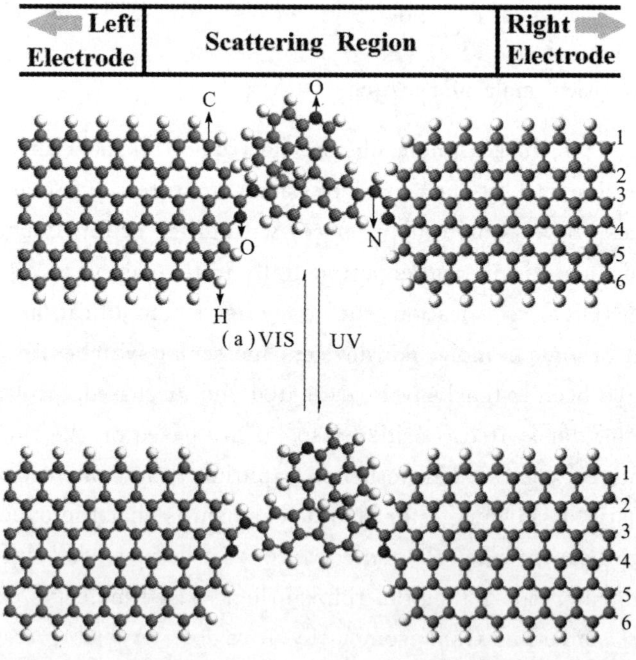

(b)

Fig. 3. 6. 1 Models of the molecular device with different chirality
(a) cis-isomer; (b) trans-isomer

The geometric optimizations and the electronic transport properties of the molecular devices are calculated by a developed first-principles software package Atomistix ToolKit, which is based on the DFT combined with self-consistent NEGF formalism. The main feature of the computational package is to model a nanostructure coupled to external electrodes with different electrochemical potentials and to realize the transport simulation of the whole two-probe system without inducing phenomenological parameters. In the electronic transport calculations, the exchange-correlation potential is described by Ceperley-Alder local density approximation (LDA). The core electrons are modeled with Troullier-Martins nonlocal pseudopotential, while the valance electrons wave functions are expanded by a SIESTA basis set. The double-zeta plus polarization (DZP) basis set is adopted for all atoms. The Brillouin zone is set to be $5\times5\times100$ points following the Monkhorst-Pack *k*-point scheme. The cut-off energy and the iterated convergence criterion for total energy are set to 150 Rydberg and 10^{-5}, respectively.

The calculated current-voltage (*I-V*) curves of the molecular device with different isomers under the bias voltage varying from 0 to 2.0 V are given in Fig. 3.6.2. It should be pointed out that at each bias, the current is determined self-consistently under the non-equilibrium condition. As shown in Fig. 3.6.2, the calculated *I-V* curves clearly demonstrate two following important features: ① Although the two isomers only differ by the position of upper half, their conductance properties are drastically different. The current through the cis-isomer is evidently larger than that through the trans-isomer over the entire bias range. For example, the calculated current at 1.4 V is about 3.45 μA and 0.61 μA for the cis-isomer and trans-isomer, respectively. Thus, when the molecule in the device changes from the cis-isomer to trans-isomer under photoexcitation, the device is predicted to switch from on (high conductance) to off (low conductance), and vice versa. The remarkable difference of current between the cis-isomer and trans-isomer under the applied bias can be quantified by the on-off ratio of the current defined as Ratio$=I_{\text{cis-isomer}}/I_{\text{trans-isomer}}$. The calculated on-off ratio varies from around 2.3 to 9.4 in the bias range from 0 to 2.0 V. The maximum on-off ratio can reach 9.4 at 1.2 V, suggesting potential applications of this type of junctions in future design of light-driven molecular switches. ② The currents of two isomers with 6-ZGNRs electrodes both vary with the bias voltage in a complicated manner. The negative differential resistance (NDR) behaviors can be observed both in the cis-isomer and trans-isomer. For example, in the bias voltage region [1.4 V, 1.7 V], the current of cis-isomer decreases quickly as the bias voltage increases, leading to significant NDRs at these bias voltages. The maximum current of the cis-isomer is up to be about 3.4 μA at the peak position($V_{\text{bias}}=1.4$ V), while the current reaches its minimum value 1.1 μA at the valley site($V_{\text{bias}}=1.5$ V).

Meanwhile, NDR also can be observed in the bias range of 1.1 to 1.8 V for the trans-isomer. The maximum peak-to-valley ratio (PVR) in is about 5 and 2 for cis-isomer and trans-isomer, respectively. The observed sequential NDR behaviors in the molecular device have important application in amplifier and logic gate.

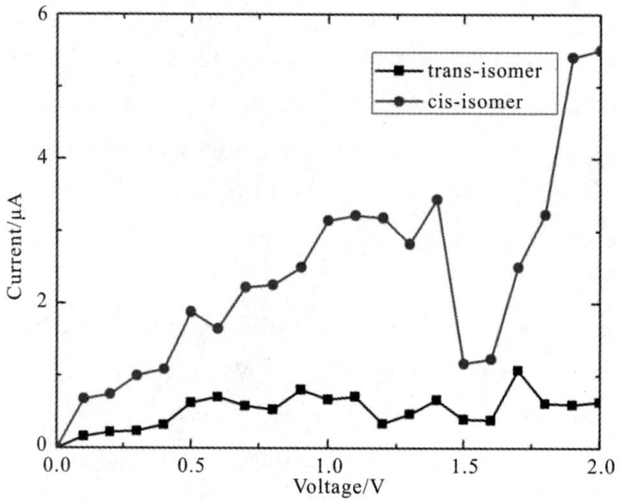

Fig. 3.6.2 The calculated *I-V* characteristics of the molecule junction with 6-ZGNRs electrodes

To understand the nature of the observedswitching behaviors, we calculate the zero-bias transmission spectra of the cis-isomer and trans-isomer with 6-ZGNRs electrodes in Fig. 3.6.3. In our calculation, the average value of the chemical potential of the left and right electrodes named average Fermi level is set as zero. The short vertical bars near the energy axes stand for the energy levels of the highest occupied molecular orbital (HOMO) and the lowest unoccupied molecular orbital (LUMO), respectively. It is clear that the transmission spectra of the cis-isomer and trans-isomer show remarkably different behavior around the Fermi level. The cis-isomer presents an overall much bigger transmission coefficient than the trans-isomer. Meanwhile, for the cis-isomer, there are two narrow peaks near the Fermi level, which dominates the conductive behavior of the molecular junction under the small bias. However, these narrow peaks become very weak and shifts toward the higher energy when switch changes from the cis-isomer to the trans-isomer under photoexcitation. Furthermore, the barrier for the electronic transport is intensively relevant to the HOMO-LUMO gap in such an open system, which the Fermi level E_F of the electrode aligns between HOMO and LUMO. From Fig. 3.6.3, we also can see that the HOMO-LUMO gap increases from 0.56 to 0.66 eV, when the cis-isomer is switched to the trans-isomer. As a result, the low transmission coefficient and the lager HOMO-LUMO gap due to the molecular structure changes accounts for the low conductivity in the trans-isomer.

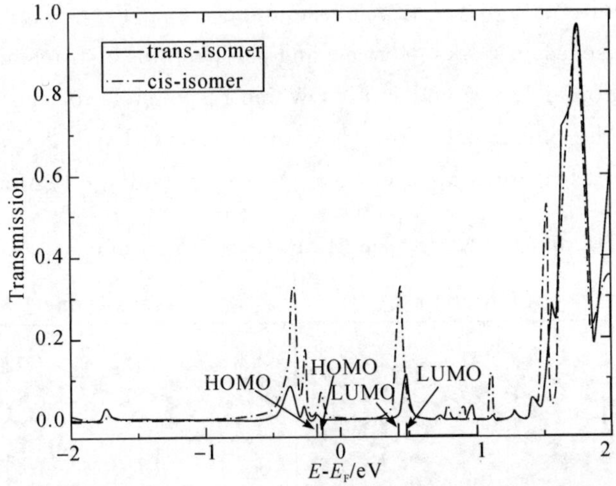

Fig. 3. 6. 3 Transmission spectra of the molecule junction with 6-ZGNRs electrodes at zero bias

Generally, the transmission coefficient can be related to the molecular orbitals, which have been modified by the electrodes. These modified molecular orbitals can be obtained from the molecular projected self-consistent Hamiltonian (MPSH). From Fig. 3. 6. 3, it is notable that the HOMO and LUMO are the main transmission channel because it is close to the Fermi energy in both of these two isomers. Therefore, to get a further insight about the origins of the peaks in the transmission spectra and of the different transmission characteristics for two isomers, we analyze the HOMO-MPSH and LUMO-MPSH on the molecular junction. As shown in Table 3. 6. 1, it is found obviously that the density of MPSH is more delocalized in the cis-isomer, which leads to low barrier and provides good transport channels for electron transport. On the contrary, these two transmission channels are localized orbital for the trans-isomer leading to high barrier for electron transport which accounts for the low transmission strength in Fig. 3. 6. 3. As a result, the current is decreased when the molecule switch from cis to trans isomer upon photoexcitation, which is in good agreement with the trends of I-V curves shown in Fig. 3. 6. 3.

Another interesting feature in Fig. 3. 6. 2 is the NDR behavior. As shown in Fig. 3. 6. 2, there exist sequential distinct negative differential resistances both in the cis-isomer and trans-isomer at the bias voltage region from 1. 0 to 1. 7 V. To investigate the mechanism responsible for NDR behavior, we calculated the transmission spectra under bias of 1. 0 to 1. 7 V as shown in Fig. 3. 6. 4. Seen from the Eq. (3. 6. 3), we can expect that only electrons with energies within a range near the Fermi level E_F contribute to the total current. Therefore, a good approximation with the range of the bias window, i. e. $[-V/2, +V/2]$ is enough to analyze a finite part of the transmission spectrum. From Fig. 3. 6. 4, we can see that the value of transmission peaks decrease and the position of transmission peak shift toward high-energy region at 1. 3 V and 1. 5 V, which responsible for the decrease of the current for cis-isomer. The similar behaviors also can be observed at 1. 2 V and 1. 5 V for trans-isomer. Thus the NDR feature is apparent over these bias voltage regions.

Table 3. 6. 1 The MPSH of HOMO and LUMO for cis-isomer to trans-isomer at zero bias

	Cis-isomer	**Trans-isomer**
MPSH-HOMO	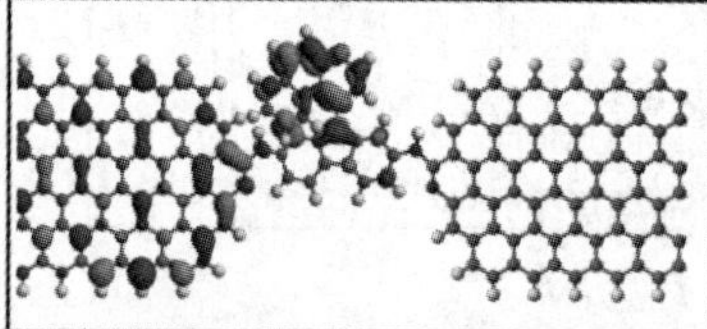	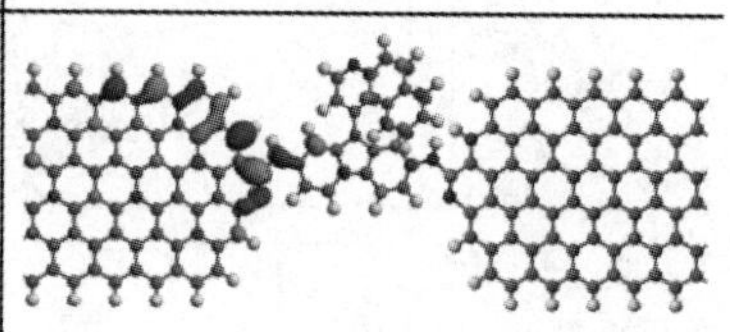
MPSH-LUMO	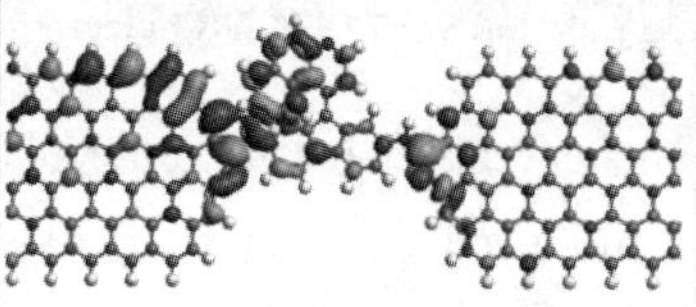	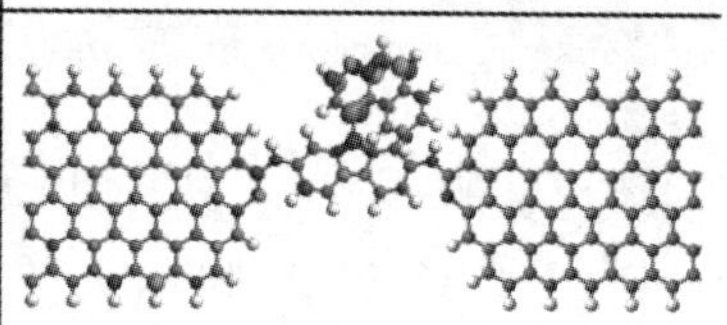

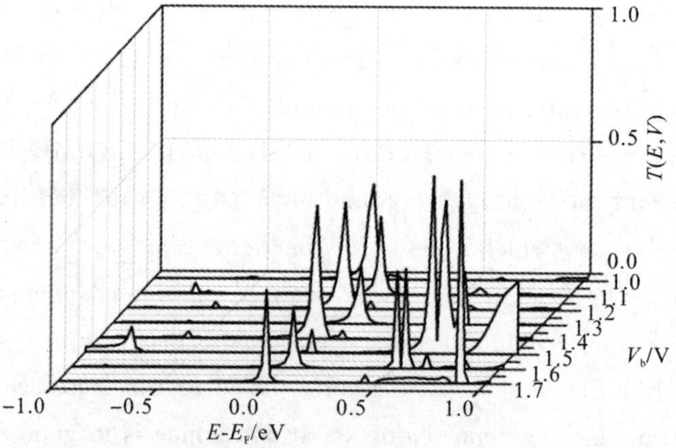

(a)

Fig. 3. 6. 4 The transmission spectrum under various bias voltages (1. 0 to 1. 7 V) for (a) cis-isomer

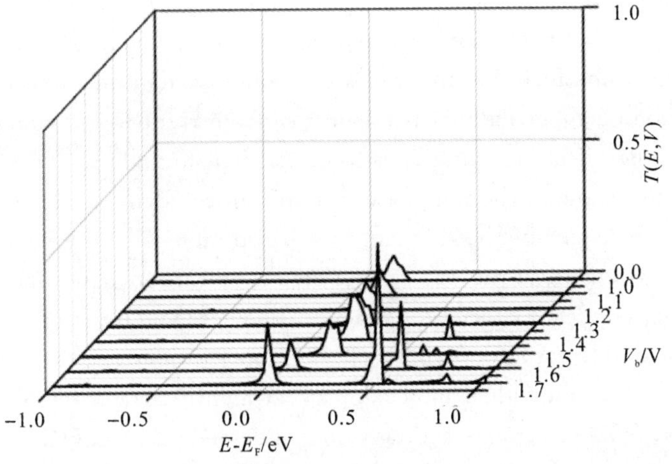

(b)

Continued Fig. 3. 6. 4 The transmission spectrum under various bias voltages (1. 0 to 1. 7 V) for (b) trans-isomer

In conclusion, the electronic transport properties ofthioxanthene-based molecule with two 6-ZGNRs electrodes have been investigated by using the DFT+NEGF first-principles method. An obvious switching behavior can be observed when the molecule convert between cis-isomer and trans-isomer. The physical origin of the switching behavior is interpreted based on the transmission spectra and molecular projected self-consistent Hamiltonian. The maximum on-off ratio and peak-to-valley ratio can reach 9. 2 and 5, which suggests that this system has attractive potential application in future molecular circuit.

3. 7 Thioxanthene-based molecular switch: effect of chirality

Following the miniaturization of electronic device, molecular devices have been considered as one of the most promising candidates for semiconductor-based devices in the future. Several molecular devices with different functionalities have been designed and measured theoretically and experimentally in the past years, such as negative differential resistance (NDR), rectification, memory effects, amplification, switch and others. As the basic element of logic in molecular devices, molecular switches, especially optical molecular switches, have been extensively investigated and discussed. In this regard, the design of optical molecular switches using photochromic molecules which can change their isomer type upon light irradiation is the subject of intensive studies. However, most mechanisms for optical molecular switch considered so far are based on the ring-opening reactions of the molecular bridge, such as diarylethene, naphthopyrans, dihydroazulenes, benzochromenes and azobenzene. The mostly disadvantage for this mechanism is that the overall length of molecule are changed significantly when the molecule convert between two different isomers. Recently, some experiments report that thioxanthene-based molecule can exhibit different chirality namely cis-form and trans-form by ultraviolet or visible irradiation. These two isomers only differ by the position of upper half, which can turn from right to left upon irradiation. This switching between the different chirality is very efficient and shows excellent reversibility, which make it usable as one of the candidates for chiroptical switch. However, to our knowledge, the transport behavior of the thioxanthene-based molecule is not reported so far.

Moreover, following the progress in fabrication of carbon nanotubes (CNTs), CNTs have become a potential ideal material for functional devices and interconnect in nanoelectronics because of their stable structures and rich electronic properties. In particularly, Guo et al. have reported that an individual molecule can covalently attach to the single-walled carbon nanotube (SWCNT) electrodes through the amide endgroup (CONH), which has proven to be very robust comparing with the common metallic electrodes materials. In this Letter, we investigate the effect of chirality on the electronic transport properties of thioxanthene-based molecular switch by applying the nonequilibrium Green's function (NEGF) method and density functional theory (DFT) calculations.

The molecular device for our theoretical study is illustrated schematically in Fig. 3. 7. 1. The pre-optimized thioxanthene-based molecule with different chirality, namely cis-form and trans-form, are sandwiched between two SWCNT electrodes. The molecular junctions can be divided into three regions including the left electrode, central scattering region and right electrode. For each electrode, eight layers of carbon atoms are included into the central scattering region (region B in Fig. 3. 7. 1) to screen the perturbation effect from the central scattering region and they are denoted as surface-atomic layers. All edge carbon atoms are saturated with hydrogen atoms. The optimized central double bond between lower half and upper half is a normal bond length (1. 383 Å), which is similar to what was found before. In our calculations, the amide endgroups were used as linkages between the molecule and SWCNT electrodes. All the configurations are relaxed until their force tolerance is smaller than 0. 05 eV/Å.

All the geometrical optimizations and the electronic transport properties of the molecular junctions are calculated by a fully self-consistent NEGF formalism combined with first-principles DFT, which is implemented in Atomistix Tool Kit (ATK) software package. The main feature of the computational package is to model a nanostructure coupled to external electrodes with different electrochemical potentials and to realize the transport simulation of the whole two-probe system without inducing phenomenological parameters. In the electronic transport calculations, the exchange-correlation potential is described by Ceperley-Alder local density approximation (LDA). The core electrons are modeled with Troullier-Martins nonlocal pseudopotential, while the valance electrons wave functions are expanded by a

SIESTA basis set. The double-zeta plus polarization (DZP) basis set is adopted for all atoms. The Brillouin zone is set to be $5\times5\times100$ points following the Monkhorst-Pack k-point scheme. The cut-off energy and the iterated convergence criterion for total energy are set to 150 Rydberg and 10^{-5}, respectively.

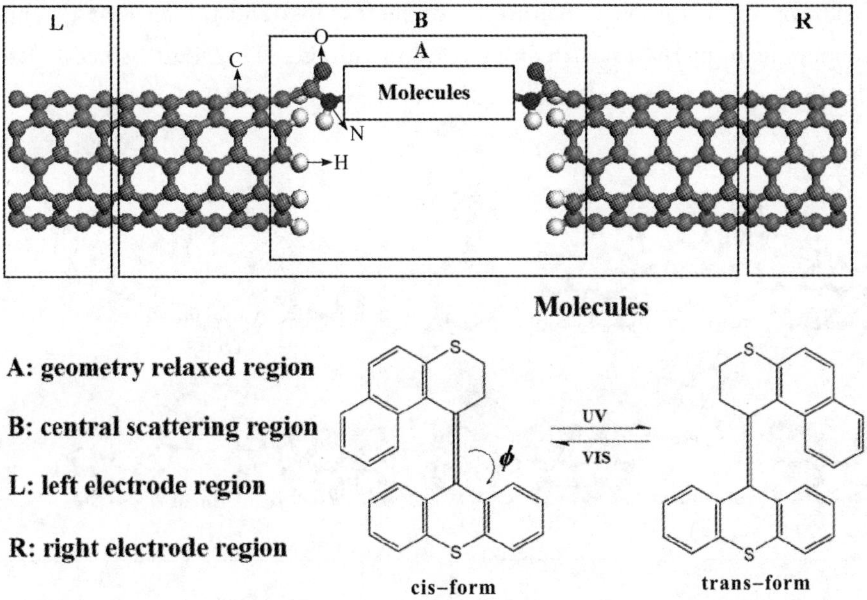

Fig. 3. 7. 1 Schematic illustration of the structure of the thioxanthene-based molecular switch

The optimized structures and self-consistently calculated current-voltage (I-V) characteristics of the junction with different SWCNT electrodes are shown in Fig. 3. 7. 2. The I-V curves show obvious switching behavior in bias voltage region [−1. 0 V, 1. 0 V]. Although the two isomers only differ by the position of upper half, their conductance properties are drastically different. The current of cis-form is evidently larger than that of trans-form over the entire bias range regardless of the electrode type. Therefore, when the molecule in the switch changes from cis-form to the trans-form under photoexcitation, one can predict that there is a switch from on (high conductance) to off (low conductance), and vice versa. This remarkable difference of current between the cis-form and trans-form under the applied bias can be quantified by the on-off ratio of the current defined as $\text{Ratio} = I_{\text{cis-form}} / I_{\text{trans-form}}$. The calculated on-off ratio of varies from around 21 to 104 and 29 to 76 for the cis-form and trans-form, respectively. The maximum on-off ratio can reach 104 at 0. 4 V for the armchair junction. Such a significant on-off ratio is desirable for the real applications in future molecular switch technology. Furthermore, an apparent NDR effect appears in cis-form with armchair

SWCNT electrodes. It is quite clear from Fig. 3.7.2(c) that initially the current of cis-form increases with the increases in external positive voltage. But beyond 0.8 V, there is a decrease in current with the increase in bias voltage. This decrease in current is the manifestation of the NDR feature. NDR behavior is quite prominent up to a bias voltage of 0.9 V. However, above 0.9 V the NDR feature is completely lost and the current through the molecular junction again increases with an increase in voltage. The NDR behavior also can been observed in the negative bias voltage region. However, no NDR behavior is seen in other models.

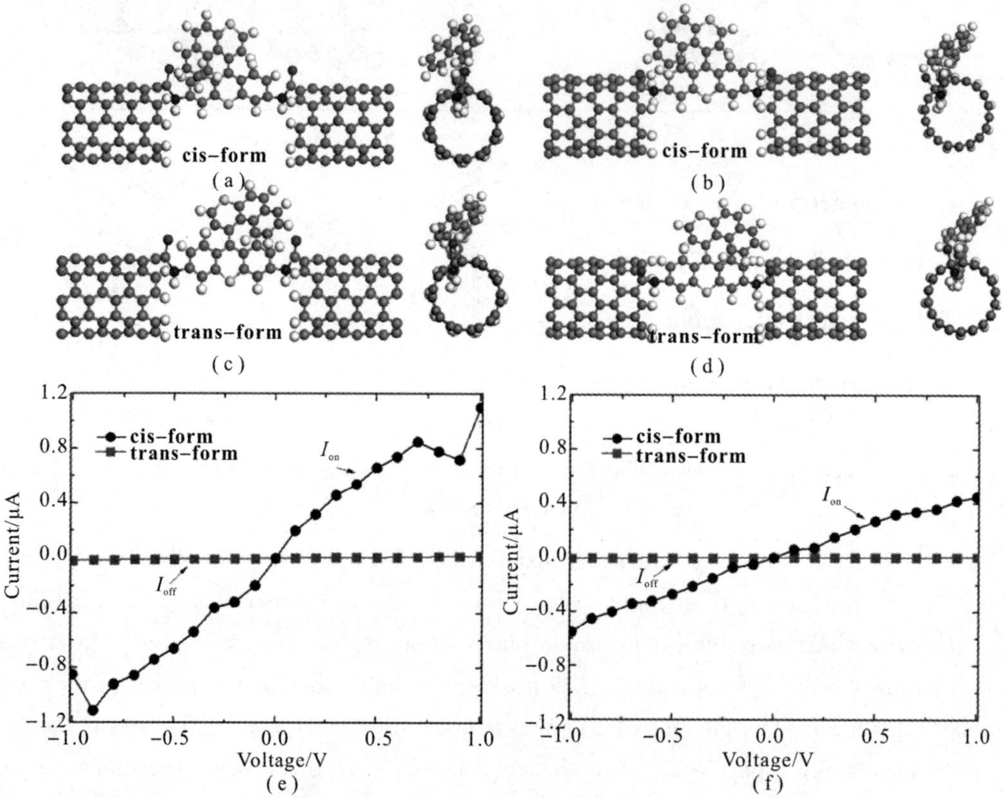

Fig. 3.7.2 The optimized structures and *I-V* characteristics of the thioxanthene-based molecular switch with
(a)(5,5) armchair; (b) (9,0) zigzag SWCNT electrodes;
(c)(5,5) armchair; (d) (9,0) zigzag SWCNT electrodes;
(e)cis-form and trans-form with (5,5) armchair;
(f)cis-form and trans-form with (9,0) zigzag SWCNT electrodes

From Fig. 3.7.2(a) and (b), we can see that the optimized structures are varying when the molecule is attached to the different SWCNT electrodes. The lower half , namely three phenyl rings, are coplanar. Meanwhile, there is a dihedral angle ϕ (see Fig. 3.7.1) between lower half and upper half of the molecule. Here, we focus on the change of the dihedral angle ϕ. Because the variety in dihedral angle ϕ is related to the intermolecular interaction, which can strongly affect the electronic transport properties of the molecular device. When the dihedral angle ϕ is bigger, the intermolecular interaction will be more weaker, and vice versa. The dihedral angle ϕ in the optimized molecules and in the relaxed armchair/zigzag junction is listed in Table 3.7.1, respectively. It is clear from Table 3.7.1 that the dihedral angle ϕ decreases when the cis-form molecule is attached to the SWCNT electrodes, whereas the reverse is true for the trans-form. It means when the molecule changes from cis-form to the trans-form, the effect of intermolecular interaction become weaker, which is corresponding to the lower current of trans-form in Fig. 3.7.2 (c) and (d).

Table 3.7.1 Dihedral angles ϕ between lower half and upper half in the optimized molecules and in the relaxed zigzag/armchair junction, respectively

System	cis-form/(°)	trans-form/(°)
Molecules	21.153	26.224
Zigzag junction	20.726	28.362
Armchair junction	19.181	29.235

To get a further insight about the origins of the difference in I-V characteristics for two forms, we analyze the transmission spectra and the energy gap of the highest occupied molecular orbital (HOMO) and the lowest unoccupied molecular orbital (LUMO) for the cis-form and trans-form with different SWCNT electrodes under zero voltage. In NEGF theory, the current of the system is calculated by the Landauer-Bütiker formula $I = \frac{2e}{h}\int [f(E-\mu_L) - f(E-\mu_R)]T(E,V)dE$, which is transmission spectra dependent. In our calculation, the average Fermi level, which is the average value of the chemical potential of the left and right electrodes, is set as zero. The mechanism of switching behavior for the molecule with different electrodes is shown clearly in the transmission spectrum from Fig. 3.7.2. The transmission coefficients of cis-form are always bigger than those of trans-form. Meanwhile, for the cis-form, there is a narrow peak above the Fermi level, i.e., LUMO transmission peak, which dominates the conductive behavior of the molecular junction. However, this narrow peak becomes very weak and shifts toward the higher energy when switch changes from the cis-form to the trans-form under photoexcitation. Furthermore, the barrier for the electronic transport is intensively relevant to the HOMO-LUMO gap in such an open system, which the Fermi level E_F of the electrode aligns between HOMO and LUMO. As shown in Fig. 3. 7. 3, the positions of the HOMO and LUMO are marked with squares for the cis-form and with circles for the trans-form, respectively. From Fig. 3. 7. 3, we also can see that the HOMO-LUMO gap increases when the cis-form is switched to the trans-form both for the armchair and zigzag junction. As a result, the low transmission coefficient and the lager HOMO-LUMO gap due to the molecular structure changes accounts for the low conductivity in the trans-form.

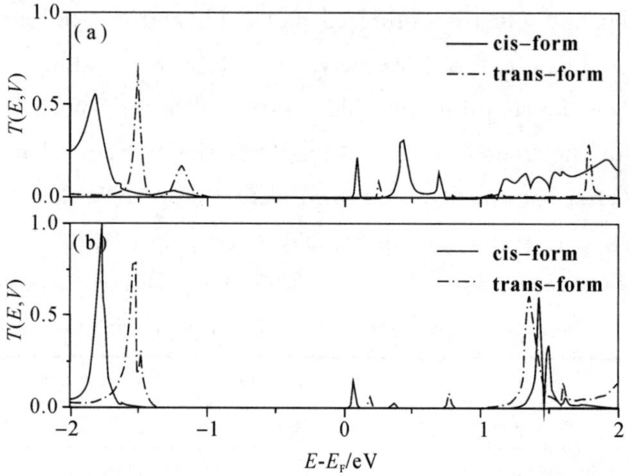

Fig. 3. 7. 3 Transmission spectra of the molecular switch with (a) (5,5) armchair; (b) (9,0) zigzag SWCNT electrodes at zero bias

Generally, the transmission coefficient can be related to the molecular orbitals, which have been modified by the electrodes. These modified molecular orbitals can be obtained from the molecular projected self-consistent Hamiltonian (MPSH). MPSH is the self-consistent Hamiltonian of the isolated molecule in the presence of the electrodes, namely the molecular part is extracted from the whole self-consistent Hamiltonian at the contact region. From Fig. 3. 7. 3, it is notable that the LUMO is the main transmission channel because it is close to the Fermi energy in both of these two isomers. Therefore, to get a further insight about the origins of the peaks in the transmission spectra and of the different transmission characteristics for two isomers, we analyze the LUMO-MPSH on the molecular junction. As shown in Table 3. 7. 2, it is found obviously that the density of MPSH is more delocalized in the cis-form, which leads to low barrier and provides good transport channels for electron transport. On the contrary, this transmission channels is localized orbital for the trans-form leading to high barrier for electronic transport which accounts for the low transmission strength in Fig. 3.7.3. As a result, the current is decreased when the molecule switch from cis to trans form upon photoexcitation.

Table 3.7.2 The MPSH of LUMO for the cis-form and trans-form at zero bias

LUMO MPSH	**Armchair junction**	**Zigzag junction**
cis-form	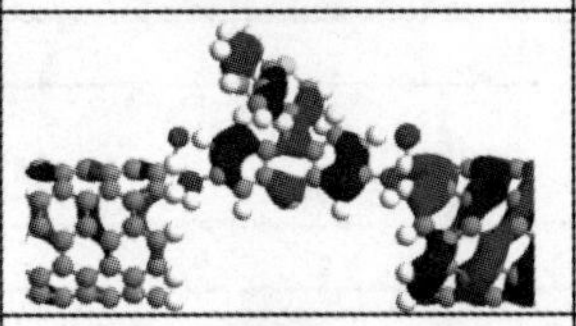	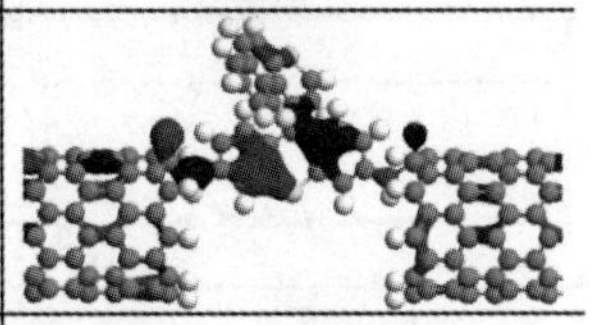
trans-form	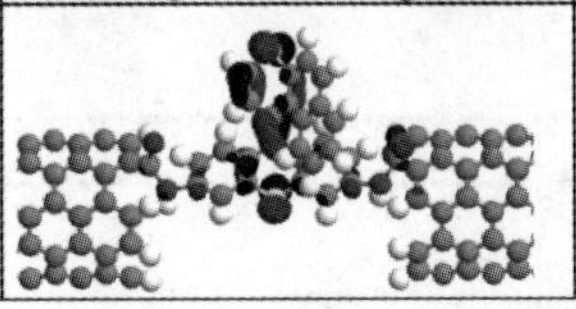	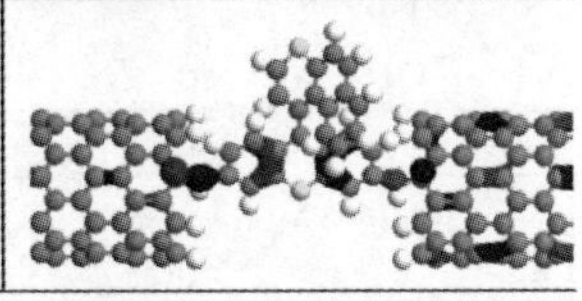

Another interesting feature in Fig. 3.7.2(c) is the NDR behavior in cis-form with armchair SWCNT electrodes. Here, we focus on the NDR behavior in the positive bias voltage region. The current through the cis-form decreases significantly when the bias exceeds 0.8 V. To investigate the mechanism responsible for NDR behavior, we calculated the transmission spectra under bias voltage from 0.8 to 1.0 V as shown in Fig. 3.7.4. Seen from the Eq. (3.7.3), we can expect that only electrons with energies within a range near the Fermi level E_F contribute to the total current. Therefore, a good approximation with the range of the bias window, i.e. $[-V/2, +V/2]$ is enough to analyze a finite part of the transmission spectrum. From Fig. 3.7.4, we can see that the positions of transmission peak shift toward low-energy region with the increase in bias. Meantime, the value of transmission peaks in the bias window decrease with the increase in bias voltage at the range of 0.8 to 0.9 V. Meanwhile, the LUMO transmission peak becomes very weak at this range. As a result, current through the molecular device diminishes with the increase in bias voltage above 0.8 V and below 1.0 V. Thus the NDR feature is apparent over the bias voltage above 0.8 V and below 1.0 V.

In conclusion, the effect of chirality on the electronic transport properties of thioxanthene-based molecule with two SWCNT electrodes have been investigated by using the DFT+NEGF first-principles method. An obvious switching behavior can be observed both in armchair and zigzag junctions. The armchair junction shows better switching characteristics as compared with the zigzag junction for the chosen system. The maximum on-off ratio reaches 109 at 0.4 V for the armchair junction, which is useful for the design of functional molecular devices. The physical origin of the switching behavior is interpreted based on the transmission spectra and the MPSH. A reversible change between the two isomers would realize a molecular nanoswitch, which suggests that this system has attractive potential application in future molecular circuit.

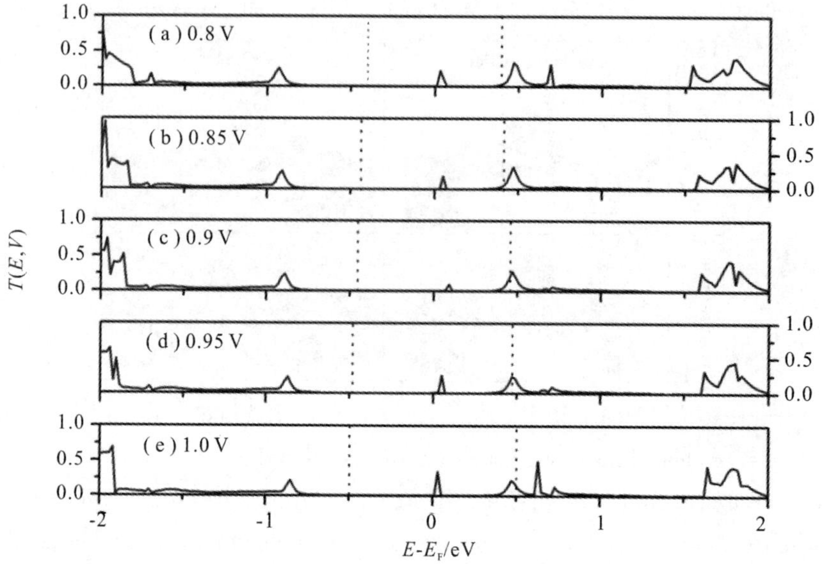

Fig. 3. 7. 4 The transmission spectra of the molecular switch under biases (a) 0. 8 V; (b) 0. 85 V; (c) 0. 9 V; (d) 0. 95 V; (e) 1. 0 V

3.8 The switching behavior of the dihydroazulene/vinylheptafulvene molecular junction

Developing electronic devices at single molecular scale has become one of the most active research fields in nanoscience, with the main aim to achieve further miniaturization compared to microelectronic devices. Molecular devices with specific functions, such as negative differential resistance (NDR), molecular rectification, field-effect characteristics and molecular switch, have been implemented and studied both experimentally and theoretically. As a critical element of design in future logic and memory circuits, molecular switches play a central role for the development of molecular electronics. In this regard, the design of molecular switches using photochromic molecules which can change their isomer type upon light irradiation has triggered a massive interest. Many photochromic molecules including diarylethene, spiropyran and azobenzene have been widely investigated and discussed. Dihydroazulene molecule (DHA), first introduced by Daub et al. in 1984, is also a photochromic molecule, which can be reversibly switched between the dihydroazulene (DHA) / vinylheptafulvene (VHF). The DHA can undergo a ring-opening to VHF irradiated by light, and the VHF can undergo a ring-closure back to the initial DHA via a thermal relaxation. The main advantage of the DHA-VHF system is that only the DHA to VHF conversion is light-induced. It means that a broad spectrum of light can be used without triggering the VHF to DHA back-reaction, which makes it usable as one of the candidates for light-driven molecular switches and useful in molecular electronic applications.

Recently, one-dimensional graphene nanoribbons (GNRs) which cut off from the novel two-dimensional material-graphene have been employed as the electrodes for several molecular devices due to its unique electronic structures and transport characteristics. The GNRs' electrical properties were found extremely sensitive to their edge geometries, namely, zigzag edged graphene nanoribbons (zGNRs) are shown to be metallic, whereas the armchair edged graphene nanoribbons (aGNRs) are semiconductors with energy gaps scaling with the inverse of the ribbon width. Moreover, when the molecule adhered to the electrodes, the nature of the anchoring groups is found to be crucial for achieving the unique functions of molecular devices. Motivated by the discussion above, we design a molecular junction constructed by DHA / VHF molecule capped with different anchoring groups, namely carbon atom and amide endgroup which are usually used as linkages between a molecule and GNRs, between two zGNR electrodes in this work. The effects of the anchoring groups on the electronic transport properties and switching behavior of the molecular junctions are investigated by applying the nonequilibrium Green's function (NEGF) formalism combined with density functional theory (DFT).

The schematic structures of the molecular devices are shown in Fig. 3. 8. 1. The DHA and VHF molecules are bonded to two zigzag-graphene nanoribbon (ZGNRs) electrodes by either carbon atom or amide anchoring groups, respectively. The molecular device can be divided into three parts: central scattering region(C), left electrode (L), and right electrode (R). For simplicity, the DHA and VHF molecules switching between the ZGNRs with carbon atom anchoring group are named M1 and M2, respectively. The DHA and VHF molecules switching between the ZGNRs with amide anchoring group are named M3 and M4.

The structural relaxation and the electronic transport properties of the molecular junctions are calculated by a fully self-consistent NEGF formalism combined with first-principles DFT, which is implemented in Atomistix ToolKit (ATK) software package. In the electronic transport calculations, the exchange-correlation potential is described by Ceperley-Alder local density approximation (LDA). The core electrons are modeled with Troullier-Martins nonlocal pseudopotential, while the valance electrons wave functions are expanded by a SIESTA basis set. The double-zeta plus polarization (DZP) basis set is adopted for all atoms. The mesh cutoff energy of 150Ry is selected to achieve a balance between the calculation efficiency and the accuracy.

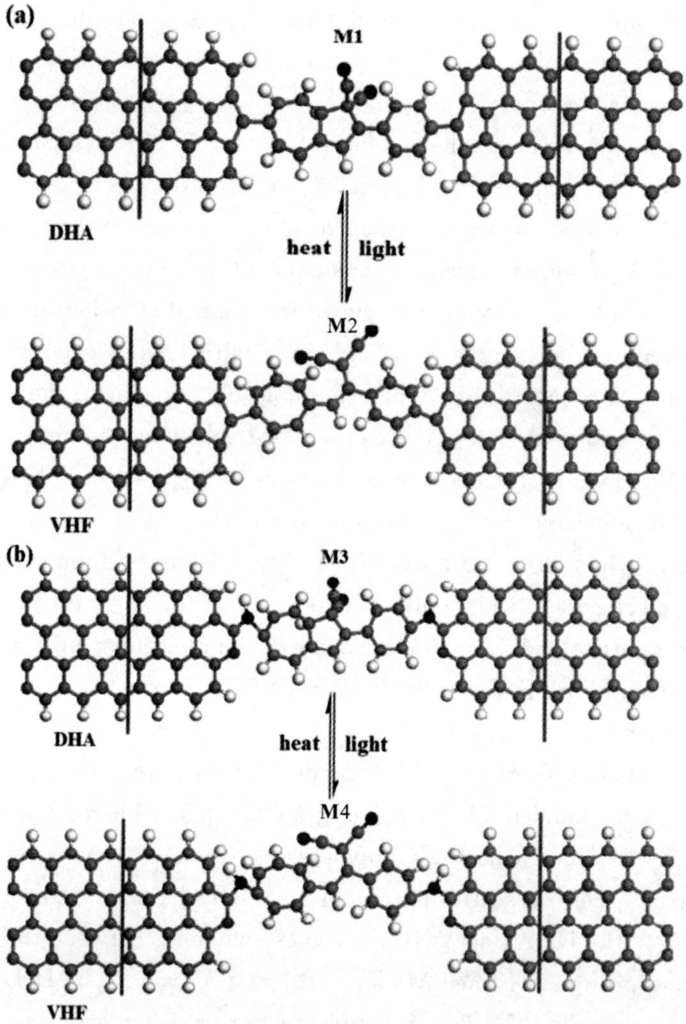

Fig. 3. 8. 1 Schematic diagrams of the devices: the DHA / VHF molecule with different anchoring group between metallic zigzag GNR electrodes, where (a) M1 and M2 correspond to the DHA/ VHF junction with carbon atom anchoring group; (b) M3 and M4 correspond to the DHA/ VHF junction with amide anchoring group

The currents through the DHA/ VHF molecular junctions with different anchoring groups under the bias voltage varying from 0 to 1. 0 V are plotted in Fig. 3. 8. 2 (a) and (b). The current switching ratios of molecular on/off states as a function of bias, namely *Ratio* = I_{DHA}/I_{VHF} , are plotted in Fig. 3. 8. 2(c). Most strikingly, the electronic transport properties of the molecular junctions with different anchoring groups are different from each other. Several important features in the evolution of current are clearly visible: ① the current through DHA is always larger than that through VHF at the same bias no matter what kind of anchoring group is. Thus, when DHA is shifted to VHF under photoexcitation, the current through the circuit can switch from on state (high conductance) to the off state (low conductance), and vice versa. ② the amide anchoring group on the molecule will hugely enlarge the currents. The currents through the molecular junction with amide anchoring group, namely M3 and M4, are larger than that through the molecular junction with carbon atom anchoring group, namely M1 and M2. ③ in Fig. 3. 8. 2(a) and (b), the currents of M1 and M3 increase with bias gradually. The current of M2 is close to zero in a bias range from 0 to 1. 0 V. The current of M4 increases with bias gradually in a bias range from 0 to 0. 7 V. Then, it slightly decreases to 4. 2 μA at 0. 8 V, which revealing a NDR behavior. ④ in order to compare the switching behaviors of molecular devices with different anchoring groups, we calculate the current switching ratios as shown in Fig. 3. 8. 2(c). The figure indicates that the current switching ratio of M1/M2 increases quickly from about 5 to 30 with the increases of the bias without any oscillation. In contrast, the switching ratio of M3/M4 is oscillating and tiny in the entire bias range, which is just around 1. 4. In some previous reports, the current switching ratios of molecular switches are small and also oscillated with applied bias, which means these switch behaviors of molecular devices are easily affected by the bias voltage. However, the switch behavior of the DHA/ VHF molecular device with carbon atom anchoring group in our study is obvious and stable in the bias region, which makes it have a broader application in future logic and memory devices.

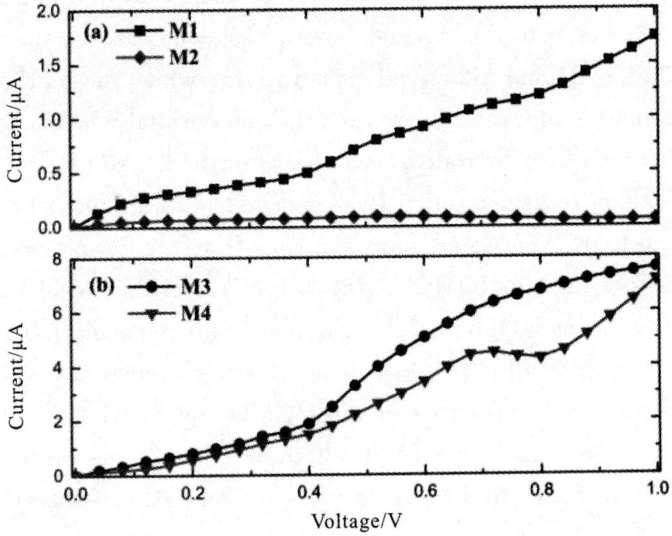

Fig. 3. 8. 2 The curreat-voltage curves and current switching ratio

(a) (b) the current-voltage curves for M1, M2 and M3, M4, respectively

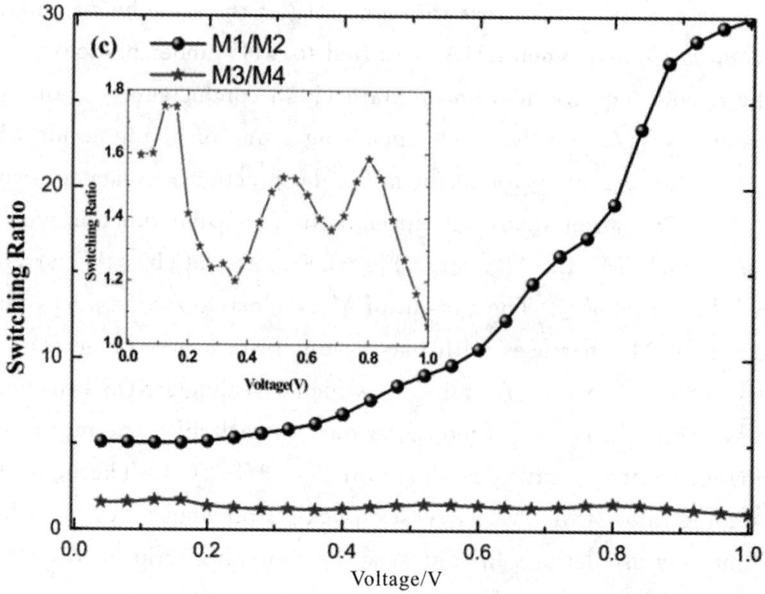

Continued Fig. 3. 8. 2 The curreat-voltage curves and current switching ratio

(c) The current switching ratio as a function of applied bias

(The insert figure is the current switching ratio of M3/M4)

To understand theelectronic transport mechanism of the molecular junctions, we calculate the electronic transmission spectra, the highest occupied molecular orbital (HOMO) and the lowest unoccupied molecular orbital (LUMO) of M1, M2, M3 and M4 at zero bias, shown in Fig. 3. 8. 3, respectively. In Fig. 3. 8. 3 (a), one can see that the transmission peaks of the HOMO for M1 and M2 are all close to zero, which means the electronic transport through this orbital is blocked. Therefore, the main orbital which accounts for conductivity is the LUMO. Although the transmission peaks of the LUMO for M1 and M2 are both broad, the corresponding transmission coefficient of M2 is lower than that of M1. In addition, the HOMO and LUMO of M2 are more far away from the Fermi energy level compared with M1, which leads to the big HOMO-LUMO Gap(HLG). As a result, the low transmission coefficient and the lager HOMO-LUMO gap due to the molecular structure changes accounts for the low conductivity in M2. As compared with the coefficients of the HOMO and LUMO for M1 and M2, the coefficients of the HOMO and LUMO for M3 and M4 all increase, especially the transmission coefficient of the HOMO is increased observably. Furthermore, when M3 transforms into M4, the transmission peaks of the HOMO and LUMO changes little.

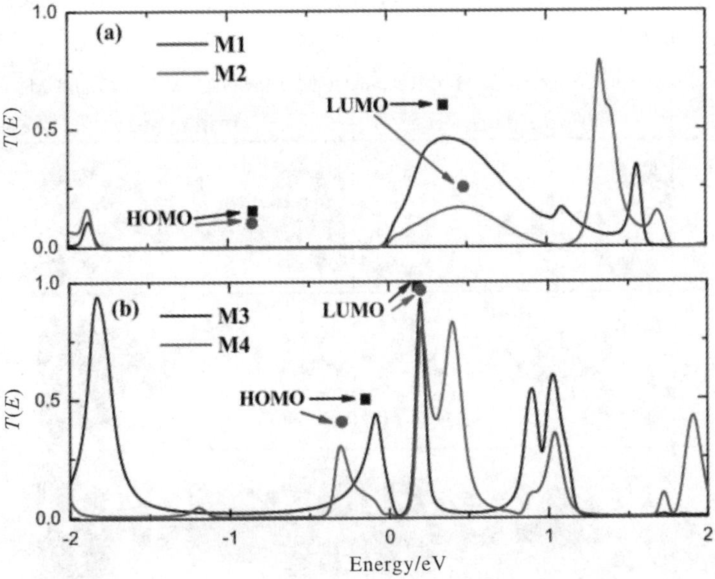

Fig. 3. 8. 3 Transmission spectra for M1, M2, M3 and M4, respectively (The HOMO and LUMO of DHA and VHF are indicated by the solid square and the solid round, the energy origin is set to be the average Fermi level E_F)

The varying of the current between the DHA and VHF with different anchoring groups can be further analyzed by the molecular projected self-consistent Hamiltonian (MPSH) eigenstates. In Fig. 3. 8. 4, we calculate the MPSH of the HOMO and LUMO to give a visual description of the electronic transport. From Table 3. 8. 1, one can see that the MPSH-HOMO of M1 and M2 are both localized, which induce the low transmission coefficients of the HOMO in Fig. 3. 8. 3(a). Meanwhile, the MPSH-LUMO of M1 is more delocalized than that of M2, which results in a large transmission coefficient and current for M1. Therefore, when the molecule switch from M1 to M2 upon photoexcitation, a higher current switching ratio can be observed. Compare with M1 and M2, the MPSH-HOMO and MPSH-LUMO are fully delocalized both in M3 and M4, due to the presence of nitrogen atom and oxygen atom which could enable larger couplings between the molecular bridge and the electrodes. These overlaps of the orbitals will translate in more delocalized molecular orbitals for the overall system. Therefore, electrons can easily transfer between the molecular orbitals and electrode, which results in the currents through M3 and M4 larger than that through M1 and M2. Furthermore, the spatial distributions of MPSH-HOMO and MPSH-LUMO for M3 and M4 are similar to each other. They are both delocalized, which leads to a lower current switching ratio of M3 and M4. In addition, the open molecular form of VHF breaking the aromatic character of the bridge leads to less delocalized molecular orbitals and conductance of the VHF in Fig. 3.8.2.

Table 3.8.1 Spatial distribution of HOMO and LUMO for M1, M2, M3 and M4 at zero bias

	MPSH-HOMO	MPSH-LUMO
M1	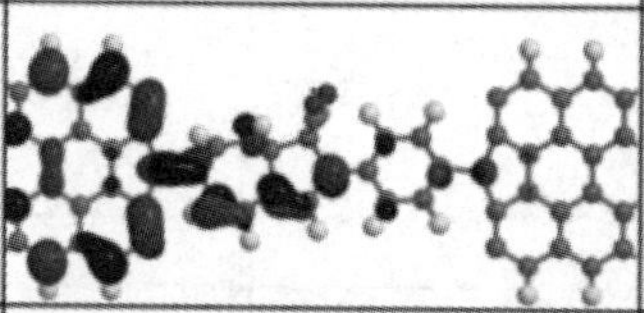	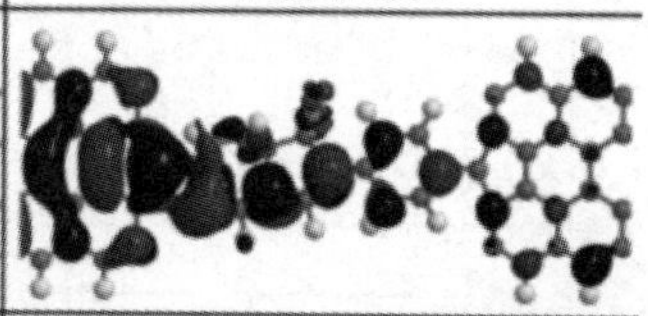
M2	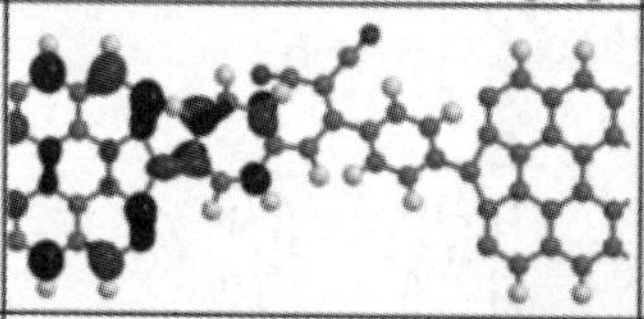	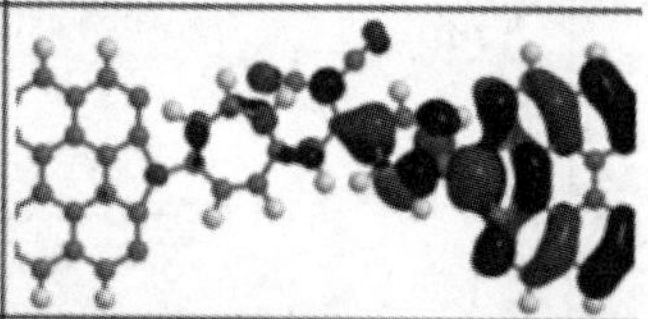
M3	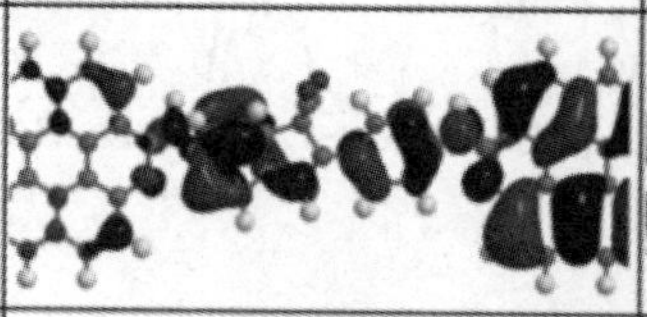	
M4	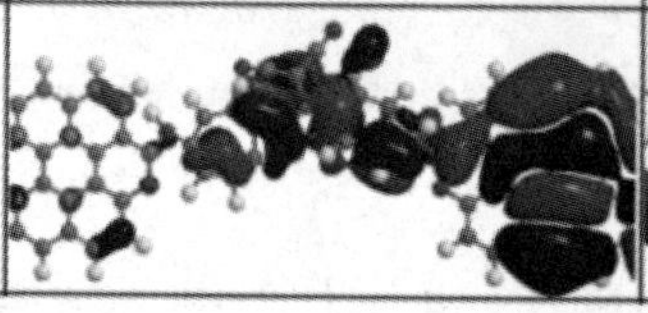	

In order to explain the origin of the switching behavior in our models, we calculate the $T(E, V_b)$ as a function of the injection energy, as shown in Fig. 3.8.4. Seen from the Eq. (3.1.3), we can expect that only electrons with energies within a range near the Fermi level E_F contribute to the total current. Therefore, a good approximation with the range of the bias window, i.e. $[-V/2, +V/2]$ is enough to analyze a finite part of the transmission spectrum. When the bias voltage is applied, the system is driven out of equilibrium. Therefore, one can see that the peaks of LUMO and HOMO shift to higher or lower energy with an increase in bias. For example, the LUMO peak of M1 shifts to higher energy for a bias of 0.2 V, and then tends again towards lower energies at higher biases. Furthermore, as shown in Fig. 3.8.4(a), when the bias voltage is increased, more and more part of the transmission peaks of the LUMO and HOMO for M1 come into bias window leading to the high current in Fig. 3.8.2(a). Meanwhile, the transmission coefficient of the LUMO for M2 gets smaller obviously and almost disappears with an increase in bias, which lead to the low current in Fig. 3.8.2(a). Contrary to Fig. 3.8.4(a), the part of the transmission peaks of the LUMO and HOMO for M3 and M4 come into bias window are almost the same leading to the similar currents in Fig. 3. 8. 2(b). Therefore, there is no significant change between the current of M3 and M4, which results in lower switching ratio when molecular switch with amide anchoring group.

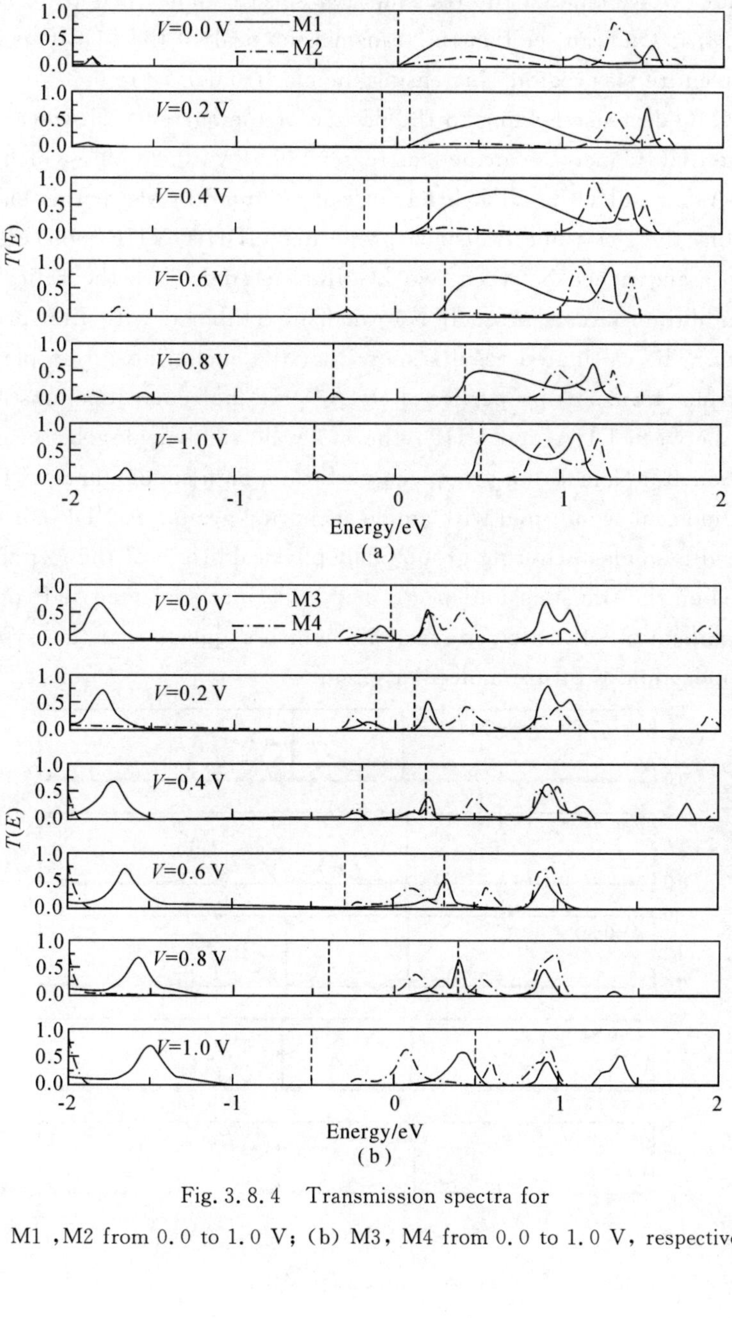

Fig. 3. 8. 4 Transmission spectra for

(a) M1 ,M2 from 0. 0 to 1. 0 V; (b) M3, M4 from 0. 0 to 1. 0 V, respectively

Another interesting feature in Fig. 3. 8. 2 is the NDR behavior in M4. The current of M4 increases with bias gradually in a bias range from 0 to 0. 7 V. The current through the M4 slightly decreases when the bias exceeds 0. 8 V. To investigate the mechanism responsible for NDR behavior, we calculated the transmission spectra under bias of 0. 72 to 0. 88 V as shown in Fig. 3. 8. 5. One can see that the transmission peak of the LUMO is in the bias window through the entire bias region. Increasing the bias from 0. 72 to 0. 8 V, the transmission peaks of the LUMO decrease leading to the decline of the current. However, the transmission peaks of the LUMO increase at the bias region [0. 84 V, 0. 88 V], which lead to the rise of the current. As a result, the NDR behavior can be found in M4 at this bias region.

In conclusion, the electronic transport properties ofDHA/VHF molecule with different anchoring groups sandwiched between two ZGNRs electrodes are theoretically investigated by using nonequilibrium Green's function formalism combined with first-principles density functional theory. The calculated results show that the anchoring group plays a significant role on the electronic transport properties of DHA/VHF molecular junctions. When the molecule translates between DHA and VHF, the molecule sandwiching between ZGNRs electrodes with carbon atom anchoring group can perform a high conductance switching behavior than that of the molecule connected with amide anchoring group. A NDR behavior can be observed in VHF with amide anchoring group. The physical origin of the switching behavior is interpreted based on the transmission spectra and molecular projected self-consistent Hamiltonian. The maximum on-off ratio can reach 30, which suggests that this system has attractive potential application in future molecular circuit.

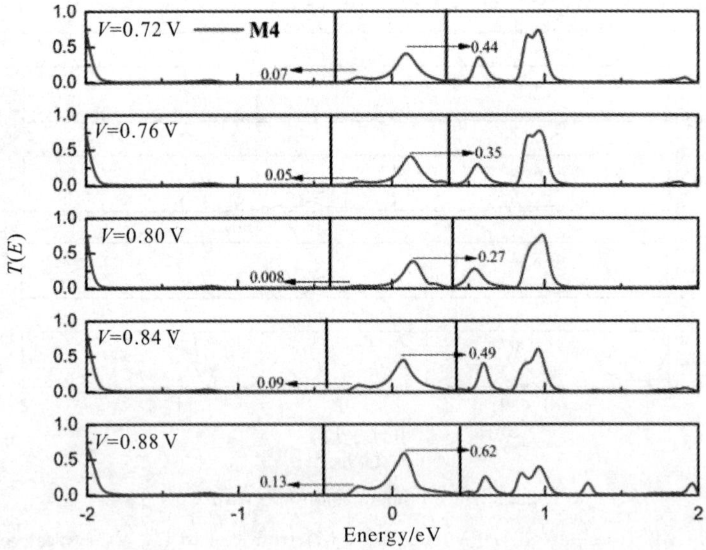

Fig. 3. 8. 5 Transmission spectra for M4 from 0. 72 to 0. 88 V

3.9 Switching behaviors of butadienimine molecular devices

Due to the experimental advances in microscale fabrication technology, designing functional devices by usingsingle molecules have become one of the most promising candidates for the next generation of electronic devices. As the basic components of almost any electronic device, molecular switches have become a research hotspot in molecular electronics. So far, various molecular switches have been intensively investigated both in experiment and theory, but some show poor stability because of weak contacts to metal electrodes. Recently, one-dimensional graphene nanoribbons (GNRs) which cut off from the novel two-dimensional material-graphene have been employed as the electrodes for several molecular devices due to its unique electronic structures and transport characteristics. Compared with metallic electrodes materials like Au, they have less serious contact problems. Many molecular devices based GNRs electrodes with interesting physical properties such as negative differential resistance (NDR), rectifying behaviors and spin-filter effects have been reported.

In this work, we propose using GNRs to fabricate single butadienimine molecular device. Butadienimine is a photochromic molecule which can convert between the keto and enol tautomeric forms activated by photoinduced hydrogen atom transfer. In contrast to most other optical molecular switch mechanisms, in particular cis-trans isomerization reactions, hydrogen translocation within the molecular bridge has the advantage that the overall length is not changed significantly. However, the research of butadienimine molecule connecting with GNRs electrodes as a switchable molecular device has rarely reported until now. Furthermore, the GNRs' electrical properties were found extremely sensitive to their edge geometries, namely, zigzag edged graphene nanoribbons (zGNRs) are shown to be metallic, whereas the armchair edged graphene nanoribbons (aGNRs) are semiconductors. Therefore, we explore the electronic transport properties of a butadienimine molecule bonded to different GNRs electrodes by applying nonequilibrium Green's function (NEGF) formalism combined with first-principles density functional theory (DFT). The results show that the current-voltage characteristics of the devices deeply depend on the edge geometry of GNRs. A higher current switching ratio can be found in butadienimine molecular device between the zGNRs.

The schematic structures of the molecular devices are illustrated in Fig. 3.9.1, the butadienimine molecules embedded in a carbon chain between two zGNRs or aGNRs electrodes. For simplicity, the keto form and the enol form of butadienimine molecules switching between the aGNRs are named M1 and M2, respectively. The keto form and the enol form of butadienimine molecules switching between the zGNRs are named M3 and M4, respectively. The molecular device can be divided into three parts: central scattering region(C), left electrode (L), and right electrode (R). The central scattering region contains three units of the GNRs on each side, thereby establishing the bonding between the molecules and the electrodes, the common Fermi level, and charge neutrality at equilibrium.

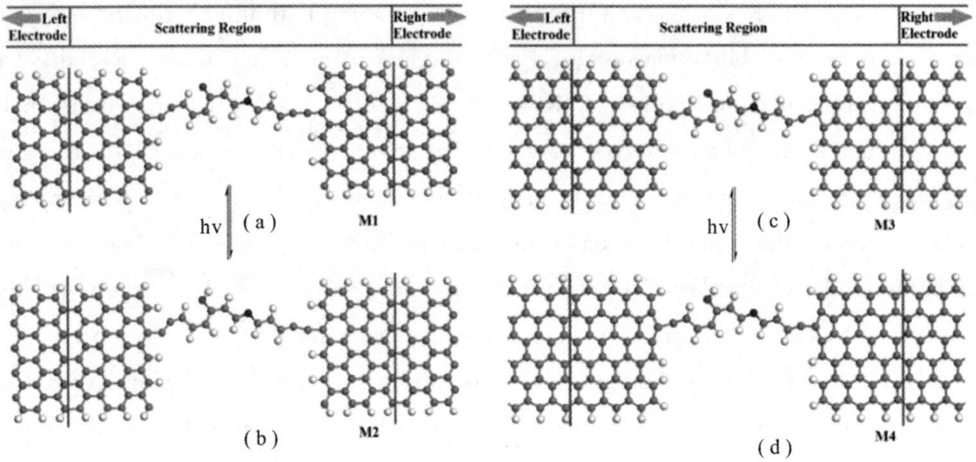

Fig. 3. 9. 1 Schematic diagrams of the devices: the butadienimine mole cule with two tautomeric forms between zGNRs or aGNRs electrodes

(a) keto-aGNRs; (b) enol-aGNRs; (c) keto-zGNRs; (d) enol-zGNRs

The geometric optimization and the electronic transport properties are calculated by a fully self-consistent NEGF formalism combined with first-principles DFT, which is implemented in Atomistix Tool Kit (ATK) software package. The Perdew-Burke-Ernzerhof (PBE) formulation of the generalized gradient approximation (GGA) is used as the exchange-correlation functional. The mesh cutoff energy of 150Ry is selected to achieve a balance between the calculation efficiency and the accuracy. The NEGF model allows the calculation of the drain current self-consistently considering the nonequilibrium and open boundary conditions within the device. The nonlinear current is calculated by using the Landauer-Bütiker formula, $I = \frac{2e}{h}\int_{\mu_L}^{\mu_R}(f_L - f_R)T(E,V)\mathrm{d}E$, where $f_{L/R}$ is the Fermi distribution function and $\mu_{L/R}$ is the electrochemical potential of the left/right electrode. The transmission coefficient of the device can be calculated using the formula, $T(E,V) = \mathrm{Tr}[\Gamma_L(E,V)G^R(E,V)\Gamma_R(E,V)G^A(E,V)]$, where $G^{R/A}$ are the retarded and advanced Green's functions, and coupling functions $\Gamma_{L/R}$ are the imaginary parts of the left and right self-energies, respectively.

The current-voltage characteristics of two tautomers with different GNRs electrodes in a bias range from 0 to 1.0 V are plotted in Fig. 3.9.2. Most strikingly, the electronic transport properties of butadienimine molecules sandwiched between the zGNRs or aGNRs electrodes are different from each other. A obvious switching behavior can be observed in Fig. 3.9.2(b). The current of M3 is evidently smaller than that of M4 in the whole bias range. When M3 is shifted to M4 under photoexcitation, the current through the circuit can switch from the low conductance to the high conductance, and vice versa. However, in Fig. 3.9.2 (a), this switching behavior is almost disappeared. The changes of switching behaviors can be clearly reflected by the current switching ratios (SR) of molecular ON/OFF states as a function of bias in Fig. 3.9.2(c). It is seen from the figure that SR2 is larger than SR1. The maximal current switching ratio of SR2 is up to 8.7at 0.35 V and then drops gradually with the bias increasing. In contrast, the switching ratio of SR1 is tiny in the entire bias range, which is just around 1.7.

To understand the electronic transport mechanism of the molecular junctions, we calculate the electronic transmission spectra, the highest occupied molecular orbital (HOMO) and the lowest unoccupied molecular orbital (LUMO) of M1, M2, M3 and M4 at zero bias, shown in Fig. 3.9.3, respectively. In Fig. 3.9.3(a), one can see that the positions of the HOMO and LUMO for M1 and M2 have no significant change. In addition, the coefficient of HOMO and LUMO for M1 is a little lower than that for M2. Therefore, the current of M2 is just a little smaller than that of M1 in the whole bias range, which induce a small switching ratio. When the electrodes change to the zGNRs, the HOMO and LUMO for M4 all shift close to Fermi energy level leading to the decrement of HOMO-LUMO Gap(HLG). It is well known that the smaller the HLG, the more easily it is to rearrange its electron density under the presence of an external electron. Furthermore, for M3, one can see that the transmission peaks of the LUMO is almost close to zero, which means the electronic transport through this orbital is blocked. As a result, an obvious switching behavior can be observed when M3 is shifted to M4 under photoexcitation.

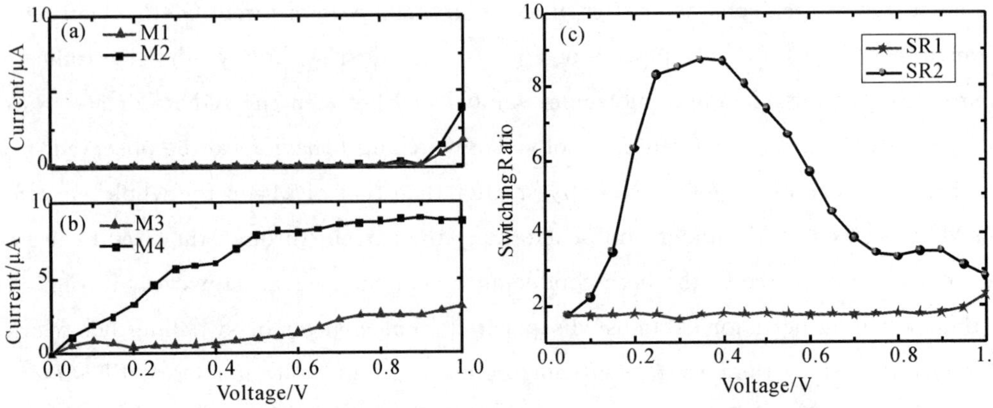

Fig. 3. 9. 2 The current-voltage curves though the molecular junctions with (a) aGNRs; (b) zGNRs electrodes, respectively; (c) The current switching ratio as a function of applied bias

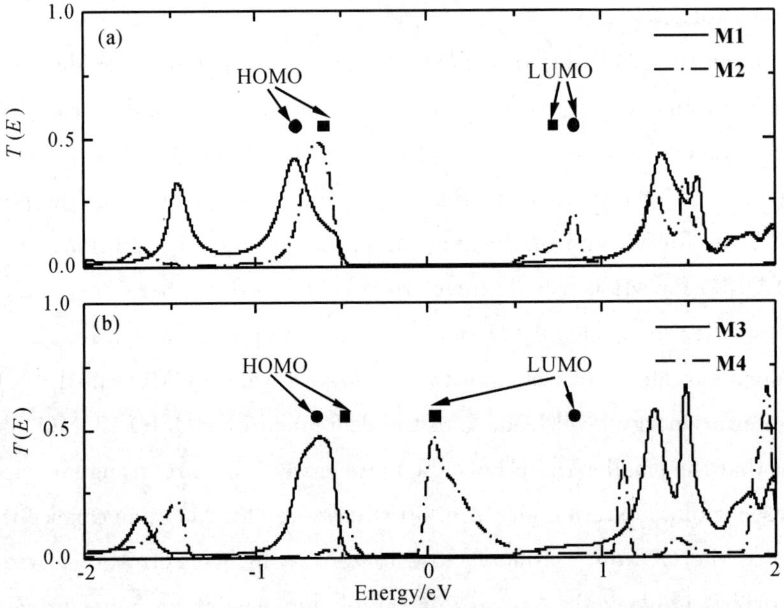

Fig. 3. 9. 3 Transmission spectra for butadienimine molecular device with (a) aGNRs electrodes; (b) zGNRs electrodes, respectively

Generally, the transmission coefficient can be related to the molecular orbitals, which have been modified by the electrodes. These modified molecular orbitals can be obtained from the molecular projected self-consistent Hamiltonian (MPSH) eigenstates. From Fig. 3. 9. 3, it is notable that the LUMO and HOMO are the main transmission channel because it is close to the Fermi energy. In order to explain the origin of the difference in conductance and switching behavior in Fig. 3. 9. 2, we calculate the LUMO-MPSH and HOMO-MPSH as shown in Table 3.9.1. It is found obviously that the spatial distributions of the LUMO and HOMO for M2 and M4 are delocalized, which leads to low barrier and provides good transport channels for electron transport. On the contrary, in M1 and M3, the LUMO and HOMO orbitals are more localized resulting in a smaller current. As a result, the current is decreased when the molecule switch from enol to keto isomer upon photoexcitation, which is in good agreement with the trends of *I-V* curves shown in Fig. 3.9.2.

Table 3.9.1 Spatial distribution of the HOMO and LUMO for butadienimine molecular device with (a) aGNRs electrodes; (b) zGNRs electrodes, respectively

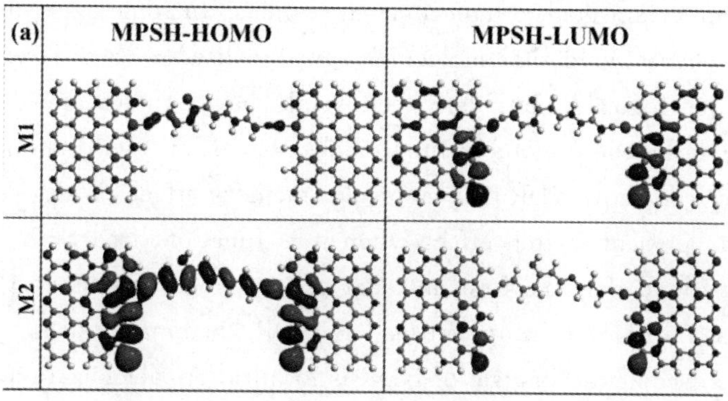

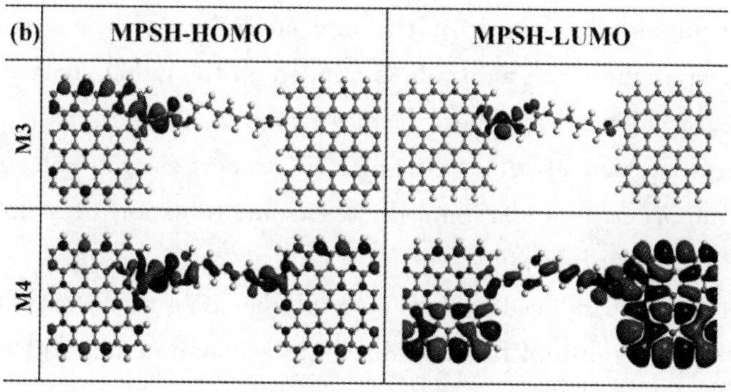

In conclusion, we investigate the electronic transport properties of butadienimine molecule sandwiched between two different GNRs electrodes by applying the DFT+NEGF first-principles method. The results show that the edge geometry of GNRs electrodes plays a significant role on the electronic transport properties of butadienimine molecular device. The molecule sandwiching between zGNRs electrodes can perform a high conductance switching behavior, which is 5 times larger than that of the molecule connected with aGNRs electrodes. The physical origin of the switching behavior is interpreted based on the transmission spectra and molecular projected self-consistent Hamiltonian. The maximum on-off ratio can reach 8.7, which suggests that this system has attractive potential application in future molecular circuit.

3.10 Effect of torsion angle in 4,4′-biphenyldithiol functionalized molecular junction

In recent years, molecular devices have attracted more and more attention because of their novel physical properties and potential for device application, including single-electron characteristics, negative differential resistance (NDR), electrostatic current switching, memory effects, Kondo effects, etc., which made them possible to realize the elementary functions in electronic circuits. In particular, the studies about molecular logical functions have become the most important goal in molecular electronics. In some experiments and theoretical studies, people have found that molecules functionalized with different side groups can perform some unique logical functions. Reed et al. have studied the electronic transport properties of phenylethylene oligomers functionalized with different side groups, and found that such molecules can show NDR or a molecular memory effect. Weiss et al. have reported that a molecule can present switch effect when it is functionalized with alkyl side groups. Cornil et al. have found that the first and last phenyl rings of twisted phenylethylene oligomers in the presence of side group can lead to NDR. Seminario et al.. theoretically proposed that the microscopic mechanism of NDR is localization/deloclization of a molecular orbital induced by the side group substitution. But in a real experiment, the structure of the molecular device including the length of the molecular bonds, the torsion angle between inter-molecule, the site where the molecule is bonded on the metal surface, etc., may be influenced by some external factors. These factors, such as temperature, pressure or electric field and so on, may also have effects on the electric properties of a molecular device. Therefore it is also very important to understand the electronic transport of a molecular junction by studying the molecular conformation.

In this work, the effect induced by torsion angle between two phenyl rings on electronic transport of a 4,4′-biphenyldithiol functionalized molecular junction will been investigated. We choose this molecular junction to demonstrate the effect since it is one of the most intensely studied molecules as a prototype of molecular transport theory, and the torsion angle between two phenyl rings is sensitive to the change of the external field, especially when it is functionalized with side groups. Furthermore, the theoretical studies of the 4,4′-biphenyldithiol functionalized molecule especially with different torsion angle have so far still lacking to the best of our knowledge. Therefore, we consider various possible molecular conformations for the lead-molecule-lead (LML) system composed of a 4,4′-biphenyldithiol functionalized molecule and semi-infinite Au leads with different torsion angle between two phenyl rings in the range 0° to 90° in steps of 30°. The main task in this work first is to study the effects of side group and torsion angle on electronic transport properties of molecular junction, and then to search for the effect of the torsion angle on the logical functions of this functionalized molecular junction. Here, we consider three cases for the LML system as following: ① the molecule without side group; ② the molecule substituted by NO_2 side group and ③the molecule substituted by NH_2 side group. Effect of different adsorption position of sulfur atom on the molecular memory characteristics is investigated finally.

As illustrated in Fig. 3.10.1, we use Au(111) surface in a 4×4 unit cell as the leads and assume that a 4,4′-biphenyldithiol molecule functionalized with side groups (NO_2, NH_2), which is chemisorbed to the surfaces through strong Au-thiol bonds. The 4×4 unit cell for Au lead is large enough to avoid the finite size effect. The structure of free molecular wire with side groups(NO_2, NH_2) is first optimized. Then the optimized molecule wires are adsorbed between two Au(111) leads. The optimized distance between molecule and metal surface is 1.813 Å, 1.847 Å, 1.854 Å separately for three systems, i.e., ① the molecule without side group; ② the molecule substituted by NO_2 side group and ③ the molecule substituted by NH_2 side group. The free molecular geometrical optimizations and the electronic transport properties of the molecular junctions are calculated by a fully self-consistent NEGF formalism combined with first-principles DFT, which is implemented in ATK2.0 package.

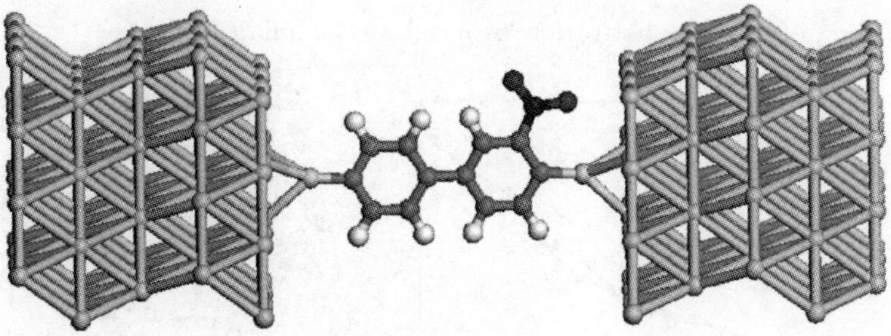

Fig. 3.10.1 Geometry of the LML system

Firstly effects of side groups on the electrical properties of three systems when torsion angle is equal to 0° are investigated. The calculated *I-V* characteristics are plotted in Fig. 3.10.2(a). We can note that the system is asymmetrical when molecule is substituted by side group (see Fig. 3.10.1), but this asymmetry in molecular conformation does not strongly affect the *I-V* characteristics. Because there is just a little asymmetry in *I-V* curve when molecule is substituted by NO_2 and NH_2 side groups [see Fig. 3.10.2(a)]. Then it can be found that the curve of molecule without side group is similar with that of NH_2 side group substitution, which means that NH_2 side group has no nearly effect on the electronic transport. However, the curve of molecule with NO_2 side group substitution presents a little difference. The current is increased larger than that of other two systems at the range of

−1.8 to 1.8 V. But when $V_b \leqslant -1.8$ V and $V_b \geqslant 1.8$ V, the value of current becomes small. To understand these differences, the transmission coefficients of three systems when V_b at the range of 0.0 to 2.4 V are shown in Fig. 3.10.2 (b) (c) and (d). The current, I is obtained from Landauer formula $I = (2e/h)\int_{\mu_L}^{\mu_R} T(E, V_b)\mathrm{d}E$, where $\mu_L(V_b)/\mu_R(V_b)$ are the electrochemical potentials of the left/right leads. The region between μ_L and μ_R is called the bias window or integral window, as shown in Fig. 3.10.2(b)(c) and (d) with the doted lines. Therefore, the current is mainly determined by $T(E, V_b)$ in the bias window, namely only determined by the transmission regions in the bias window because $T(E, V_b)$ is zero in the transmission interval and has no contribution to the current. Therefore, it can be understood that the current of molecule without side group is similar with that of NH_2 side group substitution, because there is no additional transmission region will come into the region of bias window with V_b increasing when molecule is substituted by NH_2 side group [see Fig. 3.10.2(b) and (d)]. In contrast NH_2 side group substitution, the molecule with NO_2 group substitution lifts up $T(E, V_b)$ apparently, which increases the transmission falling into the bias window compared with the case of molecule without side group [see Fig. 3.10.2(b) and (c)]. Therefore, the molecule with NO_2 side group substitution induces to a larger current than that of NH_2 side group substitution especial at the small bias.

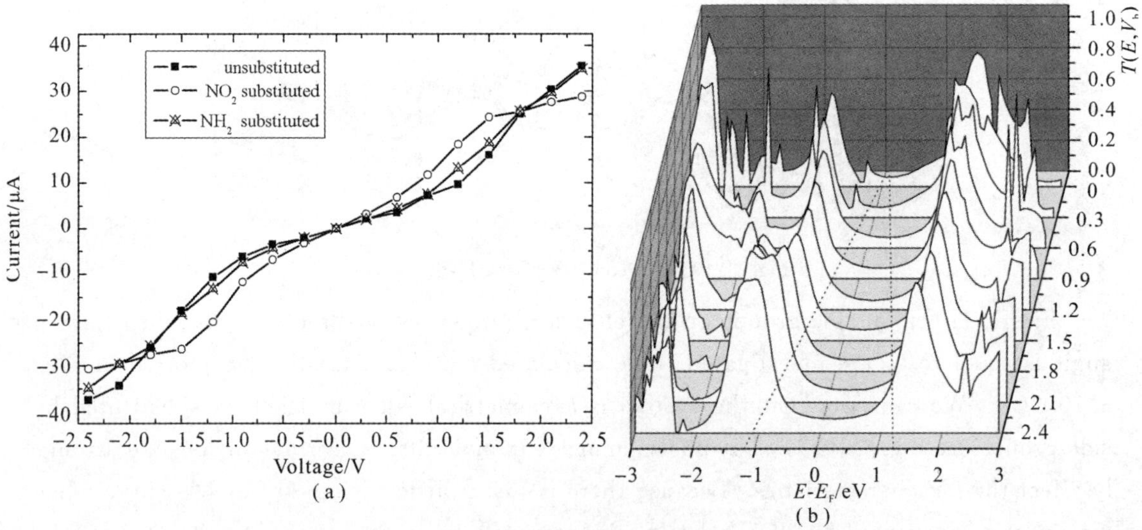

Fig. 3.10.2 The I-V characteristics and transmission coefficients for molecular device
(a) I-V_b curves for three systems; (b) the three-dimensional plot of the bias dependence of the transmission spectra for molecule without side group, with NO_2 and NH_2 side group respectively

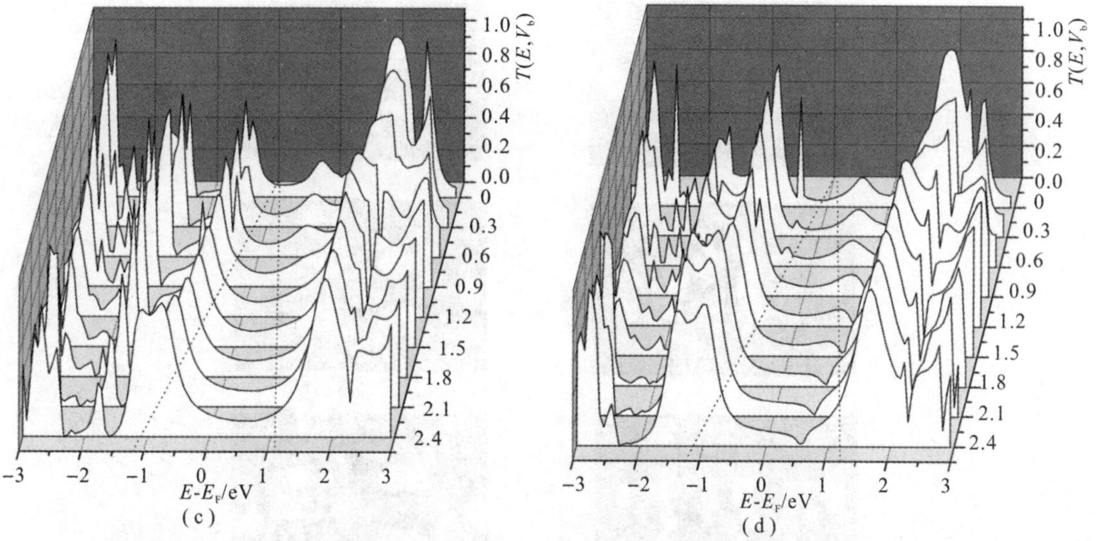

Continued Fig. 3.10.2 The *I-V* characteristics and transmission coefficients for molecular device (c) (d) the three-dimensional plot of the bias dependence of the transmission spectra for molecule without side group, with NO_2 and NH_2 side group respectively

To get a further insight, the molecular projected self-consistent Hamiltonian (MPSH) of these three systems is analyzed, as shown in Fig. 3.10.3. The molecular part is extracted from the whole self-consistent Hamiltonian at the contact region which contains the molecule-lead coupling effect. In Fig. 3.10.3(a) and (b) for the molecule without side group, we can find that the highest occupied molecular orbital (HOMO) and the lowest unoccupied molecular orbital (LUMO) are both delocalized in two phenyl rings. When molecule with NH_2 side group substitution is similar with the situation of molecule without side group, the delocalization of HOMO and LUMO are not changed essentially [see Fig. 3.10.3 (e) and (f)]. Therefore, the conducting current is similar for these two systems. However, when molecule is substituted by NO_2 side group, not only the HOMO and LUMO are fully extended, but also a high density is on each side of sulfur atom, as shown in Fig. 3.10.3(c) and (d). The high density on sulfur atom means that there is much stronger molecule-lead coupling when molecule is substituted by NO_2 side group. Therefore, the current of NO_2 side group substitution increases large more than other two systems.

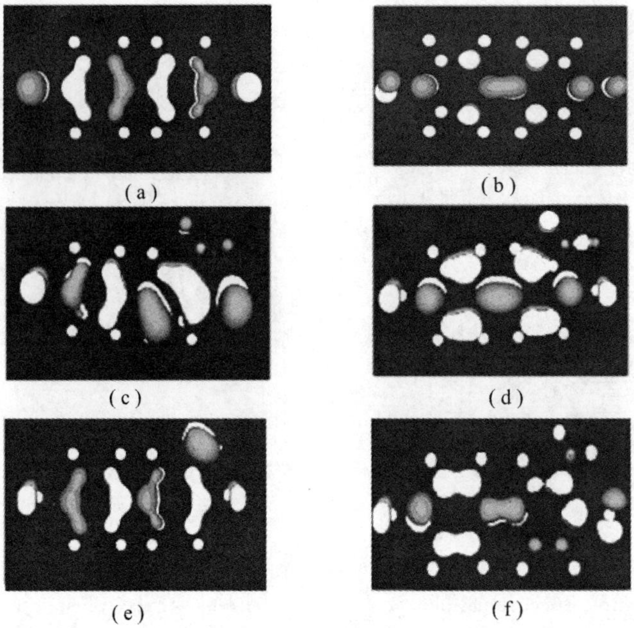

Fig. 3.10.3 Orbitals of the MPSH

(a) (b) the HOMO and LUMO respectively for molecule without side group substituted;

(c) (d) the HOMO and LUMO respectively for molecule with NO_2 side group substituted;

(e) (f) the HOMO and LUMO respectively for molecule with NH_2 side group substituted

The analysis of *I-V* characteristics, transmission coefficients and MPSH suggest that the transmission peaks in 4,4′-biphenyldithiol functionalized molecular junction are related to the delocalized nature of the π orbitals, and the functionalized molecule does not seem to perform some special logical functions except the current is increased a little with the substitution of NO_2 side group. It means the side group does not strongly affect the current in 4,4′-biphenyldithiol molecular junction when torsion angle is equal to 0°, i.e., the two phenyl rings are coplanar. But when 4,4′-biphenyldithiol molecule is functionalized with side groups, the torsion angle between two phenyl rings is sensitive to the change of the external field. Therefore, it is necessary to investigate the effect of different torsion angle between two phenyl rings on the electronic transport through 4,4′-biphenyldithiol functionalized molecular junction.

Firstly we calculate the *I-V* curve with different torsion angle between two phenyl rings for these three systems under the biases of −2.4 to 2.4 V, and the results are shown in Fig. 3.10.4. By changing the torsion angle between two phenyl rings, namely changing the magnitude of the intermolecular interaction, we can find that the current is decreased obviously when the torsion angle increases, and reached the minimum with the torsion angle increasing to 90° in all three systems. It also can be found that when torsion angle is equal to 0°, the molecular junction is in a higher conductivity state, whereas it changes to a lower conductivity when the torsion angle is equal to 30°,60°and 90°, as shown in Fig. 3. 10. 4. The conducting behavior of torsion angle equal to 120° , 150° or 180° is similar as that of 60°, 30° or 0°, and is not shown here. This change in molecular conducting behavior induced by different torsion angle can be used as molecular switch under an applied bias potential, which control the conversion between state "0" and the state "1" by the reversible field or the current pulse. When the torsion angle is equal to 90°, it will cause the switch into the "off" state, and when the torsion angle is equal to 0°,it will cause the switch into the "on" state.

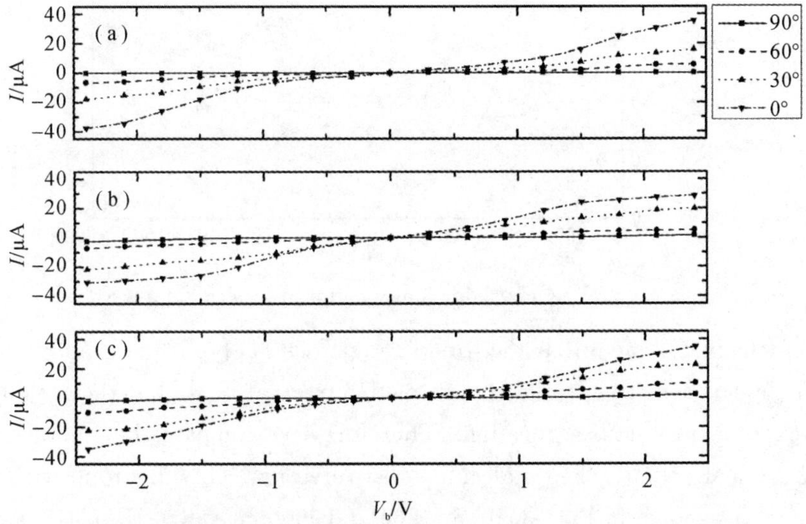

Fig. 3. 10. 4 The *I-V* curves for different torsion angle in molecular junctions

(a) the molecule without side group substitution; (b) molecule with NO_2 side group substituted; (c) molecule with NH_2 side group substituted

Then we will calculate the total energy of these three systems with different torsion angle, and compare the stability of this molecular switch with changing the torsion angle. The differences in total energy which is calculated within the Perdew, Burke, and Ernzerhof approximation for the exchange-correlation functional, i. e. ΔE ,as a function of the torsion angle α of the functionalized phenyl ring, are shown in Fig. 3. 10. 5. It is found that the intermolecular interaction energy of these three systems show a quite different behavior. With α increasing, the energy of system without substitution first increases, then decreases after α = 90°. The change in energy of molecule with NH_2 side group substitution is similar with the situation of molecule without substitution, but there is a little change nearby α = 30° and

150°. While in the system substituted by NO_2 side group, the energy emerges a local minimum at $\alpha = 60°$ and 120° separately. It can be seen that the local minima are stabilized by the NO_2 side group (see Fig. 3. 10. 5). This memory effect is a special unique logical function induced by torsion angle between two phenyl rings of 4,4′-biphenyldithiol molecule substituted by NO_2 side group.

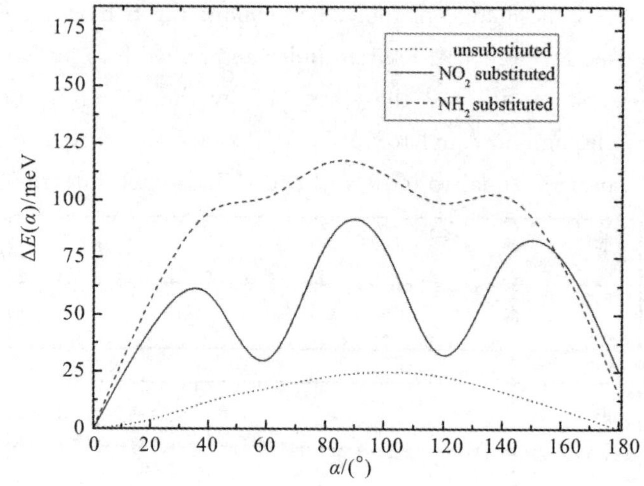

Fig. 3. 10. 5 Energy vs. torsion angle for molecules in three systems

Despite it is well known that the torsion angle of 4,4′-biphenyldithiol molecule between two phenyl rings is sensitive to the change of the external field, but the $\alpha = 60°$ whether exists in the experiment is still a question. Therefore, we compare the stability for the molecule with $\alpha = 0°$ and $\alpha = 60°$ when molecule substituted by NO_2 side group with the bias voltage increasing, as shown in Fig. 3. 10. 6. The stability of $\alpha = 120°$ is similar as that of $\alpha = 60°$ and is also not shown here. It can be found that the current of $\alpha = 60°$ is approximately two times smaller than that of $\alpha = 0°$ at $V_b = 2.4$ V [see Fig. 3. 10. 4(b)]. But the total energy of $\alpha = 60°$ is lower than that of $\alpha = 0°$ when $V_b < -1.4$ V and $V_b > 1.4$ V (see Fig. 3. 10. 6). This change in total energy means that the state of $\alpha = 60°$ could be more stable than that of $\alpha = 0°$ at the range of $V_b < -1.4$ V and $V_b > 1.4$ V, which leads to an expected conformational change as we simulating. Namely, there is a switching voltage around 1. 4 V under the bias increasing where a higher conductivity state will change to a lower conductivity state, i. e. , molecular switch.

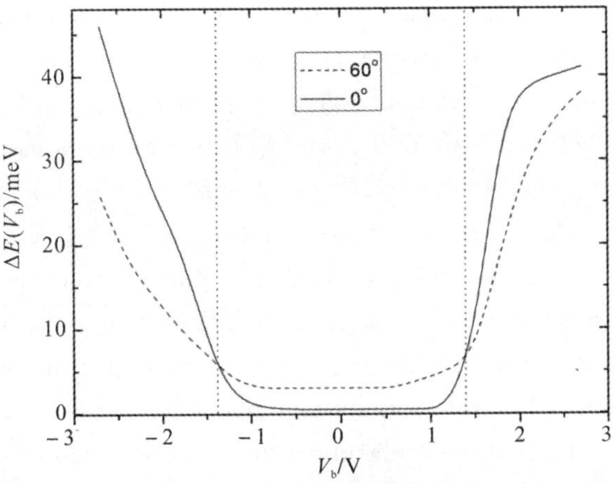

Fig. 3.10.6 Total energy for system with NO_2 side group substituted at $\alpha=60°$ and $\alpha=0°$

Finally, we discuss the effect of different adsorption position of sulfur atom on molecular memory effect. The thiol anchoring group is connected to Au surface through the hollow site in our work. It is a standard model, but it may be not adequate to describe the experimental details. The surface of Au lead in a break-junction experiment may be far away from clean (111) surface. So we construct two more anchoring geometrical models, i.e., the top and bridge adsorption positions. The total energy calculations of NO_2 side group substitution with torsion angle at different adsorption positions are given in Fig. 3.10.7. One can clearly see that the two local minima also appear at $\alpha=60°$ and $\alpha=120°$ in spite of different anchoring geometries. It means that the memory effect induced by NO_2 side group is independent of the adsorption position of sulfur atom on the molecule/lead interface. Therefore, it can be concluded that the memory effect is an intrinsic property of the molecule.

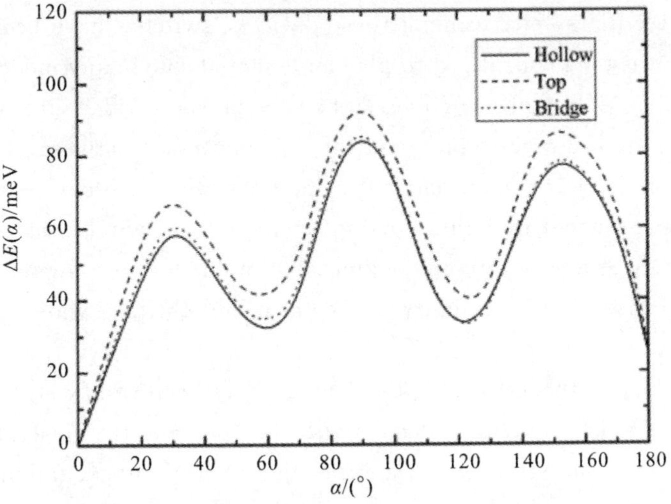

Fig. 3.10.7 Energy vs. torsion angle for system substituted by NO_2 side group with hollow, top and bridge adsorption positions, respectively

In conclusion, by using density functional theory and nonequilibrium Green's functions method, we calculate the effects of different side groups (NH_2, NO_2) and torsion angle between two phenyl rings on the electronic transport properties of 4,4′-biphenyldithiol molecule and its stability. The results show that the NO_2 side group is more effective to enhance the molecular conductance than that of NH_2 side group. It is because the molecular HOMO and LUMO become more delocalized, and the molecule-lead coupling is much stronger under NO_2 side group substitution. Furthermore, 4, 4′-biphenyldithiol molecule can exhibit a switching behavior by changing the torsion angle. Especially, the molecule of NO_2 side group substitution with different torsion angle induces two local minima at $\alpha = 60°$ and $\alpha = 120°$, which correspond to a low conductance and give a molecule memory effect. This memory effect of 4,4′-biphenyldithiol molecular junction is not only dependent on the character of side group, but also on the torsion angle between the two phenyl rings deeply. It is also found that the adsorption position of sulfur atom on Au leads does not change the memory effect of 4,4′-biphenyldithiol functionalized molecular junction. The results will be helpful to understand further the real situation in experiments and design molecular electronic devices with specific properties.

3.11 First-principles study of dihydroazulene as a possible optical molecular switch

Inrecent years, with the advancement of techniques for manipulating individual molecules, electronic devices based on single molecules have been considered as one of the most promising candidates for today's silicon-based devices both for their novel physical properties and potential for device application, such as negative differential resistance (NDR), switches, latches and rectifiers, etc. Among these devices, switches have been attracting great attention, because they are considered to play an essential role in any modern design of logic and memory circuits. Molecular switch devices can be mainly divided into two categories, including electronic switch devices which control the conversion between state "0" and state "1" by the reversible field or the current pulse, and optical switch devices which achieve the open or close state of current by light. Among them, optical switches have been extensively studied, because light is a very attractive external stimulus for such switches including short response time, the ease of addressability and compatibility with a wide range of condensed phases.

In this regard, diarylethene-based optical switch have been widely studied. This molecular switch comprises a diarylethene molecule with the open and the closed forms, which can be easily reversed from one structure to another one upon photoexcitation. Theoretical results show that the current through the closed form is significantly larger than that through the open form. Dihydroazulene molecule which also exhibit photoinduced reversible open and closed isomerism has been studied by Waele et al. in experiment. It means this molecule also can be considered as one of the better candidates for light-driven molecular switches. However, in contrast to the case of diarylethene molecules, little attention was paid to the dihydroazulene molecules. Therefore, it is very essential to offer a theoretical investigation that will help to predict the *I-V* characteristics of this molecule which may have a real application as a possible optical switch in the molecular circuit.

In this work, we carry out nonequilibrium Green's function formalism combined with first-principles density functional theory to investigate the electronic transport properties of the dihydroazulene optical molecular switch with three typical adsorption sites, namely hollow, top and bridge site, which are simple but important situations in break-junction experiments.

Firstly, the structure of freethiol(SH) capped molecule with the above-mentioned two forms was first optimized. Based on experimental issues about self-assembled monolayers (SAMs), it is generally accepted that hydrogen atoms are dissociated upon adsorption to metal surfaces. So we construct a two-probe system, which the two terminal hydrogen atoms bonded to the sulfur atom are eliminated from the optimized structure, and the remained part is sandwiched between two parallel Au(111) surfaces that correspond to the surfaces of the Au electrodes(see Fig. 3. 11. 1). The Periodic boundary conditions for the DFT calculation in the contact region have been used. A supercell consists of two layers of 4×4 Au and the molecule. It has been reported that taking two layers of Au in the self-consistent cycle is enough to avoid the finite size effect. Furthermore, in some literatures, a 3×3 supercell has been used. Our 4×4 supercell is large enough to avoid any interaction with molecules in the next supercell. As a result, there is no effect of the surface on the geometry of the contact region, while the remainders of electrodes are described by bulk parameters. These are determined from separate calculations for the bulk phase of Au employing the same model.

Furthermore, since the interface structures between the electrodes and the molecule, such as the anchoring configuration, cannot be characterized clearly in the experiments. Therefore, to obtain the most preferable configuration, different adsorption sites, including top, bridge, and hollow sites are considered. The atomic structures of the molecule structures of the molecule and the distance between two electrodes are also optimized and the Au-S bond length is about 1.9, 2.07 and 2.42 Å for hollow, top, and bridge sites, respectively. The structure of the Au(111)-molecule-Au(111) molecular junction has not been relaxed due to the limitation of computational resources. Since the switching unit is sufficiently far from the Au electrode region, the effect of geometrical optimization is minor. In NEGF theory, the molecular wire junction is divided into three regions: left electrode (L), contact region (C), and right electrode (R). The contact region includes extended molecule and two layers of Au from each electrode (see Fig. 3. 11. 1). The contact region contains parts of the electrodes include the screening effects in the calculations. The semi-infinite electrodes are calculated separately to obtain the bulk self-energy. The quantum transport calculations have been carried out by an ab initio code package, ATK2. 0.

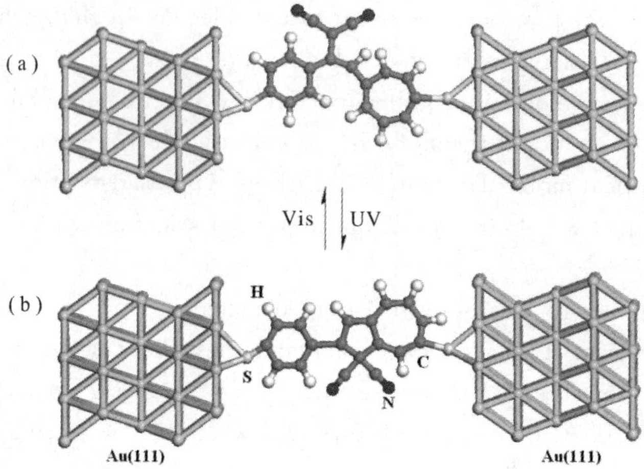

Fig. 3. 11. 1 Geometry for the molecular switch with the hollow site

(a) open-ring form; (b) closed-ring form

The calculated *I-V* characteristics of molecular switches with three different adsorption sites in the range of [0 V, 1 V] are plotted in Fig. 3. 11. 2. The switching behavior can be clearly seen from Fig. 3. 11. 2, the closed form shows good conductivity even for a small bias, which increases as stronger bias is applied. On the contrary, the open form shows very low conductivity for all three adsorption sites. One can predict that there is a switch from on (high resistance) state to the off (low resistance) state, when the closed molecule changes to the open form under photoexcitation. It also can be seen from Fig. 3. 11. 2 that, for the hollow site, the *I-V* curve has similar behavior as that of bridge site. The *I-V* curves of hollow and bridge sites increase slowly and are fairly linear in the range of voltage. But the *I-V* curve with top site increases rapidly, and the current value is larger than other sites in the range of [0 V, 1 V].

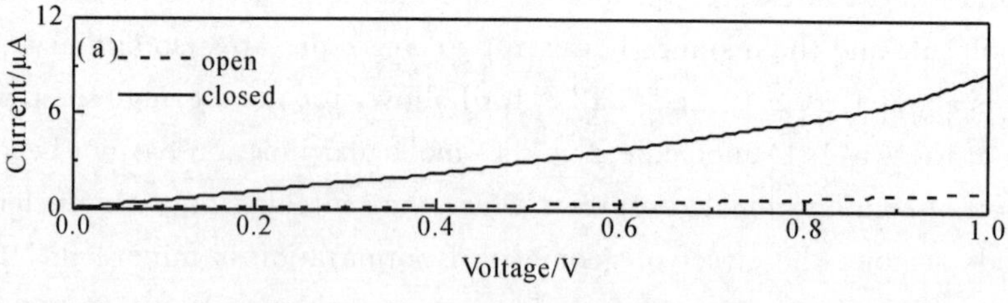

Fig. 3. 11. 2 The *I-V* characteristics of the molecular switch with

(a) hollow

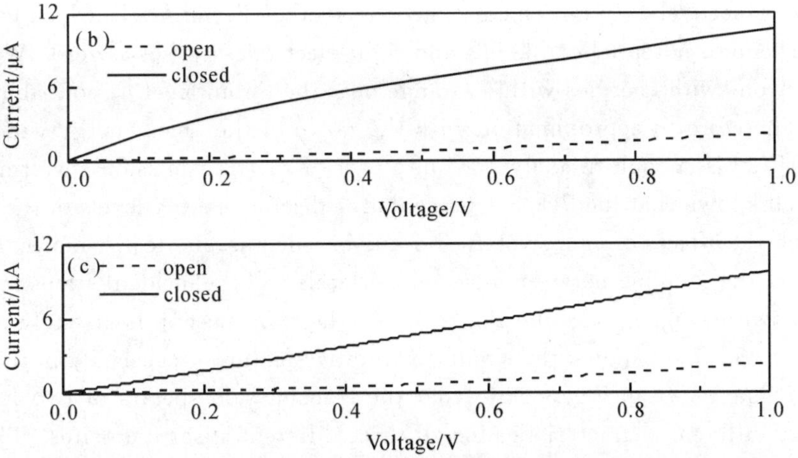

Continued Fig. 3. 11. 2 The *I-V* characteristics of the molecular switch with (b) top; (c) bridge sites

To get a further insight about the conduction change in closed and open form of dihydroazulene molecule, we define the conduction enhancement R, $R = \frac{G_{\text{closed}}}{G_{\text{open}}}$, where G_{closed} and G_{open} are the low voltage conductance of the closed and the open form, respectively. The G_{closed}, G_{open} and the R of the molecular wires adsorbed at different sites on Au(111) electrodes are given in Table 3. 11. 1. As shown in Table 3. 11. 1, there is no obvious change in the conductance of open form with three adsorption sites. But the closed form of top site adsorbed molecular wire shows more conductance than others. The R of top site is 7. 82, which is the large one compared with hollow and bridge sites, being 2. 67 and 3. 15, respectively. Thus we predict a 3 – 8 times conduction enhancement for the closed form over the open form.

Table 3. 11. 1 Conductance enhancement with various adsorption sites

Adsorption site	Molecular wire	Conductance	
		$G/\mu s$	R
Hollow	Open	0. 87	2. 67
	Closed	2. 33	
Top	Open	0. 94	7. 82
	Closed	7. 36	
Bridge	Open	0. 60	3. 15
	Closed	1. 89	

The switching characteristics of thedihydroazulene system and the origins of conductance enhancement of closed form with top adsorption sites can be understood from the transmission spectra and projected density of states (PDOS), which are shown in Fig. 3. 11. 3 and Fig. 3. 11. 4, respectively. In our calculation, the average Fermi level, which is the average value of the chemical potential of the left and right electrodes, is set as zero. We can expect that only electrons with energies within a range near the Fermi level E_F contribute to the total current. Therefore, a approximation with the range of the "bias energy window", i. e. , $[-V/2, +V/2]$ is enough to analyze a finite part of the transmission spectrum. Furthermore, it is well known that the PDOS represents the discrete energy levels of the isolate molecule including the effects of energy shift and line broadening due to the molecule-electrodes coupling, i. e. , the coupling between molecular orbitals and the incident states from the electrodes. As shown in Fig. 3. 11. 3 and Fig. 3. 11. 4, the transmission is strongly correlated to PDOS spectra, which are shown the similar qualitative features, especially in the location of their peaks. It is clear from Fig. 3. 11. 3 that the transmission spectra of two forms display extraordinarily different characteristics for all three different adsorption sites. The transmission coefficients of closed form are always significantly larger than those of open form over the entire bias range with three adsorption sites. Furthermore, the occupied and unoccupied molecular orbitals(RMO) with the vertical solid lines at zero bias ($V_b = 0$) for two forms with three different adsorption sites are also plotted in Fig. 3. 11. 3. We can clearly see that the HOMO-LUMO gaps of closed form and open form are 1. 7 eV and 2. 2 eV for the hollow site, 1. 4 eV and 2. 1 eV for the top site, 1. 5 eV and 1. 9 eV for the bridge site, respectively. As a result, the HOMO-LUMO gap of closed form is smaller than that of open form for all three adsorption sites. In such an open system, the Fermi level E_F of Au electrode aligns between HOMO and LUMO, and, therefore, the barrier for the electron transport is intensively relevant to the HOMO-LUMO gap. Consequently, the current through the closed form is always larger than that though the open form.

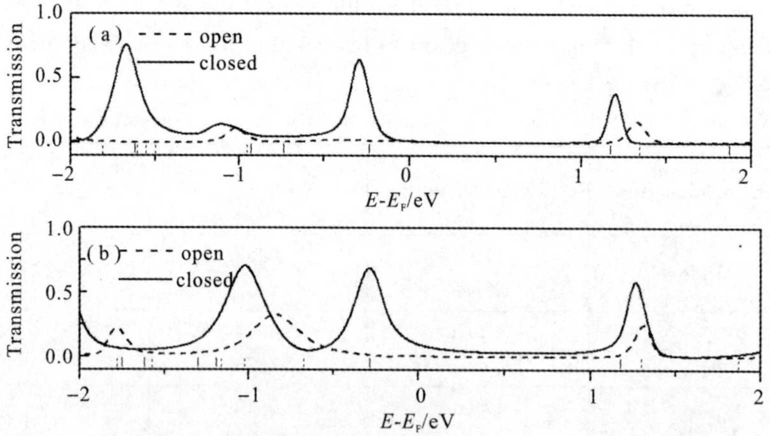

Fig. 3. 11. 3 The transmission spectra of the molecular switch at zero bias

(a)hollow; (b) top

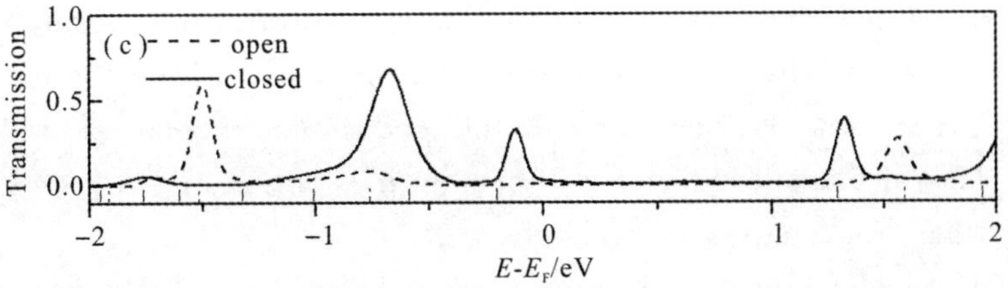

Continued Fig. 3. 11. 3 The transmission spectra of the molecular switch at zero bias

(c) bridge sites

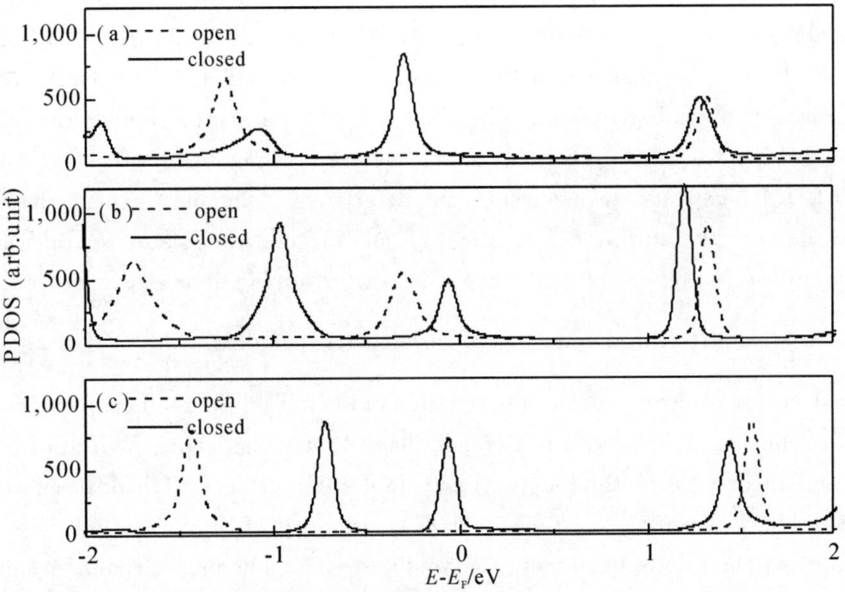

Fig. 3. 11. 4 The PDOS

(a)hollow; (b) top; (c) bridge sites

Furthermore, the conductance enhancement of closed form for top site is also shown clearly in the transmission spectra. For the closed form, there are four transmission peaks for all three adsorption sites: three transmission peaks are below the electrode Fermi energy E_F, and one is above it. However, just one transmission peak is within the bias energy window[-0.5 eV, $+0.5$ eV], and the other three transmission peaks are out of the bias energy window. That means there are four scattering channels, but only one of the channels makes its contribution to conductance. It is evident from Fig. 3. 11. 3 that the transmission peak and the value of the transmission coefficient of top site is more broader and larger than others in the energy window. This broad and large peak results in great transmission probability at the Fermi energy and is responsible for the conductance enhancement for the top site of closed form observed in Table 3. 11. 1. On the other hand, for the open form, there are three transmission peaks for all three adsorption sites, but no one is within the bias energy window. Therefore, there is no obvious change in the conductance of opened form with three adsorption sites. The calculated conductance of open form is only 0. 87 μs, 0. 94 μs and 0. 60 μs for hollow, top, and bridge sites, respectively.

To conclude, the electronic transport properties ofdihydroazulene (open form and closed form) optical molecular switch with three typical adsorption sites have been investigated by using the DFT + NEGF first-principles method. The electronic transmission coefficients, projected density of states, and *I-V* characteristics corresponding to different forms are calculated and analyzed. The dramatic difference in conductivity appearing in two different forms can be observed. A 3 – 8 times conduction enhancement is predicted for the closed form over the open form after optical switching. The main contribution to the conduction enhancement is due to the different molecular geometry and π-electron conjugation, which results in different HOMO-LUMO gap and transmission characteristics. One may expect that dihydroazulene molecule with two different forms can be one of good candidates for optical switches due to this unique advantage, and may have some future applications in the molecular circuit.

3. 12 Negative differential resistance by asymmetric couplings

In recent years, electronic devices based on single molecules have been considered as one of the most promising technologies to extend silicon-based electronics. With advances in experimental and theoretical methodology, many molecular devices with different functionalities like negative differential resistance (NDR), memory effects, molecular rectification, and electronic switching have been measured and designed. The most prominent among these is negative differential resistance (NDR), which is the basic principle of several electronics components. NDR have already been observed or predicted in some molecular devices. Chen et al. reported NDR behavior in molecules containing a nitroamine redox center. They suggested a possible mechanism for NDR is a two-step reduction process that modifies charge transport through the molecule. Based on quantum chemical calculations, Karzazi et al. described NDR behavior in polyphenylene-based molecular wires incorporating saturated spacers. They conjectured that resonant tunneling originating from shifting of the molecular energy level by an external electric field might lead to NDR. Taylor et al. calculated the same structure and found that the side groups play an important role in NDR. Luo et al. have also found the NDR behavior in benzene molecule junctions with donor/acceptor substitutions by applying the elastic scattering Green's function theory approach in combination with the frontier molecular orbital theory. They suggested that the one-electron reduction is responsible for the NDR.

Albeit there have been many theoretical as well as experimental investigations about NDR in various kinds of molecular devices, the number of molecular systems having the NDR feature are still very limited. Especially, the NDR performance of molecular devices induced by asymmetric coupling between molecule and electrode, up until now, has not been reported yet. In the present work, by applyingnonequilibrium Green's function (NEGF) formalism combined with first-principles density functional theory (DFT), we investigate the electronic transport properties of the pyrene-based molecular device which consists of the different molecule-electrodes contacts chemically bound to two Au electrodes. It has been found that the variation of the coupling between the molecule and the electrodes with external bias will leads to NDR. An analysis of the origin of negative differential resistance has been given by observing the shift in transmission resonance peak across the bias window with varying bias voltage. Furthermore, the influence of the spatial distributions of molecular orbitals on the quantum transport through the molecular device is also discussed in detail.

Fig. 3. 12. 1 illustrates the simulation setup. The pre-optimized pyrene-based molecule with two different forms are sandwiched between two parallel Au(111) surfaces that correspond to the surfaces of the Au electrodes. Each layer of Au electrodes is represented by a 4×4 supercell with the periodic boundary conditions so that it imitates bulk metal structures. The 4×4 supercell is large enough to avoid any interaction with molecules in the next supercell.

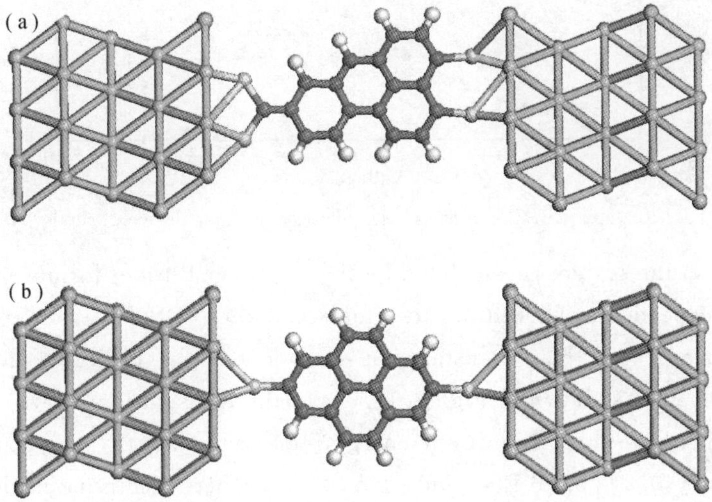

Fig. 3. 12. 1 Geometry of the molecular device in our simulation. Pyrene-based molecule with two forms are sandwiched between two Au electrodes through the different end group, which is referred to (a) model A; (b) model B

Fig. 3.12.2 shows the calculated *I-V* characteristics curve for models A and B in a bias range from 0 to 2.0 V in steps of 0.2 V. It should be pointed out that at each bias, the current is determined self-consistently under the non-equilibrium condition. It is quite clear from Fig. 3.12.2 that initially the current increases with the increases in external voltage for both models. But this increase in current of model A due to the variation in voltage is observed up to 1.2 V. Beyond 1.2 V, there is a rapid decrease in current with the increase in bias voltage. This decrease in current due to an increase in voltages is the manifestation of the NDR feature. NDR behavior is quite prominent up to a bias voltage of 1.6 V. However, above 1.6 V the NDR feature is completely lost and the current trough the molecular junction again increases with an increase in voltage. In contrast to model A, the current of model B always increase over the entire bias range.

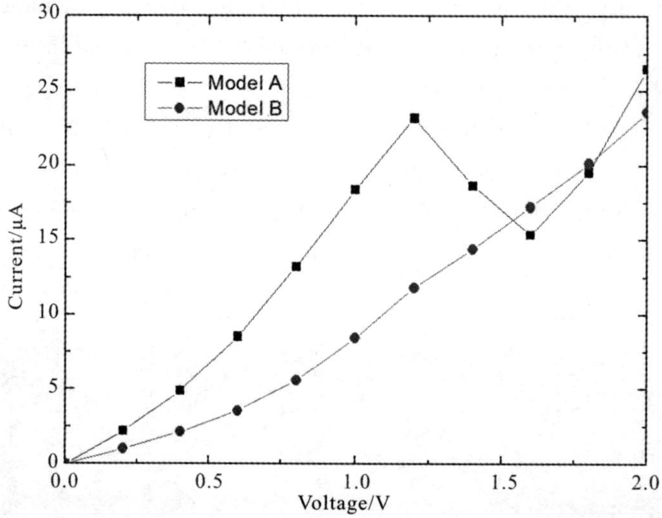

Fig. 3.12.2 The calculated *I-V* characteristics of the molecular device of model A and model B

The current in the system is calculated by the Landauer-Bütiker formula, which is transmission spectra dependent. Therefore, to understand the NDR behavior, we calculate the energy dependence of zero-bias transmission spectra and the projected density of states (PDOS) of model A. As shown in Fig. 3.12.3(a), the transmission curve consists of a series of peaks. There are three significant energy regions in the zero bias transmission spectra, i.e. [−1.8, −0.2] [0.8, 1.4] and [1.5, 2.0], where electrons incident from one of the electrodes can transmit across the molecule to the other electrode significantly. The transmission is determined by the electronic structure of the molecule and the coupling: the peak corresponds to the resonant transmission through the molecular states. To understand why incident states in these three energy regions can transmit across the molecule significantly, we calculated the projection of the density of states(PDOS), namely the density of states of the combined system onto all the molecule subspace. It is well known that the PDOS represents the discrete energy levels of the isolate molecule including the effects of energy shift and line broadening due to the molecule-electrode coupling, i. e. , the coupling between molecular orbitals and the incident states from the electrodes. The PDOS is shown in Fig. 3. 12. 3(b). It is evident from Fig. 3. 12. 3 that the transmission and the PDOS spectra are strongly correlated and show the same qualitative features, especially in the location of their peaks. A strong coupling makes incident electrons at a certain energy easily transmit across the molecule, and this will give rise to a large transmission coefficient at this energy. As a consequence, a large transmission coefficient indicates a strong coupling between the electrodes and the molecule, and the evolution of transmission curves with external biases can help us to understand how the changes of the coupling between the electrodes and molecule determines the *I-V* characteristics in the system. Therefore, to get a further insight about the mechanism responsible for NDR behavior, the voltage dependence of the transmission function will be studied next.

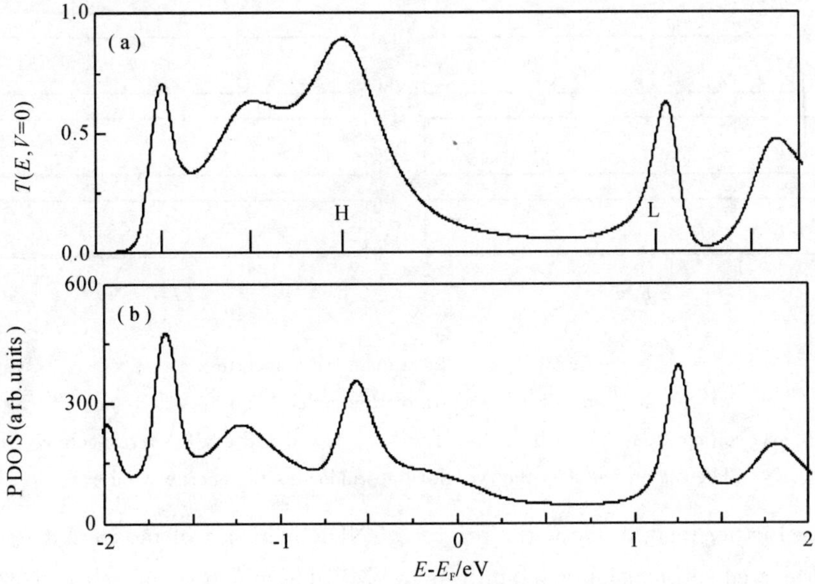

Fig. 3. 12. 3 The transmission and PDOS

(a) the transmission spectra of model A under zero bias(The short vertical bars near the enery axes stand for the energy levels of the extend molecule); (b) the corresponding PDOS

The bias dependence of the transmission characteristics at biases of 1. 2 V, 1. 4 V, 1. 6 V, 1. 8 V and 2. 0 V of the molecule device are presented in Fig. 3. 12. 4. In our calculation, a good approximation with the range of the bias window, i. e. ,[$-V/2, +V/2$] is enough to analyze a finite part of the transmission spectrum. As shown in Fig. 3. 12. 4 (a), we can see that the value of transmission peaks in the bias window decrease with the increase in bias voltage at the range of 1. 2 to 1. 6 V. As a result, current through the molecular device diminishes with the increase in bias voltage above 1. 2 V and below 1. 6 V. Thus the NDR feature is apparent over the bias voltage above 1. 2 V and below 1. 6 V. At relatively higher bias voltage, contributions from other resonance peaks enter into the bias window [see Fig. 3. 12. 4(a)d - e]. Thus the current through the molecular device increases. As a result, the NDR feature disappears at this bias voltage range. However, from Fig. 3. 12. 4 (b), we can find that the transmission area of model B in the bias window gets bigger with the increase of bias voltage, which leads the current to increase(see Fig. 3. 12. 2). Therefore, the current is enhanced over the entire bias range.

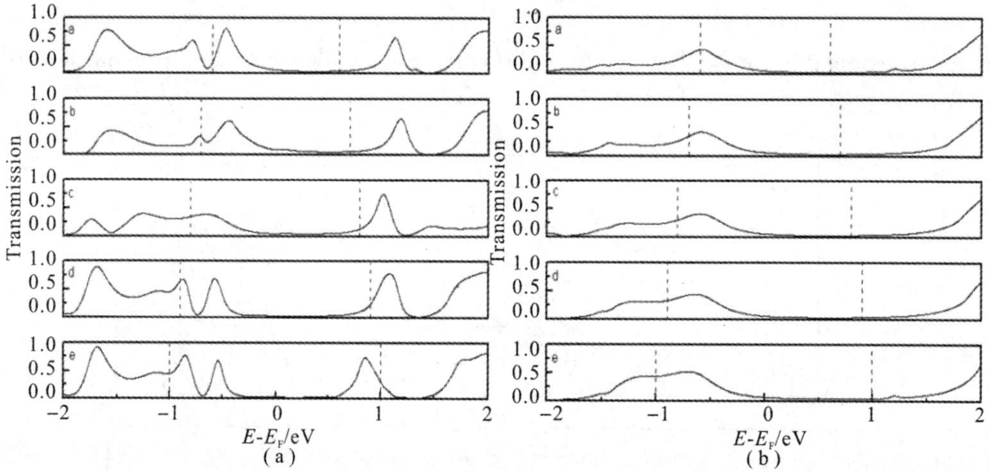

Fig. 3. 12. 4 The transmission spectra

(a) (b) the transmission spectra of the model A and model B under bias voltage: a 1. 2 V, b 1. 4 V, c 1. 6 V, d 1. 8 V and e 2. 0 V, respectively

(The region between the two dot lines indicates the energy window)

To get a further insight about the origins of NDR behavior of model A, we analyze the molecular projected self-consistent Hamiltonian (MPSH) which is the self-consistent Hamiltonian of the isolated molecule in the presence of the electrodes, namely the molecular part is extracted from the whole self-consistent Hamiltonian at the contact region. Table 3. 12. 1 presents a comparative study of MPSH-HOMO and MPSH-LUMO states which make their contribution to current within the bias voltages of 1. 2 V, 1. 4 V, 1. 6 V, 1. 8 V and 2. 0 V. From Table 3. 12. 1, we can see that although there is a complete delocalization of MPSH-

HOMO states at 1. 2 V and 2. 0 V, respective MPSH-HOMO state is completely localized at 1. 4 V, 1. 6 V and 1. 8 V. Furthermore, the HOMO orbital is the main transmission channel for the current of model A[see Fig. 3. 12. 3(a)]. Therefore, this feature clearly indicates that there will be a reduction in current at 1. 4 V, 1. 6 V and 1. 8 V because of localized MPSH-HOMO states.

Table 3. 12. 1 The MPSH to HOMO and LUMO of model A at bias voltages of 1. 2 V, 1. 4 V, 1. 6 V, 1. 8 V and 2. 6 V, respectively

MPSH	HOMO	LUMO
1. 2 V	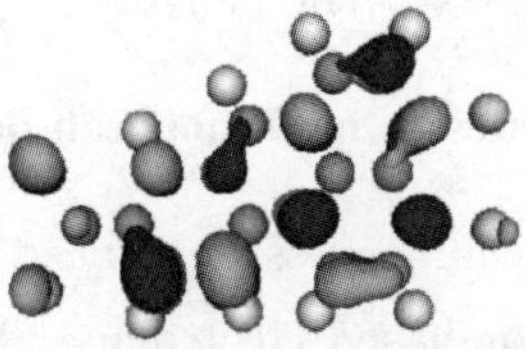	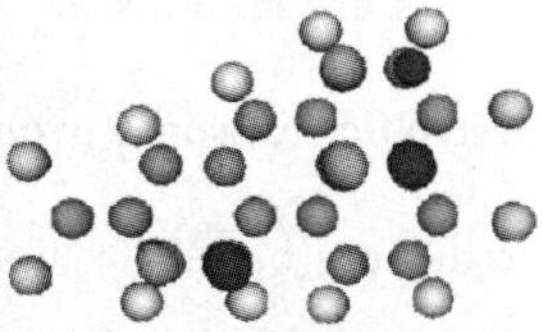
1. 4 V	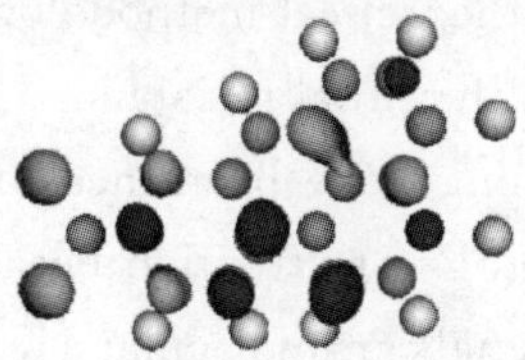	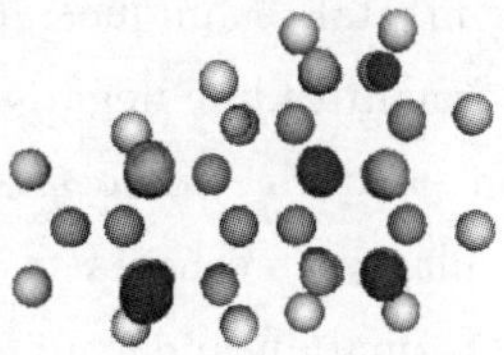
1. 6 V	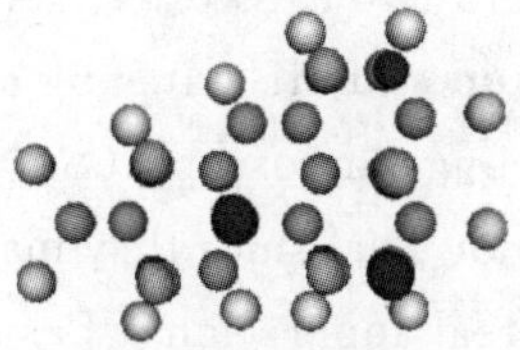	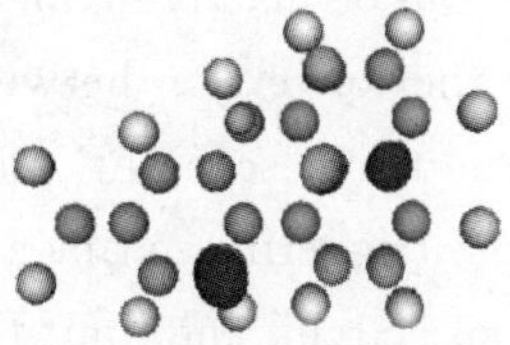
1. 8 V	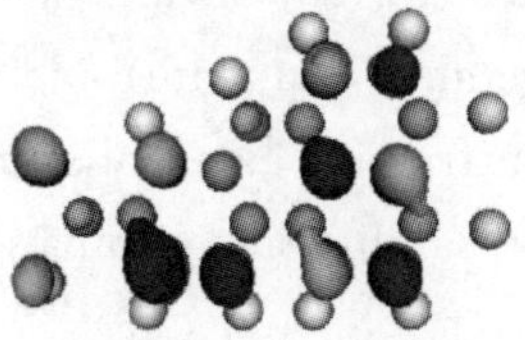	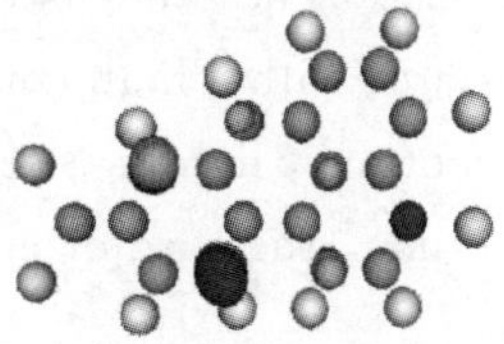
2. 0 V	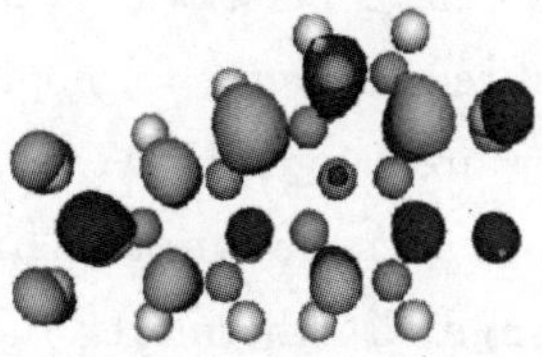	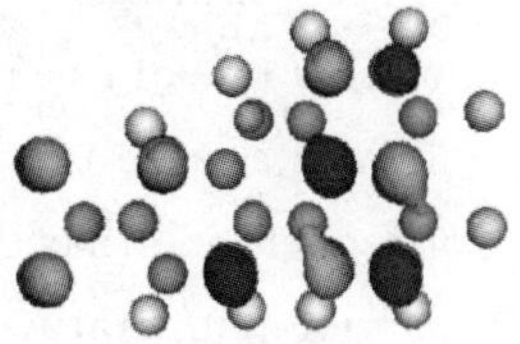

In conclusion, the electronic transport properties of the pyrene-based molecular device with different molecule-electrodes have been investigated by using the DFT+NEGF first-principles method. The electronic transmission coefficients, partial density of states, and I-V characteristics corresponding to different forms are calculated and analyzed. It is shown that the NDR behavior can be observed over a certain range of applied bias voltage(1. 2 to 2. 0 V) in the asymmetric system namely model A. This observed NDR behavior has been explained by the shift in transmission resonance peak across the bias window with varying bias voltage and the spatial distributions of molecular orbitals on the quantum transport through the molecular. It is found that the asymmetric structure of both the molecule and the molecule-electrode couplings are responsible for the NDR behavior.

3. 13 Electronic transport properties in naphthopyran-based optical molecular switch

Electronic devices based on single molecules have been widely recognized as potentially promising alternatives to semiconductor-based electronics. With the help of the rapidly developing experimental techniques and theoretical methodology, many molecular devices with different functionalities like negative differential resistance (NDR), memory effects, molecular rectification, and electronic switching have been measured and designed. Among these devices, molecular switch has attracted great attention because it is critical to the ability to store digital information and route signals in molecular electronic logic circuits. Molecular switch devices can be mainly divided into two categories. One is the electronic switch devices which control the conversion between on and off states by an external trigger such as the electric field, the tip of scanning tunneling microscopy (STM), and the redox process, etc. However, these triggering means are not ideal since they may interfere greatly with the function of a nanosize circuit and limit the real applications. Besides, all these means are relatively slow in response. The other is optical switch devices which exist in two thermally sufficiently stable states with a high conductance (on state) and a low conductance (off state) by light. On the contrary, light is a very attractive external stimulus for such switches including short response time, the ease of addressability and compatibility with a wide range of condensed phases.

Azobenzene-based optical switches have been widely studied in this regard in recent years. It is known that one of essential requirements for applications as molecular switches is the thermal stability in a large temperature range. However, the spontaneous cis-trans photoisomerzation of azobenzene that occurs at moderate temperature creates a drawback for the application over a wide temperature range. Furthermore, the on-off ratio of this molecule is about two orders fro magnitude for biases below 0. 3 V. The diarylethene-based optical switches also have been studied. However, it is just one-way switch due to the strong overlap between the excited state of the open form of the diarylethene molecule with the plasmon bands of the Au electrodes. Therefore, it is very important to search new candidates for the molecular switch with large on-off ratio and which can be used over a wider temperature. Recently, Szaciöowski et al. have reported that the naphthopyran molecule which characterized by having a pyran ring to a naphthoquinone can be reversed from open ring to closed ring form by irradiation with ultraviolet and visible light, respectively. And it was found that, in solution as well as in the solid state, they present excellent photochromic properties including short response time, heat stability, and large changes of the absorption wavalengths between the two isomers, which make it usable as one of the candidates for light-driven molecular switches. But up to now, the electrical conductivity of this molecule has not been reported both in experiment and on the theory. Therefore, it is very essential to offer a theoretical investigation that will help to predict the *I-V* characteristics of this molecule which may have a potential application as a possible optical switch in the molecular circuit. In the present work, we report the results of theoretical studies on the electronic transport properties of the naphthopyran-based optical molecular switch. It has been found that the current through the open form is significantly larger than that through the closed form which is different from other optical switches based on ring-opening reactions of the molecular bridge. Furthermore, its on-off ratio can reaches 90 at a bias of 1. 4 V.

Fig. 3. 13. 1 illustrates the simulation setup. The pre-optimized molecule with two forms are sandwiched via thiolate bonds between two parallel Au(111) surfaces that correspond to the surfaces of the Au electrodes.

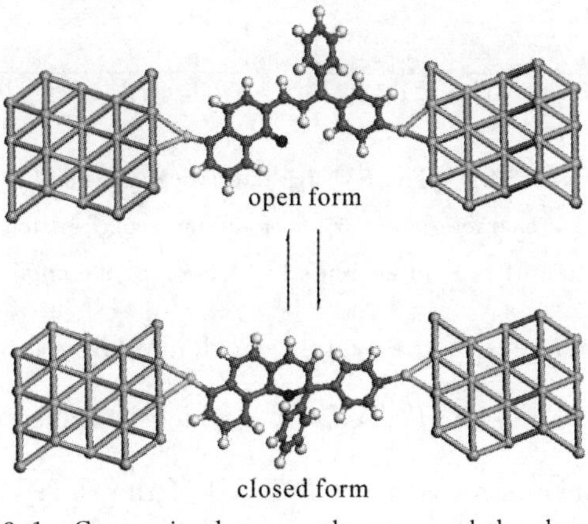

Fig. 3. 13. 1 Conversion between the open and the closed forms

The self-consistently calculated current-voltage (*I-V*) characteristic curves of the molecular switch with two forms in a bias range from 0 to 1. 4 V in steps of 0. 2 V are plotted in Fig. 3. 13. 2(a). It should be pointed out that at each bias, the current is determined self-consistently under the non-equilibrium condition. The switching behavior can be clearly seen from Fig. 3.13.2(a). In contrast to most other optical switches which are also based on the ring opening/ring closure reactions, the open form of naphthopyran exhibits a significant current and an almost Ohmic current-voltage characteristics. However, the closed form shows a rather low current. Consequently, the current through the open form is evidently larger than that through the closed form over the entire bias range, which can realize different conductance states of the molecular bridge corresponding to high conductance to low conductance. Namely, when the molecule in the switch changes from the open form to the closed form under photoexcitation, one can predict that there is a switch from on state (open form) to the off state (closed form). The on-off ratio, $R = I_{\text{open-form}} / I_{\text{closed-form}}$, versus bias V is also plotted in Fig. 3.13.2(b). It can be noted that a very large on-off ratio which varies from around 65 to 90 in the bias range under investigation, which suggests that this system has attractive potential application in future molecular switch technology.

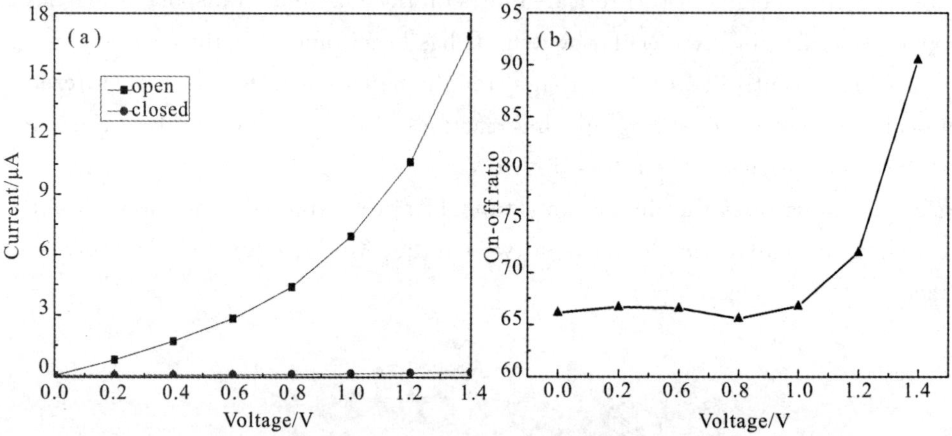

Fig. 3.13.2 The *I-V* characteristic curves and *R*

(a) the calculated *I-V* characteristic curves of the molecular switch with two forms;

(b) the on-off ratio of currents as function of the applied bias

In NEGF theory, the current of the system is calculated by the Landauer-Bütiker formula $I = \frac{2e}{h}\int[f(E-\mu_L) - f(E-\mu_R)]T(E,V)\mathrm{d}E$, which is transmission spectra dependent. Thus, we calculate the transmission spectra $T(E,V)$ of the closed and the open forms under zero voltage to understand the dramatic difference in conductance appearing in *I-V* curves. As shown in Fig. 3.13.3, the characters H and L represent the highest occupied molecular orbital (HOMO) and the lowest unoccupied molecular orbital (LUMO) and the average Fermi level E_F which is the average value of the chemical potential of the left and right electrodes is set as zero. The current at low voltage is mainly determined by the transmission function close to the Fermi energy. The mechanism of conductance enhancement for the open form is shown clearly in the transmission spectra from Fig. 3. 13. 3. For the open form, there are two significant peaks which dominate the conductive behavior under small bias voltages: one is above the Fermi level, and the other is below it. These two peaks still can be found in the transmission spectrum of closed form, but both of them become very weak and the first peak is shifted toward the higher energy. Besides the change of these two peaks, the HOMO-LUMO gap of the closed form also becomes larger than that of the open form. The results show that the HOMO-LUMO gap increases from 1. 34 to 2. 44 eV when the open form is switched to the closed form. In such an open system, the Fermi level E_F aligns between HOMO and LUMO, and the barrier for the electron transport is intensively relevant to the HOMO-LUMO gap. As a result, the large transmission strength and small HOMO-LUMO gap account for the high conductance in the open form.

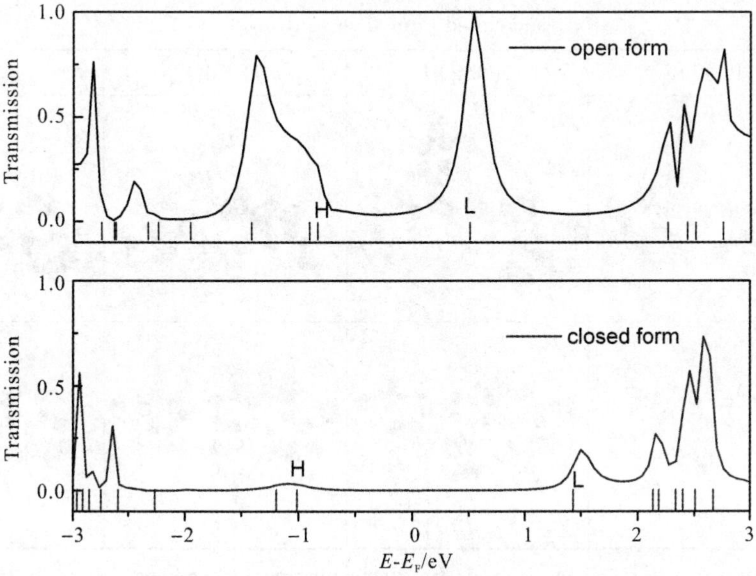

Fig. 3. 13. 3 The transmission spectra of the molecular switch at zero bias

To further elucidate the microscopic origin of the peaks in the transmission spectra and of the different transmission characteristics for two forms, we analyze the molecular projected self-consistent Hamiltonian (MPSH) which is the projected self-consistent Hamiltonian of the molecular switch on to the Hilbert space spanned by the basis functions of the molecule. The eigenstates of the MPSH can be referred to molecular orbitals which includes the modification on the electronic structure of the molecule due to the presence of the electrodes.

Table 3. 13. 1 shows the MPSH orbital near the Fermi level of the electrodes. When the energies of the resonances in the transmission spectra are compared with the eigenvalues of MPSH, we find from Fig. 3. 13. 3 that the two main transmission peaks of the open form mainly are contributed by MPSH orbitals HOMO-2, HOMO-1, HOMO and LUMO, while the peaks of closed form take their origin from MPSH orbitals HOMO-1 HOMO and LUMO. Therefore, it can be found from Table 3. 13. 1 that the main transmission channels, i. e. the HOMO-2, HOMO-1, HOMO and LUMO, are delocalized orbital for the open form which leads to low barrier and provides good transport channels for electron transport. On the contrary, these four transmission channels are localized orbitals for the closed form leading to high barrier for electron transport which accounts for the low transmission strength in Fig. 3. 13. 3. As a result, the current is decreased when the molecule switch from open to closed form upon photoexcitation, which is in good agreement with the trends of *I-V* curves shown in Fig. 3. 13. 2.

Table 3. 13. 1 The MPSH to HOMO-2, HOMO-1 HOMO and LUMO for open and closed forms, respectively

MPSH	HOMO-2	HOMO-1	HOMO	LUMO
open form	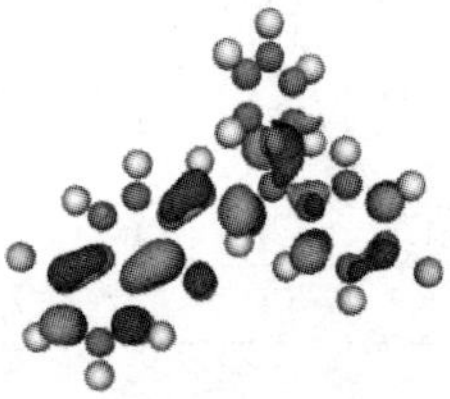	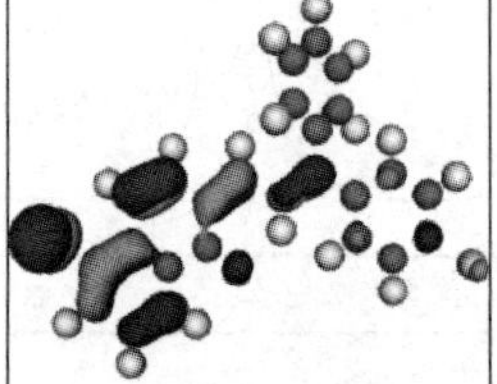	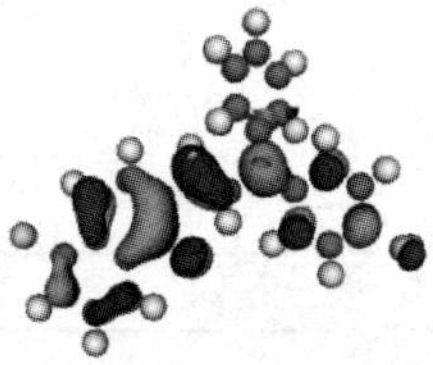	
closed form	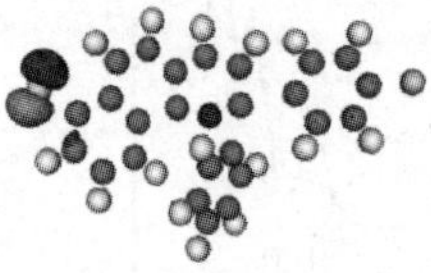	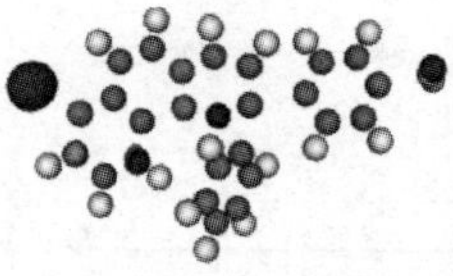	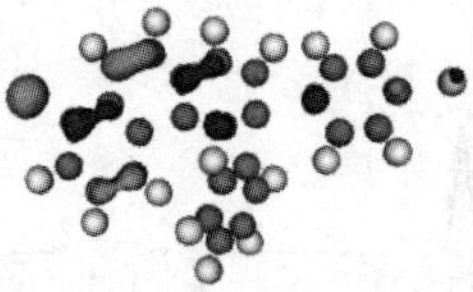	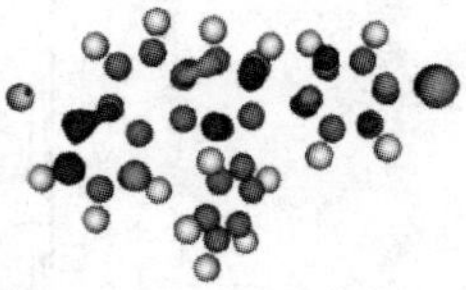

In conclusion, we have investigated the electronic transport properties of the naphthopyran-based optical molecular by using NEGF formalism combined with first-principles DFT. The dramatic difference in conductance properties appearing in the open and the closed forms is interpreted based on the transmission spectra and spatial distribution of MPSH orbitals. Theoretical results show that the current through the open form is significantly larger than that through the closed form. A reversible change between the two forms would realize a molecular nanoswitch, which suggests that this system has attractive potential application in future molecular circuit.

3.14 Pyridine-substituted dithienylethene optical molecular switch

The development of molecular electronics is a field that offers great potential to today's semiconductor-based electronics. Molecules can be developed to perform operations that are unattainable via conventional semiconductor technology. Recently, with the advent of experimental techniques for characterizing and manipulating individual molecules, such as scanning tunneling microscopy (STM), mechanically controllable break junction, atomic force microscope, creative fabricated techniques, and so forth, many molecular devices with different functionalities have been designed and measured. Meantime, the theoretical simulations based on density functional theory (DFT) in combination with the nonequilibrium Green's function formalism (NEGF) are used to help to understand the underling working principle of these molecular devices. Among these devices, molecular switch has been attracting great attention, because they are considered to play an essential role in any modern design of logic and memory circuits. Molecular switch devices can be mainly divided into two categories, including electronic switch devices which control the conversion between on and off states by an external trigger such as the electric field, the tip of scanning tunneling microscopy (STM), and the redox process, etc, and optical switch devices which exist in two thermally sufficiently stable states with a high conductance (on state) and a low conductance (off state) by light. Among them, optical switch has been extensively studied, because light is a very attractive external stimulus for such switches including short response time, the ease of addressability and compatibility with a wide range of condensed phases. In this regard, diarylethene-based photochromic switch has been investigated and discussed theoretically and experimentally. However, a few of these molecules can be found real application as a possible optical "switch" of the electric current, because majority of them cannot overcome the potential barrier between the two stable forms in the real monolayer or in the Au nanonetwork. Dulić et al. have added thiophene rings at the two ends of the basic diarylperfluorocyclopentenes. They found that the unperturbed system shows good photoreversible properties, but when it was attached to the Au electrodes, the ring-closing reaction was blocked. Furthermore, another representative of molecular photochromic switches is the group of azobenzene switches, whose two stable forms(cis- and trans-) is reversible, when it was attached to the Au electrodes. But the on-off ratio is about two orders magnitude only for biases below 0.3 V. Therefore, the challenge is to search new candidates for the molecular switch with large on-off ratio and which can operate in a real molecular circuit.

The pyridine-substituteddithienylethene molecule which is based on the basic diarylethene system with two pyridine rings attached on the terminating C atoms is one derivative of the diarylethene molecule which is considered as the most potential candidates for active materials in photoswitching devices. Recently Mendoza et al. have reported that the closed form of this molecule can be transformed into its open form upon irradiation with 420 nm visible light (VIS) when it contact with Au surfaces via thiol anchoring groups. On the other hand, upon irradiation with 313 nm UV light, the open form undergoes cycloreversion and reverse back to the closed form(see Fig. 3. 14. 1). These two forms can keep stable over a wider temperature range and reversibly switch from each other, which make it usable as one of the good candidates for light-driven molecular switches. But up to now, the electrical conductivity of this molecule has not been reported both in experiment and on the theory to the best of our knowledge. Therefore, it is very essential to offer a theoretical investigation that will help to predict the *I-V* characteristics of this molecule which may has real application as a possible optical switch in the molecular circuit. In this work, we carry out nonequilibrium Green's function formalism combined first-principles density functional theory to investigate the electronic transport properties of the pyridine-substituted dithienylethene optical molecule with open- and closed-ring forms.

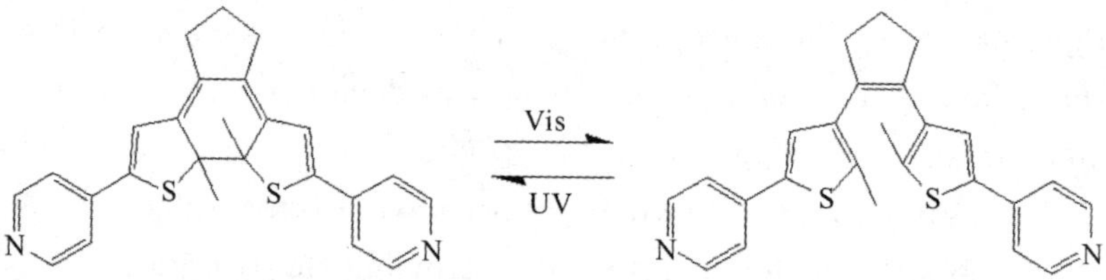

Fig. 3. 14. 1 Conversion between closed- and open-ring of molecule

First, the structure of freethiol(SH) capped molecule with the above-mentioned two forms was first optimized. Based on experimental issues about self-assembled monolayers (SAMs), it is generally accepted that hydrogen atoms are dissociated upon adsorption to metal surfaces. So we construct a two-probe system, which the two terminal hydrogen atoms bonded to the sulfur atom are eliminated from the optimized structure, and the remained part is sandwiched between two parallel Au(111) surfaces that correspond to the surfaces of the Au electrodes. In this work, we focus on the switch of conductivity through a photochemical reaction. Furthermore, many studies have predicted that the hollow adsorption site is more favorable in energy. Therefore, we investigate the sulfur atom to locate at the hollow site of each Au surface without considering other two adsorption sites, i. e. top and bridge sites. The perpendicular distance between the Au surface and sulfur atom is set to 1. 9 Å, which is a typical Au-S distance. The Au(111) surface is represented by a 4×4 supercell with the periodic boundary conditions so that it imitates bulk metal structures. The contact region includes extended molecule and two layers of Au from each electrode (see Fig. 3. 14. 2). The contact region contains parts of the electrodes include the screening effects in the calculations. The semi-infinite electrodes are calculated separately to obtain the bulk self-energy. The structure of the Au(111)-molecule-Au(111) molecular junction has not been relaxed due to the limitation of computational resources. Since the switching unit is far from the Au electrode region, we believe that the effect of geometrical optimization is minor.

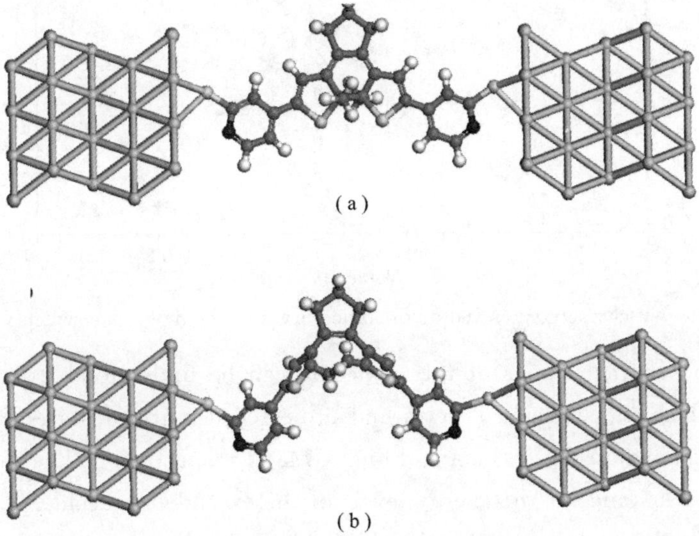

Fig. 3. 14. 2 Geometry for the molecular switch with the hollow site

(a) closed-ring form; (b)open-ring form

The calculated *I-V* characteristics of the molecular switch with two forms at a bias up to 1. 0 V are plotted in Fig. 3. 14. 3. The switching behavior can be clearly seen from Fig. 3. 14. 3, the current of closed-ring form is greater than that of open-ring form at the same bias. One can predict that there is a switch from on (low resistance) state to the off (high resistance) state, when the closed-ring molecule changes to the open-ring form under photoexcitation. The on-off ratio, $R = I_{\text{closed-form}} / I_{\text{open-form}}$, versus bias V is also plotted in Fig. 3. 14. 3. It can be noted that a large on-off ratio which varies from around 9 to 200 in the bias range is desirable for real application. Furthermore, when the bias voltage reaches 0. 7 to 2. 0 V, the current become very large, and when the bias voltage is 0. 8 V, the current decreased. Negative differential resistance(NDR) appears between 0. 7 to 0. 8 V. In the following, we will explain this switching behavior and NDR phenomena in detail.

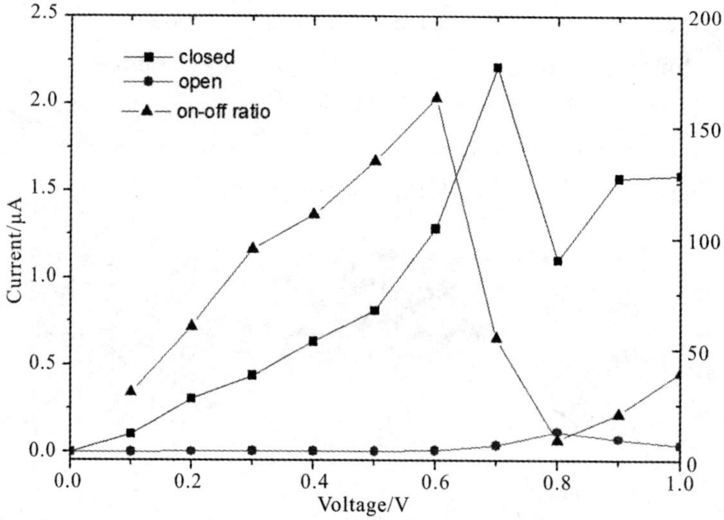

Fig. 3. 14. 3 The *I-V* characteristics and on-off ratio curves of the molecular switch with two forms

The switching characteristics of this molecule can be understood from the energy dependence of zero-bias transmission spectra and projected density of states (PDOS) for two forms, which are shown in Fig. 3. 14. 4 and Fig. 3. 14. 5, respectively. The short vertical bars near the energy axes stand for the energy levels of the extended molecule. The characters H and L represent the highest occupied molecular orbital (HOMO) and the lowest unoccupied molecular orbital (LUMO), respectively. In our calculation, the average Fermi level, which is the average value of the chemical potential of the left and right electrodes, is set as zero. A approximation with the range of the "voltage window", i. e. , $[-V/2, +V/2]$ is enough to analyze a finite part of the transmission spectrum. Furthermore, it is well known that the PDOS represents the discrete energy levels of the isolate molecule including the effects of energy shift and line broadening due to the molecule-electrodes coupling, i. e. , the coupling between molecular orbitals and the incident states from the electrodes.

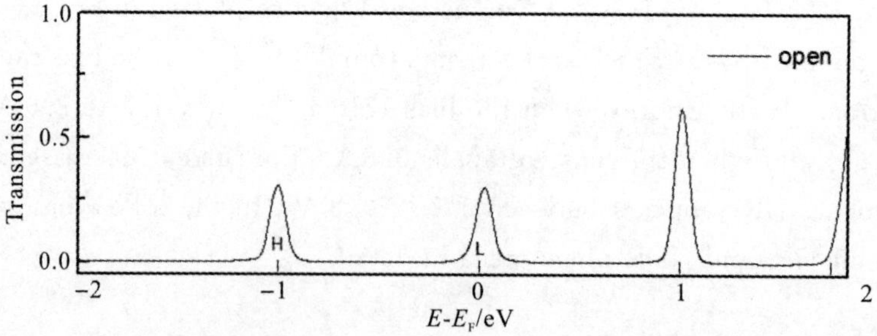

Fig. 3. 14. 4 The transmission spectra of the molecular switch at zero bias

(The short vertical bars near the enery axes stand for the energy levels of the extend molecule)

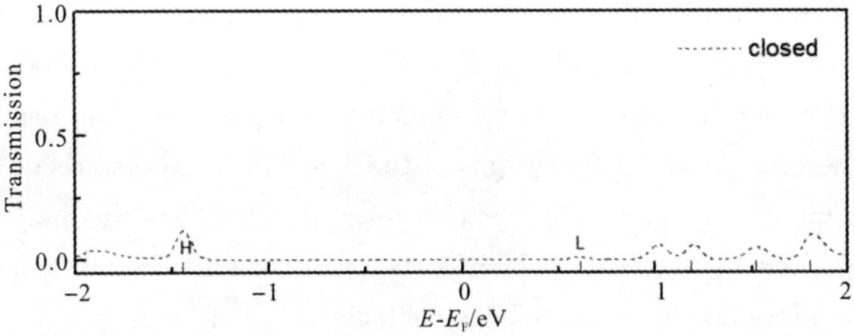

Continued Fig. 3. 14. 4 The transmission spectra of the molecular switch at zero bias

(The short vertical bars near the enery axes stand for the energy levels of the extend molecule)

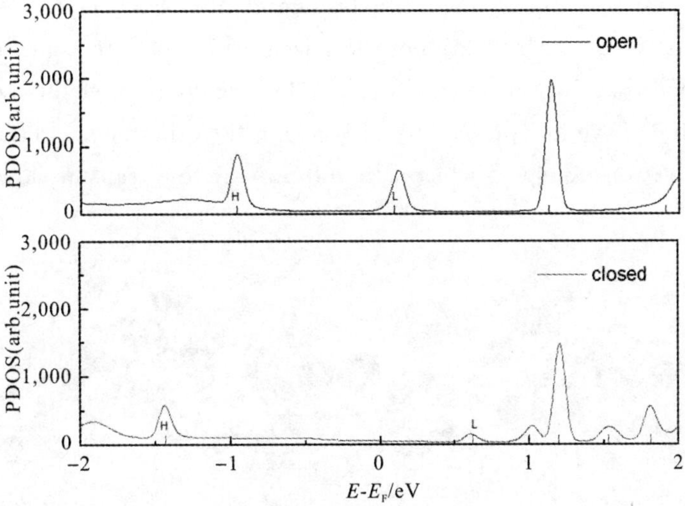

Fig. 3. 14. 5 The PDOS, i. e. , the DOS of the molecular switch projected onto the molecule subspace at zero bias

As shown in Fig. 3. 14. 4 and Fig. 3. 14. 5, the transmission is strongly correlated to PDOS spectra, which are shown the similar qualitative features, especially in the location of their peaks. It is clear from Fig. 3. 14. 4 that the transmission spectra of two forms display extraordinarily different characteristics. When the system with open-ring form, the value of the transmission coefficient is quite small in the energy region of [−2. 0 eV, 2. 0 eV]. Meanwhile, the HOMO and LUMO levels are −1. 13 eV and 0. 09 eV for the closed-ring form, −1. 43 eV and 0. 61 eV for the open-ring form, respectively. Therefore, the HOMO-LUMO gap of the closed-ring form is smaller than that of the open-ring form. In such an open system, the Fermi level E_F of the Au electrode aligns between HOMO and LUMO, and, therefore, the barrier for the electron transport is intensively relevant to the HOMO-LUMO gap. The results show that the HOMO-LUMO gap increases from 1. 22 to 2. 04 eV when the closed-ring form is switched to the open-ring form. As a result, the lack of any significant peak in the region of [−2.0 eV, 2.0 eV] and the lager HOMO-LUMO gap due to the molecular structure changes accounts for the low conductivity in the open-ring form.

To get a further insight about the origins of the peaks in the transmission spectra and of the different transmission characteristics for two forms, we analyze the molecular projected self-consistent Hamiltonian (MPSH), which is the self-consistent Hamiltonian of the isolated molecule in the presence of the electrodes, namely the molecular part is extracted from the whole self-consistent Hamiltonian at the contact region(see Fig. 3.14.6). It is found that both the HOMO and the LUMO are delocalized orbitals which provide the main electronic transport channel for the closed-ring form leading to low barrier for electron transport. However, the HOMO and LUMO are both localized orbitals for the open-ring form, which cannot provide good transport channels, because electrons that enter the molecule at the energy of these orbitals have low probability of reaching the other end. As a result, it leads to high barrier for electron transport which accounts for the low transmission strength in Fig. 3.14.4.

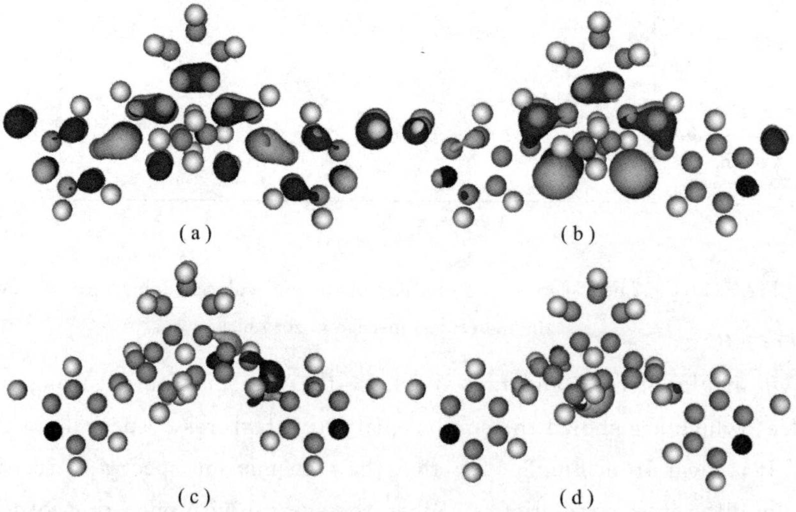

Fig. 3.14.6 The spatial distribution of the MPSH states corresponding to
(a) HOMO of closed-ring form; (b) LUMO of closed-ring form;
(c) HOMO of open-ring form; (d) LUMO of open-ring form

Another interesting feature in Fig. 3.14.3 is the NDR behavior. To investigate the mechanism responsible for NDR, we analyze the transmission coefficient spectrums of closed-ring form for 0.7 V, 0.8 V, 0.9 V and 1.0 V respectively, as shown in Fig. 3.14.7. Each transmission coefficient spectrum has four transmission peaks at the energy region [−2.0

eV,2.0 eV]: one transmission peak is within the bias energy window, and the other three peaks are out of the bias energy window. That means that there are four scattering channels, but only one of the channels makes its contribution to current. From Fig. 3.14.7, we can also see that the transmission peak for 0.8 V is much lower than the transmission peak for 0.7 V, 0.9 V and 1.0 V. As a result, the current of 0.8 V is decreased and NDR behavior appears.

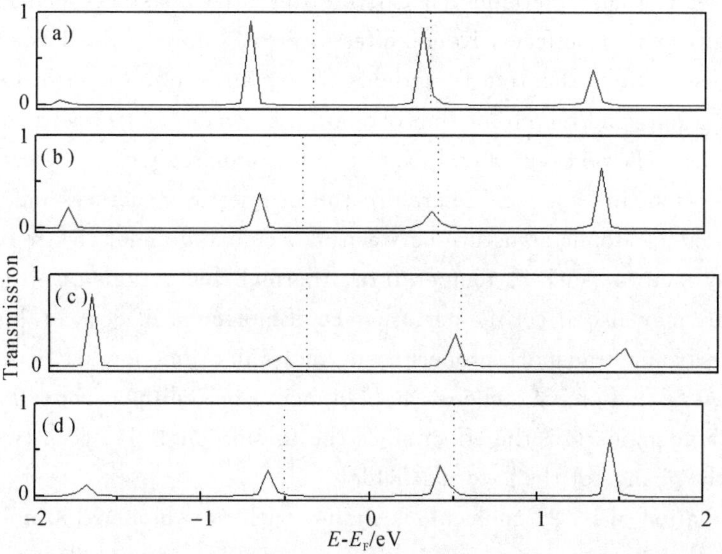

Fig. 3.14.7 The transmission spectra of the system with the closed-ring form under biases (The region between the two dot lines indicates the energy window)
(a) 0.7 V; (b) 0.8 V; (c) 0.9 V; (d) 1.0V

As a summary, the electronic transport properties of thepyridine-substituted dithienylethene optical molecular switch have been investigated by using the DFT + NEGF first-principles method. The electronic transmission coefficients, partial density of states, and *I-V* characteristics corresponding to different forms are calculated and analyzed. The dramatic difference in conductivity appearing in two different forms can be observed. The molecule that comprises the switch shares the following features: ① the two forms of this molecule can keep stable over a wider temperature range and reversibly switch from each other, when it was attached to the Au electrodes; ② there is a large current ratio in a wider bias window, these of which make it usable as one of the good candidates for light-driven molecular switches.

3.15 The field-induced current-switch by dithiocarboxylate anchoring group in molecular junction

In recent years, with the advancement of techniques for manipulating individual molecules, the electronic transport properties through molecular devices have attracted more and more attention both for their novel physical properties and potential for device application, such as single-electron characteristics, negative differential resistance (NDR), electrostatic current switching, memory effects, Kondo effects, etc. Following the development in experiments, people use various theoretical methods to explain the mechanism for molecular devices during operation, and search for the correlation between the geometric structures and electronic properties. However, in a real experiment, compared with the electrode, the molecule is a small system in the size. Therefore the geometric structures including molecular conformation, contact atomic structure between molecule and metal electrodes, etc., can be affected by many factors, such as temperature, thermal fluctuations or electric field and so on. These factors may also affect the performance of molecular device. In this article, we investigate the electronic transport properties of molecular junction with the electric field-induced geometry relaxation. We choose 4,4′-biphenyl bis (dithiocarboxylate)(BDCT) molecular junction to demonstrate the effect since the torsion angle between two phenyls could be sensitive to the change of the external field.

The conformation of BDCT molecule is similar with 4,4′-biphenyldithiolate(BDT)molecule except only the anchoring groups are different. BDT has the standard thiol groups, and BDCT is connected to Au surface through dithiocarboxylate groups. Tivanski et al. experimentally find that the molecular conductance can be enhanced with dithiocarboxylate group connection. Recently, Li et al. theoretically approve this conductance enhancement induced by dithiocarboxylate anchoring groups on investigating the coplanar BDCT and BDT molecular junctions. However, Li et al. have not considered the effect of the field-induced geometry relaxation of the molecule on the electronic transport properties in molecular junction. In this work, we carry out a theoretical simulation that combines both first-principles DFT and non-equilibrium Green's function (NEGF) formalism to discuss the electronic transport properties of a BDCT molecular junction in the presence of an external electric field. We have optimized the geometrical structure of the molecule under the external voltages, and calculated the *I-V* characteristics inclusion of electric field-induced geometry relaxation.

The structure of free molecule is first optimized. Based on the experimental study, the H atoms are dissociated upon adsorption to metal surfaces. Then the optimized molecules are adsorbed between two Au(111) electrodes bythe sulfur atoms are located above the hollow site of the Au triangle. In our calculations, the torsion angle of 37° is the equilibrium structure of optimized molecular junction, as illustrated in Fig. 3.15.1.

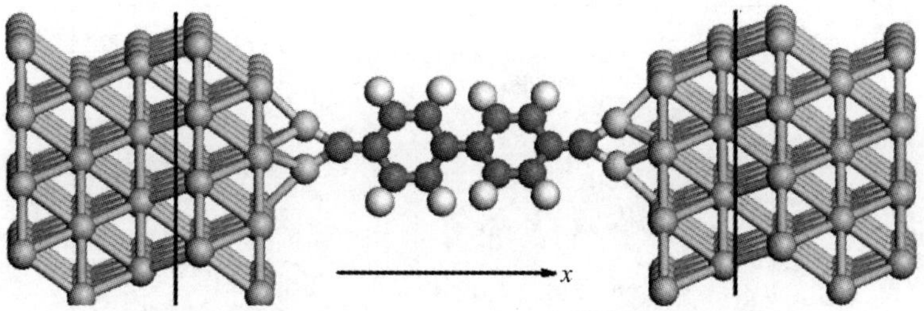

Fig. 3. 15. 1 Schematic view of BDCT molecular junction

Firstly when the voltage is turned on, there is an electric field along the molecular axis acts on the molecule (labeled as x direction in Fig. 3. 15. 1). We optimize the molecular geometrical structure under the external voltages, the bond distances between the sulfur atom and the Au electrodes and the torsion angles between two phenyl rings become field-depended (see Fig. 3. 15. 2). It can be clearly seen that the absolute change of the bond distances between the sulfur atom and the Au electrodes is very small, which is less than 0. 01 nm at the range of −2. 1 to 2. 1 V. But the bond distances between the sulfur atom and the right Au electrode is lengthened under the positive electric field, on the contrary, for the left electrode it is shortened. A reversed behavior can be found under the negative electric field. However, the change of the torsion angle is quite different from that of the bonding distances. It is noted that there is a vibration around 37° and 45° at the range of −1. 2 to 1. 2 V and 1. 5 to 2. 1 V(−2. 1 to −1. 5 V) respectively. However the vibration of torsion angle is very small, which is less than 1 degree at these three ranges(−1. 2 to 1. 2 V ,1. 5 to 2. 1 V and −2. 1 to −1. 5 V)(see Fig. 3. 15. 2(b)).

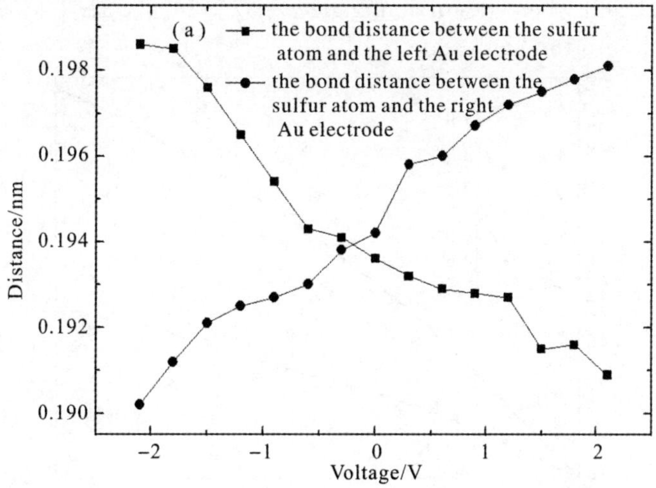

Fig. 3. 15. 2 The change of the bond distance and torsion angle
(a) the bond distance between the sulfur atom and the Au electrode on the left (square) and on the right (circle)

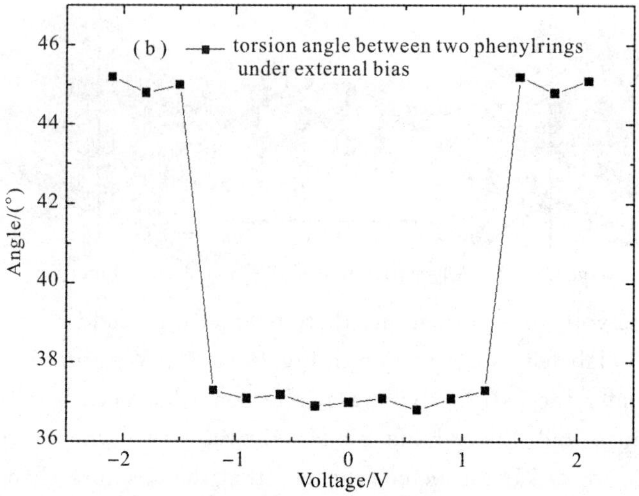

Continued Fig. 3. 15. 2 The change of the bond distance and torsion angle
(b) torsion angle between two phenyl rings under external bias

The calculated *I*-*V* characteristics of molecular junction with different torsion angles in the range of [0 V, 2. 1 V] are plotted in Fig. 3. 15. 3. The *I*-*V* curves appear quite different, though there is just 8° difference between 37° and 45°. In the figure there are eight current curves corresponding to torsion angle α = 37°, 36. 9°, 37. 2°, 37. 1°, 37. 3°, 45°, 44. 8° and 45. 2°. We can find that the molecular device presents an active switch function. The current can be controlled due to different molecular conformations induced by torsion angle in adjusting the gate bias. When the torsion angle α = 45°, 44. 8° and 45. 2°, it will cause the switch into the "off" state, and when the torsion angle α = 37°, 36. 9°, 37. 2°, 37. 1° and 37. 3°, it will cause the switch into the "on" state.

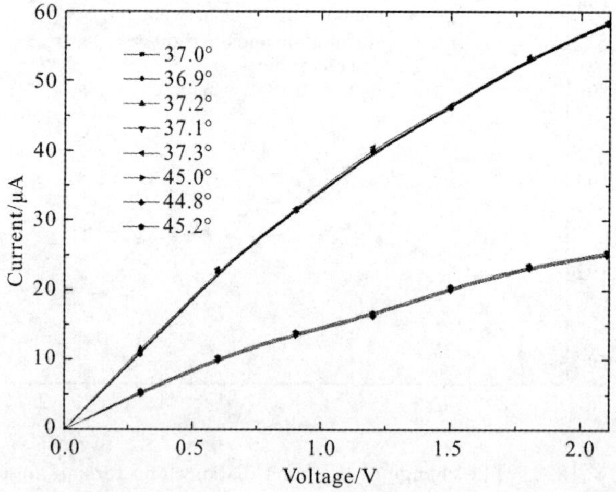

Fig. 3. 15. 3 The *I*-*V* characteristics of BDCT molecule with different torsion angles

The current in the system is calculated by the Landauer-Bütiker formula $I = G_0\int n(E)T(E,V)\mathrm{d}E$, which is transmission spectra dependent. It is well known that eigenstates of the whole system consist of scattering states, which are molecular-orbital-like in the molecule, and Bloch-wave-like in the metal electrodes. If an orbital is delocalized across the molecule, an electron that enters the molecule at the energy of the orbital has a high probability of reaching the other end, and there is a high corresponding peak in the transmission. The partial density of states (PDOS) is strongly correlated with transmission, which represents the discrete energy levels of the isolate molecule including the effects of energy shift and line broadening due to the molecule-electrode coupling, and it shows the same qualitative features as the peaks and valleys in transmission. Thus, the switching characteristics of the BDCT system can be understood from the transmission spectra and projected density of states (PDOS), which are shown in Fig. 3. 15. 4(a) and (b). In our calculation, the average Fermi level, which is the average value of the chemical potential of the left and right electrodes, is set as zero. As shown in Fig. 3. 15. 4, in order to get more clearly figures, we only give the transmission, PDOS and the molecular projected self-consistent Hamiltonian (MPSH) of $\alpha=37°$ and 45° here. Because, the torsion angle of phenyl rings by $\alpha=36.9°$, 37.2°, 37.1° and 37.3° or $\alpha=44.8°$ and 45.2° have negligible influence on the transmission spectra of $\alpha=37°$ or 45°. Therefore, an approximation with $\alpha=37°$ and 45° is enough to analyze relationship between the transmission spectrum and current. It is clearly seen that the transmission is strongly correlated to PDOS spectra, which are shown the similar qualitative features, especially in the location of their peaks. The number of peaks is almost the same for these two molecular conformations, but the peak value is different. With the increase in torsion angle, the peak of $\alpha=45°$ near the Fermi Level becomes smaller. Furthermore, the molecular projected self-consistent Hamiltonian (MPSH) of BDCT with two conformations is analyzed. MPSH is the self-consistent Hamiltonian of the isolated molecule in the presence of the electrodes, namely the molecular part is extracted from the whole self-consistent Hamiltonian at the contact region. It contains the molecule-electrode coupling effects because during the self-consistent iteration, the electron density is for the contact region as a whole. To get a further insight about the change in current of BDCT molecule with two conformations, MPSH of system is analyzed. The spatial distribution of the MPSH states corresponding to the highest occupied molecular orbital (HOMO) and the lowest unoccupied molecular orbital (LUMO) are presented in Fig. 3. 15. 4(c). It can been found that the HOMO and LUMO are both delocalized states for $\alpha=37°$, and the LUMO become localization when the torsion angle increase to 45°. Furthermore, the density on two sulfur atoms of LUMO that denotes the molecule-electrode coupling strengths is decreased with the change of torsion angle from 37° to 45°, which corresponding to the decrease of PDOS in Fig. 3. 15. 4(b). The decrease in the overlap between π electron clouds and molecule-electrode coupling strengths is the mainly reasons for current reduction. It can be found that the HOMO and LUMO are both delocalized state for 37°("on" state) and it become localized on one side of the molecule when the torsion angle approaches to 45°("off "state).

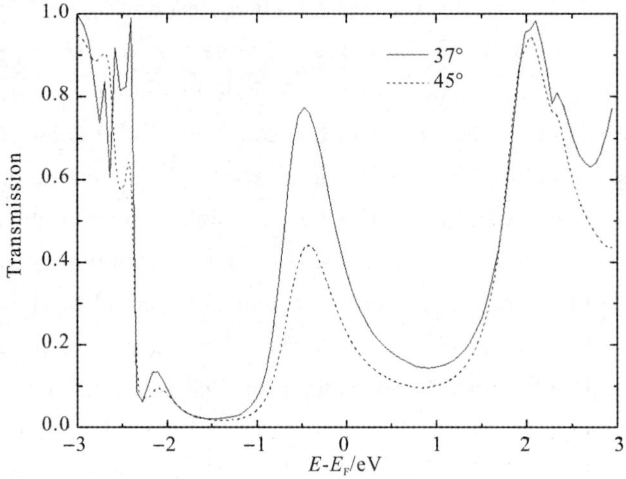

(a)

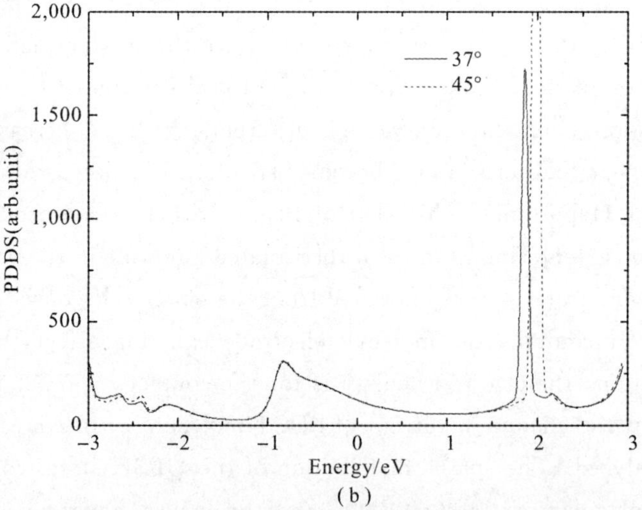

(b)

Fig. 3. 15. 4 The Transmission , PDOS and MPSH

(a) the Transmission coefficient; (b) partial density of states (PDOS)

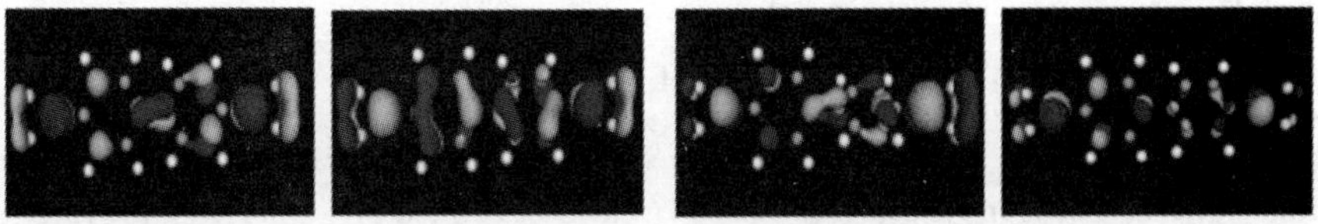

(c)

Continued Fig. 3.15.4 The Transmission, PDOS and MPSH
(c) molecular projected self-consistent Hamiltonian (MPSH) for
BDCT with $\alpha = 37°$ and 45° at zero bias voltage

We will now show the changes of the transmission function with different optimizing molecular conformations under various biases in a three-dimensional plot (see Fig. 3.15.4). The current is determined by $T(E, V_b)$ in the voltage window and is further only determined by the transmission regions in the bias window because $T(E, V_b)$ is zero in the transmission interval and has no contribution to the current. It can be seen from Fig. 3.15.4 that with the external bias increasing form 0 to 2.1 V, no additional transmission region will be include into the bias window. But the transmission near the Fermi Level becomes smaller.

In summary, the effect of field-induced geometry relaxation on the electronic transport properties of BDCT molecular device has been numerically simulated using the DFT+NEGF first-principles method. The results show that the BDCT can present a distinct switch function under the external electric field exists. The mechanism of this conformational switch is the strong intermolecular coupling between the two phenyl rings that lead to the localization of electron in the molecules. The results will be helpful to understand further the possible situation in experiments and design molecular electronic devices with specific properties.

3.16 The *I-V* characteristics of the butadienimine-based optical molecular switch

With the physical limitations of today's semiconductor-based micoelectronics fabrication in sight, molecular electronic devices are considered as appropriate candidates for nanometer electronics. During the past ten years, with the advantages of experimental techniques for characterizing and manipulating individual molecules, such as scanning tunneling microscopy (STM), mechanically controllable break junction, atomic force microscope, creative fabricated techniques, and so on, many molecular devices with different functionalities have been designed and measured. A lot of novel physical properties, such as negative differential resistance (NDR), memory effects, switching properties and so on, are found in various systems including organics, carbon nanotubes, DNA, etc. Meanwhile, the development of theoretical work has helped to understand the underling working principle of molecular devices.

Among these devices, molecular switch has been attracting great attention, because they are considered to play an essential role in any modern design of logic and memory circuits. Molecular switch devices can be mainly divided into two categories, including electronic switch devices which control the conversion between on and off states by an external trigger such as the electric field, the tip of scanning tunneling microscopy (STM), and the redox process, etc, and optical switch devices which exist in two thermally sufficiently stable states with a high conductance (on state) and a low conductance (off state) by light. Among them, optical switch has been widely studied, because light is a very attractive external stimulus for such switches including short response time, the ease of addressability and compatibility with a wide range of condensed phases. In this regard, diarylethene- and azobenzene-based photochromic switch has been extensively studied. The mechanisms for these two optical switches are mainly based on light-induced conformational changes, namely ring-opening reactions and isomerization reactions of the molecular bridge.

Recently, Benesch et al. have reported another mechanism which is based on photoinduced excited state hydrogen transfer in the molecular bridge. In contrast to most other mechanisms, hydrogen translocation within the molecular bridge has the advantage that the overall length and thus the molecule-electrode binding geometry of the junction is not changed significantly, as shown in Fig. 3. 16. 1. These two forms, namely the oxo-amine(keto) and hydroxyl-imine(enol) form, can keep stable over a wider temperature range and reversibly switch from each other, which make it usable as one of the good candidates for light-driven molecular switches and may have some future applications in the molecular circuit. Therefore, we use nonequilibrium Green's function formalism combined first-principles calculations to investigate the electronic transport properties of this optical molecular switch, which has been functionalized by replacing a hydrogen atom with a sulfur atom as the linker at both end of the molecular wire to ensure good contact with Au electrodes. In this work, we extend the previous work of Benesch's and consider a possible molecular conformation with R = benzene which with and without donor/acceptor substituent, since the electron transport through the molecular device can be controlled by the substituent ligand. The influence of the HOMO-LUMO gaps and the spatial distributions of molecular orbitals on the quantum transport through the molecular device are discussed in detail. The calculated results show that donor/acceptor substituent plays an important role in the electronic transport of molecular devices. The paper is organized as follows: First, the model and computational method are described. Then the optical molecular switch and switch with donor/acceptor substituent are studied. Finally, the conclusion is given.

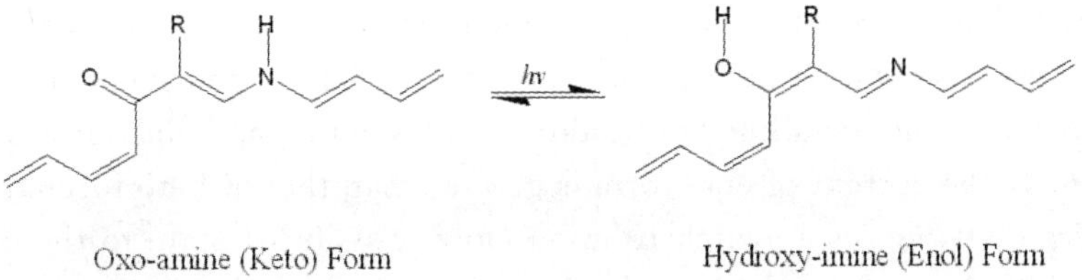

Fig. 3. 16. 1 Conversion between keto and enol form of molecule

First, the structure of freethiol(SH) capped molecule with the above-mentioned two forms was first optimized. Based on experimental issues about self-assembled monolayers (SAMs), it is generally accepted that hydrogen atoms are dissociated upon adsorption to metal surfaces. So we construct a two-probe system, which the two terminal hydrogen atoms bonded to the sulfur atom are eliminated from the optimized structure, and the remained part is sandwiched between two parallel Au(111) surfaces that correspond to the surfaces of the Au electrodes. The perpendicular distance between the Au surface and sulfur atom is set to 1.9 Å, which is a typical Au-S distance. The contact region includes extended molecule and two layers of Au from each electrode (see Fig. 3. 16. 2). The contact region contains parts of the electrodes include the screening effects in the calculations. The semi-infinite electrodes are calculated separately to obtain the bulk self-energy. The structure of the Au(111)-molecule-Au(111) molecular junction has not been relaxed due to the limitation of computational resources. Since the switching unit is far from the Au electrode region, we believe that the effect of geometrical optimization is minor.

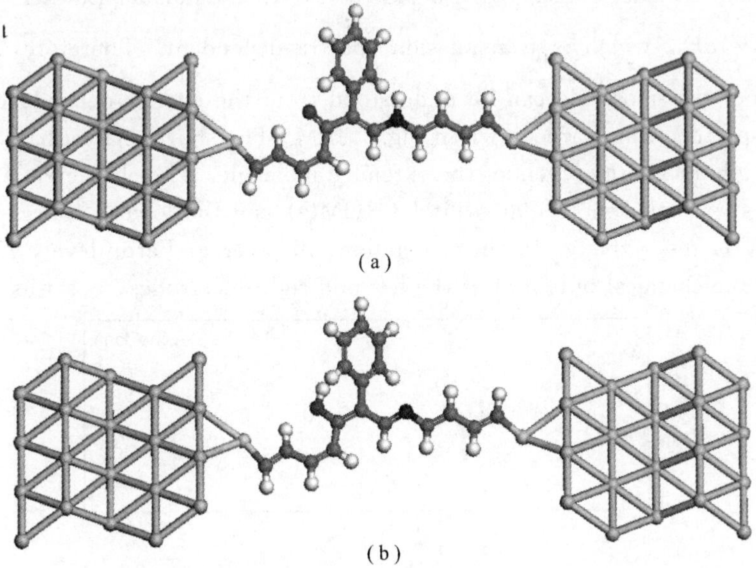

Fig. 3. 16. 2 Geometry for the molecular switch with the hollow site

(a) keto form; (b) enol form

Fig. 3. 16. 3 shows the calculated *I-V* characteristics of the system with two forms at a bias up to 1. 2 V. It should be pointed out that at each bias, the current is determined self-consistently under the non-equilibrium condition. The switching behavior can be clearly seen from Fig. 3. 16. 3, the current of enol form is greater than that of keto form at the same bias. One can predict that there is a switch from on (low resistance) state to the off (high resistance) state, when the enol form molecule changes to the keto form.

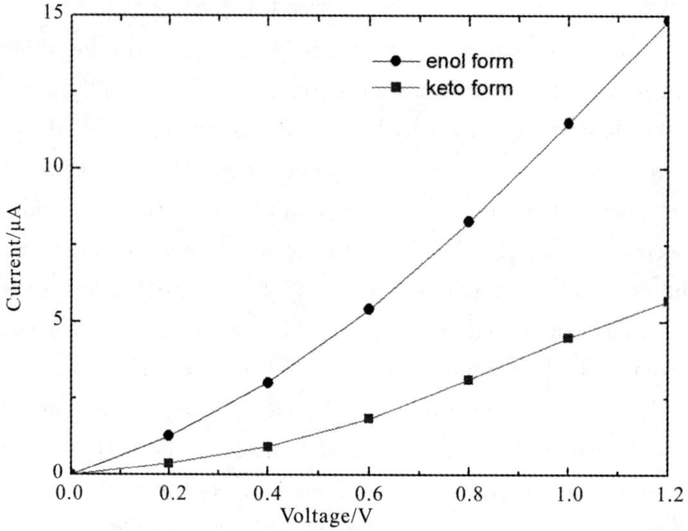

Fig. 3. 16. 3 The *I-V* characteristics of the molecular switch with two forms

The current in the system is calculated by the Landauer-Bütiker formula $I = G_0\int n(E)T(E,V)\mathrm{d}E$, which is transmission spectra dependent. Therefore, the switching characteristics of this molecule can be understood from the energy dependence of zero-bias transmission spectra, which are shown in Fig. 3. 16. 4. The short vertical bars near the energy axes stand for the energy levels of the extended molecule. The characters H and L represent the highest occupied molecular orbital (HOMO) and the lowest unoccupied molecular orbital (LUMO), respectively. In our calculation, the average Fermi level, which is the average value of the chemical potential of the left and right electrodes, is set as zero.

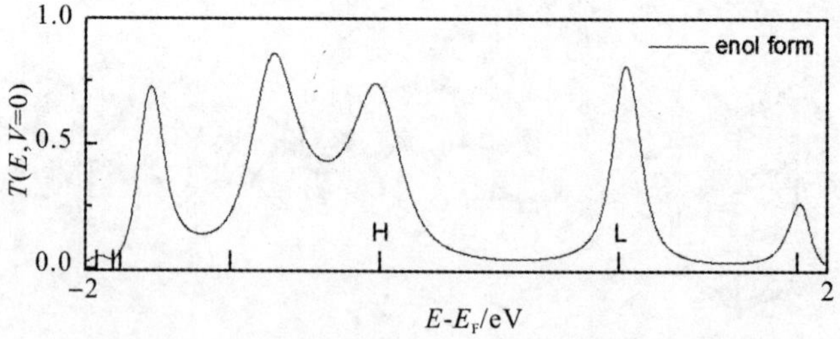

Fig. 3. 16. 4 The transmission spectra of the molecular switch at zero bias

(The short vertical bars near the enery axes stand for the energy levels of the extend molecule)

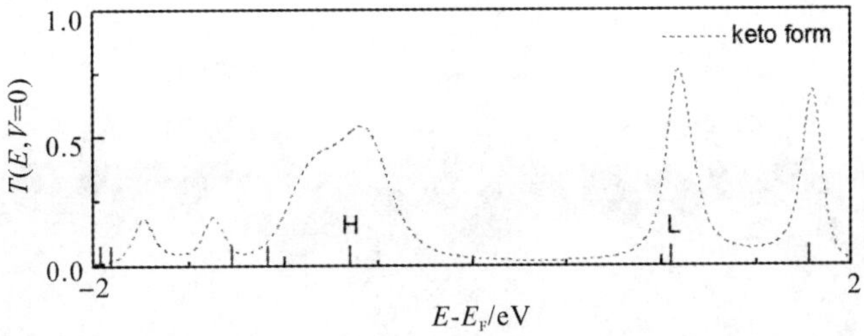

Continued Fig. 3. 16. 4 The transmission spectra of the molecular switch at zero bias

(The short vertical bars near the enery axes stand for the energy levels of the extend molecule)

It is clear from Fig. 3. 16. 4 that the transmission spectra of two forms display extraordinarily different characteristics. When the system with keto form, the value of the transmission coefficient is small than that of enol form in the energy region of [−2.0 eV, 2.0 eV]. Meanwhile, the HOMO and LUMO levels are −0.41 eV and 0.89 eV for the enol form, −0.64 eV and 1.05 eV for the keto form, respectively. Therefore, the HOMO-LUMO gap of the enol form is smaller than that of the keto form. In such an open system, the Fermi level E_F of the Au electrode aligns between HOMO and LUMO, and, therefore, the barrier for the electron transport is intensively relevant to the HOMO-LUMO gap. The results show that the HOMO-LUMO gap increases from 1.30 to 1.69 eV when the enol form is switched to the keto form. As a result, the lack of any significant peak in the region of [−2.0 eV, 2.0 eV] and the lager HOMO-LUMO gap due to the molecular structure changes accounts for the low conductivity in the keto form.

To get a further insight about the origins of the peaks in the transmission spectra and of the different transmission characteristics for two forms, we analyze the molecular projected self-consistent Hamiltonian (MPSH) and the partial density of states (PDOS) projected on the molecular junction, as shown in Fig. 3. 16. 5 and Fig. 3. 16. 6, respectively. MPSH is the self-consistent Hamiltonian of the isolated molecule in the presence of the electrodes, namely the molecular part is extracted from the whole self-consistent Hamiltonian at the contact region. It is found from Fig. 3. 16. 5 that both the HOMO and the LUMO are delocalized orbitals which provide the main electronic transport channel for the enol form leading to low barrier for electron transport. However, the HOMO and LUMO are both localized orbitals for the keto form, which cannot provide good transport channels, because electrons that enter the molecule at the energy of these orbitals have low probability of reaching the other end. As a result, it leads to high barrier for electron transport which accounts for the low transmission strength in Fig. 3. 16. 4.

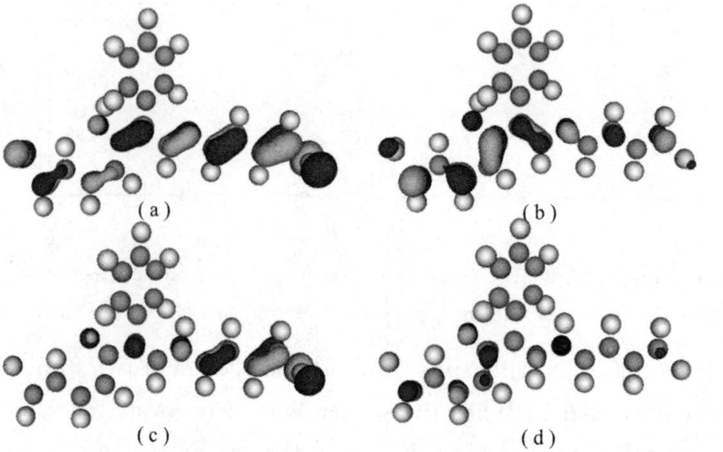

Fig. 3. 16. 5 The spatial distribution of the MPSH states corresponding to (a) HOMO of enol from; (b) LUMO of enol form; (c) HOMO of keto form; (d) LUMO of keto form

Furthermore, it is well known that the PDOS represents the discrete energy levels of the isolate molecule including the effects of energy shift and line broadening due to the molecule-electrodes coupling, i. e. , the coupling between molecularorbitals and the incident states from the electrodes. Therefore, the PDOS will give us information on how strongly the molecule couples with the electrodes at a certain energy E. As shown in Fig. 3. 16. 4 and Fig. 3. 16. 6, the transmission is strongly correlated to PDOS spectra, which are shown the similar qualitative features, especially in the location of their peaks where the PDOS takes a comparatively large value. A strong coupling makes incident electrons at a certain energy easily transmit across the molecule which will give rise to a large transmission coefficient at this energy. By comparison of transmission and PDOS spectra, it can be found that a large transmission coefficient indicates a strong coupling between the electrodes and the molecule. Furthermore, the evolution of transmission curves with external biases can help us understand how the changes of the coupling between the electrodes and molecule determines the I-V characteristics in the system. Thus, the voltage dependence of transmission function will be studied next.

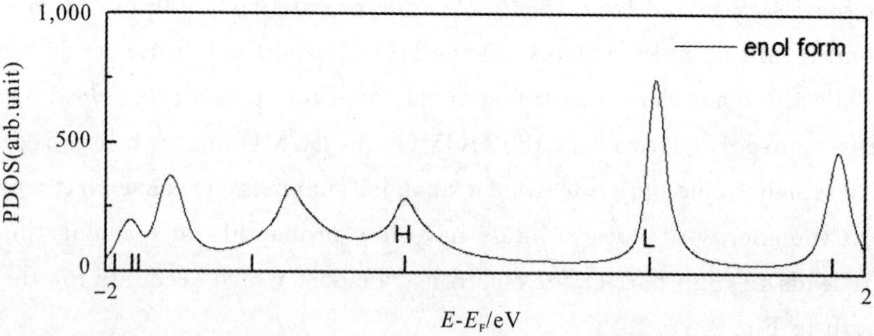

Fig. 3. 16. 6 The PDOS, i. e. , the DOS of the molecular switch projected onto the molecule subspace at zero bias

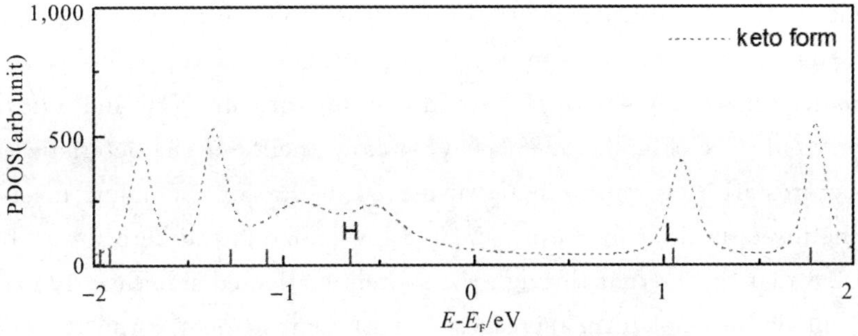

Continued Fig. 3. 16. 6 The PDOS, i. e. , the DOS of the molecular switch projected onto the molecule subspace at zero bias

The bias dependence of the transmission characteristics, $T(E,V)$, at biases of 0. 2 V, 0. 4 V, 0. 6 V, 0. 8 V, 1. 0 V and 1. 2 V of enol and keto form are presented in Fig. 3. 16. 7. As shown in Fig. 3. 16. 7, we can see that the transmission peaks are close to the Fermi level E_F upon increasing the applied bias, while more part of HOMO transmission peak enters the bias window gradually in all cases, i. e. , the integral area in the bias window gets bigger with the increase of bias. Thus, the current through the switch increases. However, the integral area of enol form is always greater than that of keto form since the peaks in the transmission spectra of the former is larger than that of the latter. Furthermore, the HOMO of enol form is more near the Fermi level E_F than that of keto form. As a result, the current of enol form is always greater than that of keto form in the presented bias window.

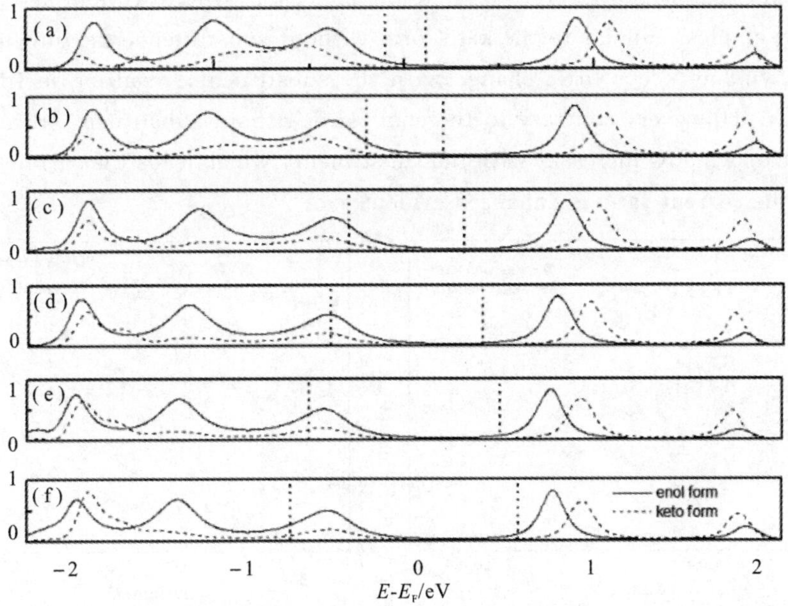

Fig. 3. 16. 7 The transmission spectra of the molecular switch under biases (a) 0. 2 V; (b) 0. 4 V; (c)0. 6 V; (d)0. 8 V; (e)1. 0 V; (f) 1. 2 V

Furthermore, the electron transport through the molecular device can be controlled by the substituentligand, in order to improve the performance of the switch, we compare the *I-V* characteristics for switches with electron donor substituent -NH_2 and -OH, and electron acceptor substituent -CN. Fig. 3. 16. 8 (a)-(c) show their *I-V* characteristic curves. Their geometry structures after optimization are displayed in the corresponding insets. The upper structure of the insets is the enol form, and the lower one is the keto form. It can be seen from Fig. 3. 16. 8 that the current through the switch is affected significantly by the substituent. The current of the enol form increases with the substituent. And the current of enol form is always greater than that of keto form no matter what the substituent is. When enol form in the switch changes to keto form under photoexcitation, the molecular switch is predicted to change from the on state to the off state. To characterize the conduction change due to photoexcitation, we define the current ratio, Ratio $= I_{\text{enol-form}}/I_{\text{keto-form}}$, where $I_{\text{enol-form}}$ and $I_{\text{keto-form}}$ are the current of enol and keto form, respectively. The current ratios vs. bias curves of these three cases are given in Fig. 3. 16. 8(d). From Fig. 3. 16. 8(d), one can see that the maximum current ratio is about 5. 3 at 0. 6 V for the switch with the substituent -CN which suggests that the switching performance can be improved to some extent. Figs. 3. 16. 9(a_1)-3. 16. 9(d_2) present the mechanism that the transmission function determines the current behaviors in Fig. 3. 16. 3 and Fig. 3. 16. 8 at 0. 6 V. The regions between the two vertical dash lines indicate the bias window. Here we take the molecular switch without and with substituent -CN as examples. Similar to the keto form without substituent, the integral area entering the bias window keeps little change with the substituent, resulting in the current to change slightly. However, contrary to the enol form without substituent, the integral area entering the bias window increases with the substituent, which leads the current to increase. As a result, the current ratio is enhanced evidently.

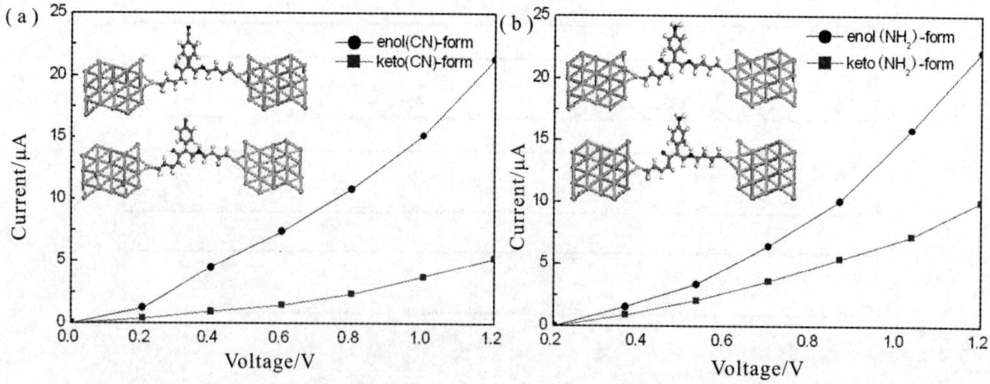

Fig. 3. 16. 8 The calculated *I-V* characteristics of the molecular switch with substituent (a) with acceptor substituent -CN; (b) with donor substituent -NH_2

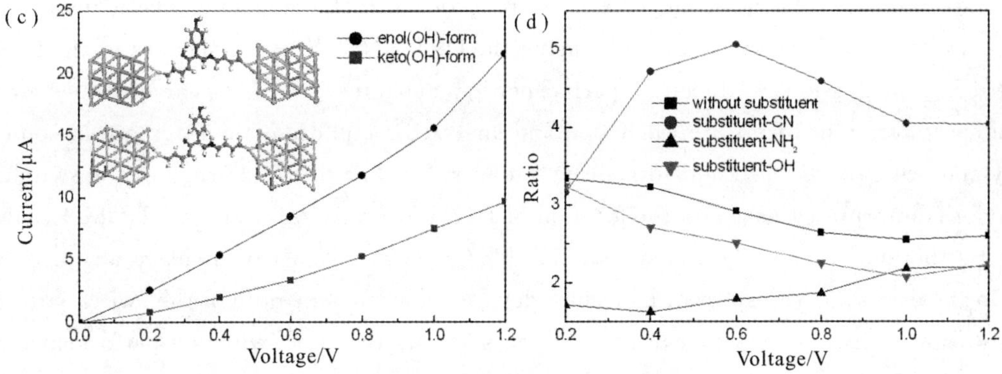

Continued Fig. 3. 16. 8 The calculated *I-V* characteristics of the molecular switch with substituent (c) with acceptor substituent -OH; (d) the calculated current ratio of the molecular switches with different substituents at different biases

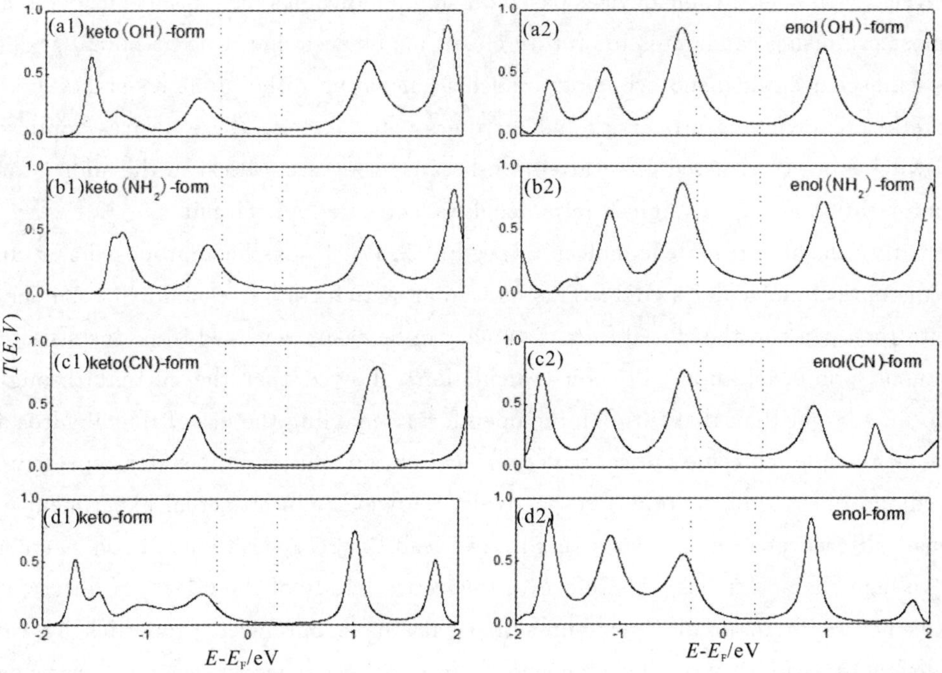

Fig. 3. 16. 9 The calculated transmission characteristics of the molecular switch with and without substituents at 0. 6 V(The regions between the two vertical dash lines indicate the bias window). The energy origin is set to be the Fermi level of the system

(a_1) (a_2) keto-form and enol-form with acceptor substituent -OH;

(b_1) (b_2)keto-form and enol-form with donor substituent $-NH_2$;

(c_1) (c_2) keto-form and enol-form with electron acceptor substituent -CN ;

(d_1) (d_2)keto-form and enol-form without substituent

As a summary, the electronic transport properties of thebutadienimine-based optical molecular switch have been investigated by using the DFT+NEGF first-principles method. The electronic transmission coefficients, partial density of states, and *I-V* characteristics corresponding to different forms are calculated and analyzed. The dramatic difference in conductivity appearing in two different forms can be observed. The physical origin of the switching behavior is interpreted based on the location of HOMO and LUMO, and the HOMO-LUMO gap. Furthermore, it can be found that the donor/acceptor substituent plays an important role in the electronic transport of molecular devices. The current ratio of the switch with the substituent - CN is enhanced evidently which suggests that the switching performance can be improved to some extent.

3.17 Effect of chemical modifications on the electron transport properties

In recent years, electronic devices based on single molecules have been considered as one of the most promising candidates for future electronic devices due to their potential applications in nanoscience and nanotechnology, such as negative differential resistance (NDR), memory effects, switching properties and rectifiers, etc. Among these devices, optical molecular switches have attracted great attention because they are critical to the ability to store digital information and route signals in molecular electronic logic circuits.

Recently, thedihydroazulene molecule (see Fig. 3.17. 1) has been proposed as a component of the optical molecular switch. This molecular switch comprises a dihydroazulene molecule with the open and the closed forms, which can be easily reversed from one structure to another upon photoexcitation. Previous calculations showed that the current through the closed form is larger than that through the open form, enabling the use of the dihydroazulene molecule as a single-molecule optical molecular switch with on and off states represented by the closed and open conformations, respectively. However, in an actual experiment condition, small differences in molecular structure can lead to great variations in conductance and current-voltage characteristics. Therefore, a theoretical study of the effects of different substituent is necessary to predict structures that may have optimized properties by varying chemical structures for a particular application. In the present work, we consider a possible molecular conformation with R=H, CH_3, NH_2, NO_2 and OH, respectively and carry out non-equilibrium Green's function formalistic study under the first-principles density functional theory to investigate the effects of chemical modifications on the electron transport properties of the dihydroazulene molecule. The influence of the transmission coefficients and the spatial distributions of molecular orbitals on the quantum transport through the molecular device are discussed in detail. Theoretical results show that the chemical modifications play an important role in determining the switching behavior of such molecular device. The current-voltage characteristics can be manipulated with the careful selection of their substituents, and such modifications become crucial in optimizing the electron transport properties of chemical structures.

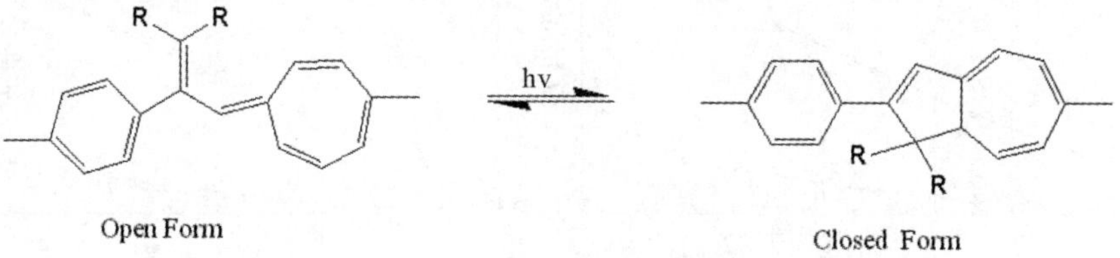

Fig. 3. 17. 1 Conversion between open and closed form of molecule

The pre-optimizeddihydroazulene molecule with two different forms are sandwiched between two parallel Au(111) surfaces that correspond to the surfaces of the Au electrodes, as shown in the insets of Fig. 3. 17. 1.

Fig. 3. 17. 2 shows the calculated *I*-*V* characteristics for the dihydroazulene molecule system with and without substituents. Their geometry structures after optimization are displayed in the corresponding insets. The upper structure of the insets is the closed form, and the lower one is the open form. It should be pointed out that at each bias, the current is determined self-consistently under the non-equilibrium condition. The switching behavior of the dihydroazulene molecule system can be clearly seen from Fig. 3. 17. 2, the current of closed form is always greater than that of open form at the same bias no matter what the substituent is. Furthermore, it also can be seen from Fig. 3. 17. 2 that the current through the switch is affected significantly by the substituent. The closed conformations of the CH_3 and NH_2 systems display current larger than that of the unsubstituted dihydroazulene system, and the open conformations of these two systems show an almost complete suppression of current. However, the OH system shows the opposite behavior. The current of closed conformation for OH system decrease in the entire bias range comparison to the unsubstituted system, and the open conformation display a current higher than that of the unsubstituted system. The NO_2 substitution displays a different *I*-*V* characteristic to that of these three systems. The closed conformation of the NO_2 system shows an enhancement of the current at bias voltages lower than 0. 8 V. Beyond 0. 8 V, the closed system begins to show a current decrease than that of the unsubstituted system. While the open conformation display a current higher than that of the unsubstituted system.

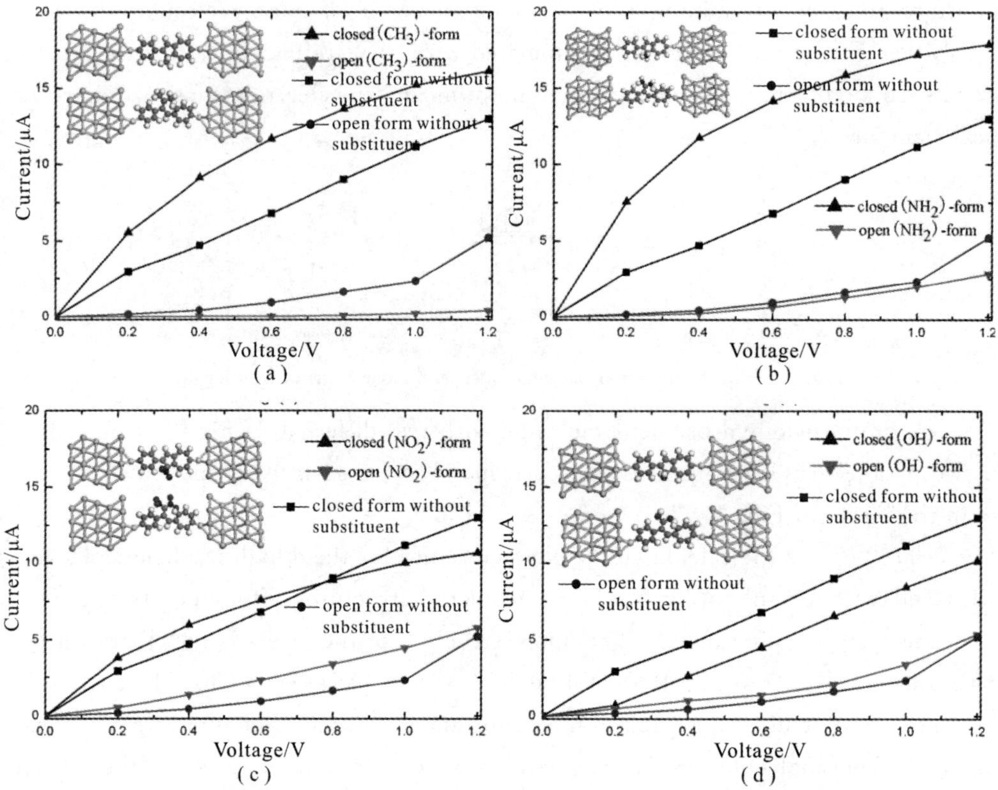

Fig. 3. 17. 2 The *I-V* characteristics for the molecular switch with and without substituents (The black lines represent the unsubstituted molecule of closed form, red lines represent the unsubstituted molecule of open form)

(a) with donor substituent -CH_3; (b) with donor substituent -NH_2;

(c) with acceptor substituent -NO_2; (d) with acceptor substituent -OH

Furthermore, to characterize the conduction change due to differentsubstituents, we define the current ratio, Ratio= $I_{closed\text{-}form}/I_{open\text{-}form}$, where $I_{closed\text{-}form}$ and $I_{open\text{-}form}$ are the current of closed and open form, respectively. As shown in Fig. 3. 17. 3, the maximum current ratio is about 32. 6 at 1. 0 V for the switch without the substituent. When the molecule switches are substituted by donor group CH_3 and NH_2, the maximum current ratio is changed to 105. 4 and 90. 0, respectively. However, when the molecule switches are substituted by acceptor group NO_2 and OH, the maximum current ratio is changed to 11. 0 and 3. 2, respectively. Although all substituents have different current ratio, in general, we observe that the electron-donating groups enhance the current ratio while the electron-accepting groups suppress it.

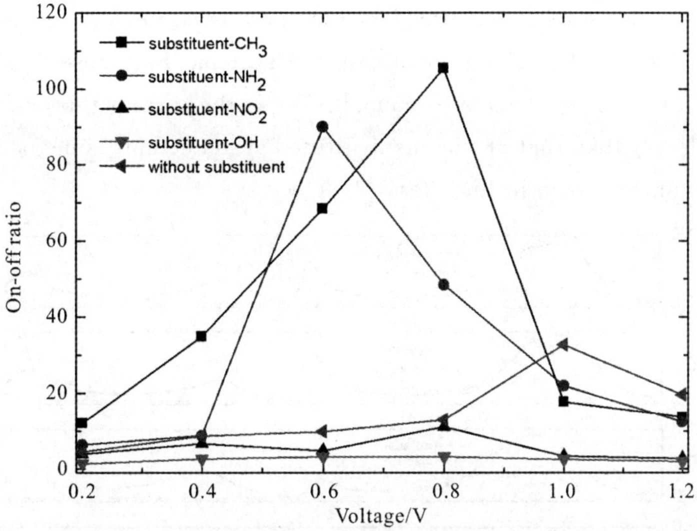

Fig. 3.17.3 The calculated current ratio of the molecular switches with and without different substituents at different biases

The current in the system is calculated by the Landauer-Bütiker formula $I = \frac{2e}{h}\int[f(E-\mu_L)-f(E-\mu_R)]T(E,V)\mathrm{d}E$, which is transmission spectra dependent. Therefore, the different switching characteristics of these molecules can be understood from the energy dependence of zero-bias transmission spectra, which are shown in Fig. 3.17.4. In our calculation, the average Fermi level, which is the average value of the chemical potential of the left and right electrodes, is set as zero.

It is clear from Fig. 3.17.4 that the transmission spectra of different substituents display extraordinarily different characteristics. From Fig. 3.17.4(a) and (b), it can be found that when the closed conformations are substituted by donor group CH_3 and NH_2, the value of the main peak near the Fermi level becomes larger than that of the unsubstituted conformation. However, for the NO_2 and OH systems, the value of the main peak near the Fermi level becomes smaller than that of the unsubstituted system. Especially, when the system is substituted by OH group, the peak near the Fermi energy nearly disappears. Therefore, comparison to the unsubstituted system, the currents of the NO_2 and OH system are decreased, and the currents of the CH_3 and NH_2 system are increased. On the contrary, the transmission spectra of open forms show the absolute different behavior comparison to the closed forms. For the open conformation the of the CH_3 and NH_2 systems, the main peak near the Fermi level is shifted toward the lower energy. While the same peak of the NO_2 system is moved to the higher energy. Moreover, for the OH system, the value of the peak near the Fermi level becomes smaller but broader, but there is no obvious change in the position of this peak compared to that of the unsubstituted system. For these reasons, the current through the CH_3 and NH_2 systems is very small. While the current through the NO_2 and OH systems is more larger than that of the unsubstituted system. This is in good agreement with the trend of *I-V* curves shown in Fig. 3. 17. 2.

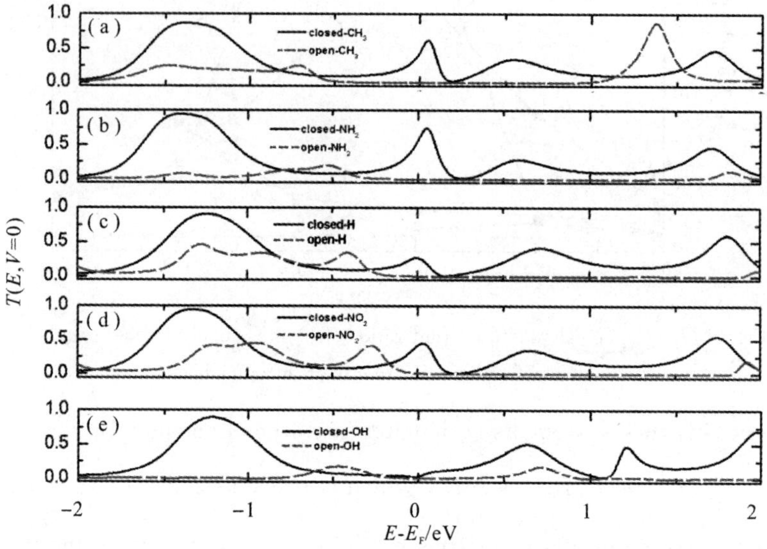

Fig. 3. 17. 4 The transmission spectra of the molecular switches with and without different substituents at zero bias

(a) with donor substituent -CH_3; (b) with donor substituent -NH_2; (c) with substituent -H; (d) with acceptor substituent -NO_2; (e) with acceptor substituent -OH

Generally speaking, the transmission coefficients can be related to the modified molecular orbitals, which can be obtained from the molecular projected self-consistent Hamiltonian (MPSH). MPSH is the self-consistent Hamiltonian of the isolated molecule in the presence of the electrodes, namely the molecular part is extracted from the whole self-consistent Hamiltonian at the contact region. Table 3. 17. 1 illustrates the MPSH orbitals of the highest occupied molecular orbital (HOMO) and the lowest unoccupied molecular orbital (LUMO) with and without substituent. We can see from Fig. 3. 17. 4 that the HOMO peak is close to the Fermi energy for the closed form, and the LUMO peak is close to the Fermi energy for the open form. It means that the electron transport of the closed form is dominated by HOMO at low bias. On the contrary, the electron transport of the open form is dominated by LUMO. It is found from Table 1that the HOMO orbital of the closed form with substituent NH_2 is more delocalized than other four systems. It is obvious that the π-overlapping and the

π-electron conjugation of HOMO orbital are increased by the substituent NH_2, which is the main reason leads the current to increase. However, for the open form, the LUMO is become more localized than that of unsubstituted system orbitals which accounts for the low transmission strength in Fig. 3.17.4.

Table 3.17.1 The MPSH of HOMO and LUMO for closed and open forms with different substituents at zero bias

MPSH	Closed-CH_3	Closed-NH_2	Closed-H	Closed-NO_2	Closed-OH
HUMO	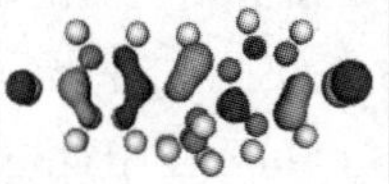	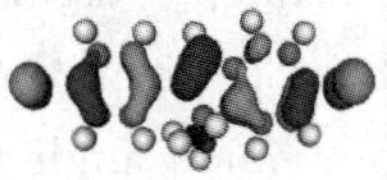	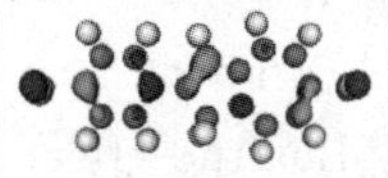	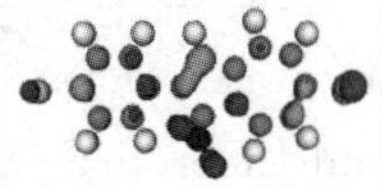	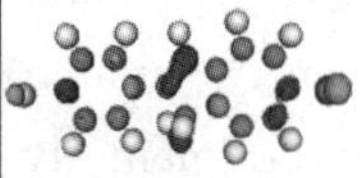
LUMO	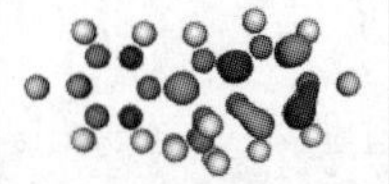	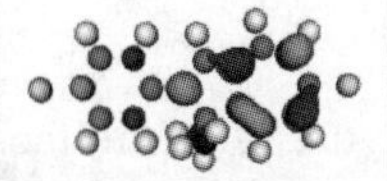	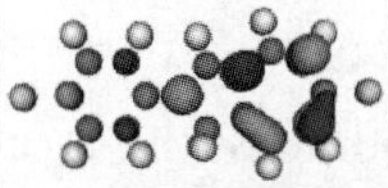	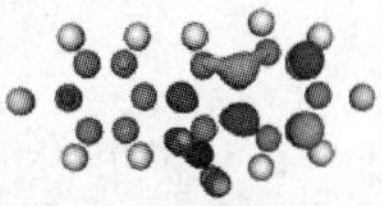	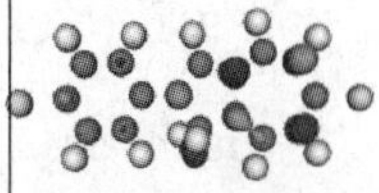
MPSH	Open-CH_3	Open-NH_2	Open-H	Open-NO_2	Open-OH
HOMO	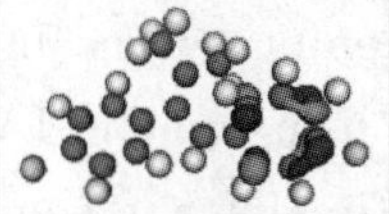	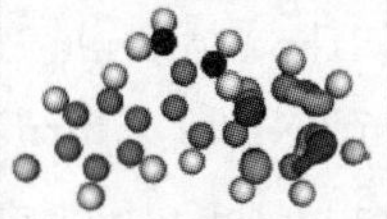	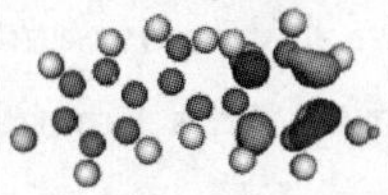	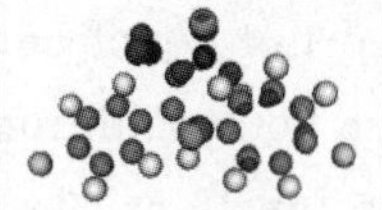	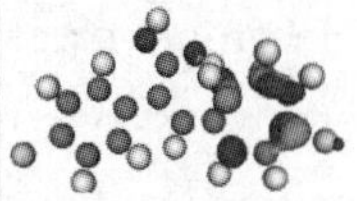
LUMO	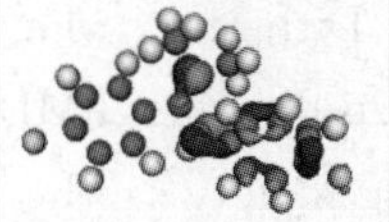	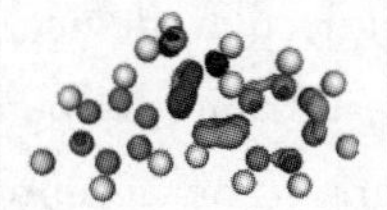	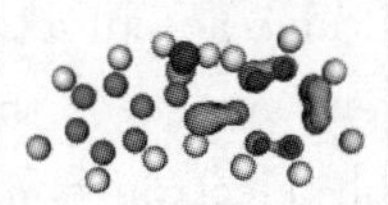	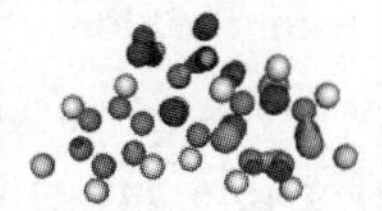	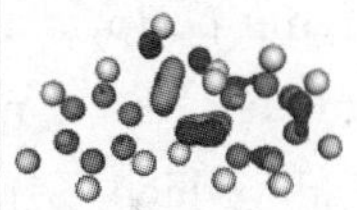

As a summary, the effect of chemical modifications on the electron transport properties of the dihydroazulene optical molecular switch has been investigated by using the DFT+NEGF first-principles method. The results indicate that the transport properties in molecular device can be manipulated, enhanced, or suppressed by the consideration of the effects of chemical modification. The current ratio can be enhanced by the electron-donating groups, while the electron-accepting groups can suppress it. This result reflects that the transport properties of molecular structures are intimately related to the chemical modification and can provide fundamental guidelines for the design of functional molecular devices.

3.18 Effect of torsion angle on the rectifying performance in the donor-bridge-acceptor single molecular device

In recent years, the research on electron transport through individual molecules is gaining a lot of attention, motivated by the prospect of using molecules as components in atomic-scale circuits. With the recent advances in experimental techniques including scanning tunneling spectroscopy, lithographically fabricated nanoelectrodes, colloid solutions, and mechanically controllable break junctions, various molecular devices with the interesting physical properties such as negative differential resistance (NDR), memory effects, molecular rectification, and current switching have been measured and designed. Among these devices, rectification has attracted great attention because it is one of the most important functions in a traditional electronic component. The first molecular rectifier was proposed by Aviram and Ratner in 1974 using D-B-A molecules, where D and A are respectively, an electron donor and an electron acceptor, and B is a covalent "sigma" bridge(insulator). Recently, numerous works of this AR model have been performed on molecular rectifier both experimentally and theoretically. But the effect of torsion angle on the rectifying performance in the donor(D)-bridge(B)-acceptor(A) single molecular device, up until now, has not reported yet.

In an actual experiment condition, the structure of a molecular device could be influenced by external factors during fabrication which may also have effects on the electronic transport properties of a molecular device. Therefore, it is very important to understand the electronic transport of a molecular device by study the molecular conformation. In the present work, we use nonequilibrium Green's function (NEGF) formalism combined with first-principles density functional theory (DFT) to examine the effect of torsion angle on the rectifying performance in the donor-bridge-acceptor molecular device. Different from other AR model, the donor moiety of present system is the tetramethylbenzo-dioxole(TMDO) molecule which is reversible and stable electron donors. The acceptor moiety is the 1,2,4,5-tetracyanobenzyl unit which is substituted by the acceptor(-CN) functional groups at the outer phenyl rings. These two moieties are bridged together by the middle phenyl ring. The influence of the HOMO-LUMO gaps and the spatial distributions of molecular orbitals on the quantum transport through the molecular device are discussed in detail. The calculated results show that the torsion angles play an important role in determining the rectifying performance of the molecular devices.

The molecular architectures we study are illustrated schematically in Fig. 3. 18. 1. The D-B-A molecule with different torsion angles are sandwiched between two parallel Au(111) surfaces that correspond to the surfaces of the Au electrodes. Each layer of Au electrodes is represented by a 4×4 supercell with the periodic boundary conditions so that it imitates bulk metal structures. The 4×4 supercell is large enough to avoid any interaction with molecules in the next supercell.

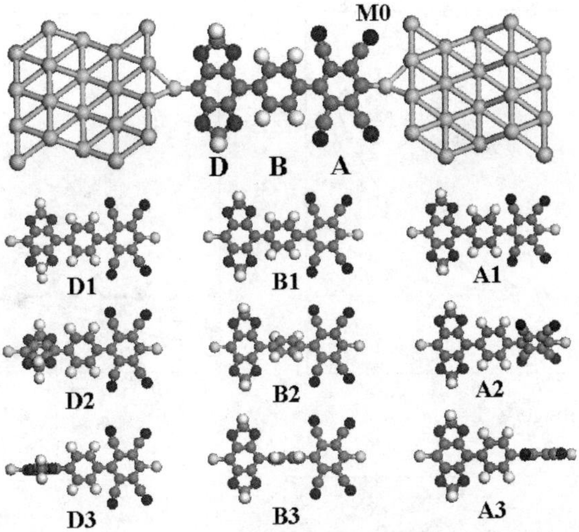

Fig. 3. 18. 1 Schematic illustration of the D-B-A molecular device in our simulation. M0, D1-D3, B1-B3, A1-A3 describe the related position of the donor, bridge and acceptor moiety with torsion angles 0°, 30°, 60° and 90° respectively

As shown in Fig. 3. 18. 2, the self-consistently calculated current-voltage (I-V) characteristics and the rectification ratio for the D-B-A molecular device with different torsion angles given in Fig. 3. 18. 1 have been plotted in the bias range from −1. 2 to 1. 2 V in steps of 0. 2 V. It is quite clear from Fig. 3. 18. 2(a)-(c) that a significant rectification in the current is observed, because the electric current under the positive bias are larger than that under the same negative bias for all molecular junctions. Furthermore, by increasing the torsion angle, the current is decreased obviously. Particularly, when the dihedral angle is equal to 90°, the current reaches its minimum. As a figure of merit, the rectification ratio of current, $R(V) = |I(V)/I(-V)|$, versus bias V is plotted in Fig. 3. 18. 2(d)-(f). From the figures, we can find that the torsion angles have an evident influence on rectifying ratio of the D-B-A molecular device. When the relative dihedral angle is equal to 60°, the rectification ratio reaches the maximum in all three figures. Especially, model B2, namely the relative dihedral angle of B is equal to 60°, shows the strongest rectifying performance compared with other models. For example, the rectification ratio reaches 11. 7 at a bias of 1. 2 V for model B2 and only 5. 2 and 8. 6 at the same bias for model A2 and D2. In Fig. 3. 18. 2(d) (e), we also can find that when the relative dihedral angle is equal to 0°, 30°, 90°, the rectification ratio is increased evidently with the increasing of the torsion angles. In contrast, when the relative dihedral angle is equal to 90°, the rectification ratio is reaches its minimum in Fig. 3. 18. 2(d).

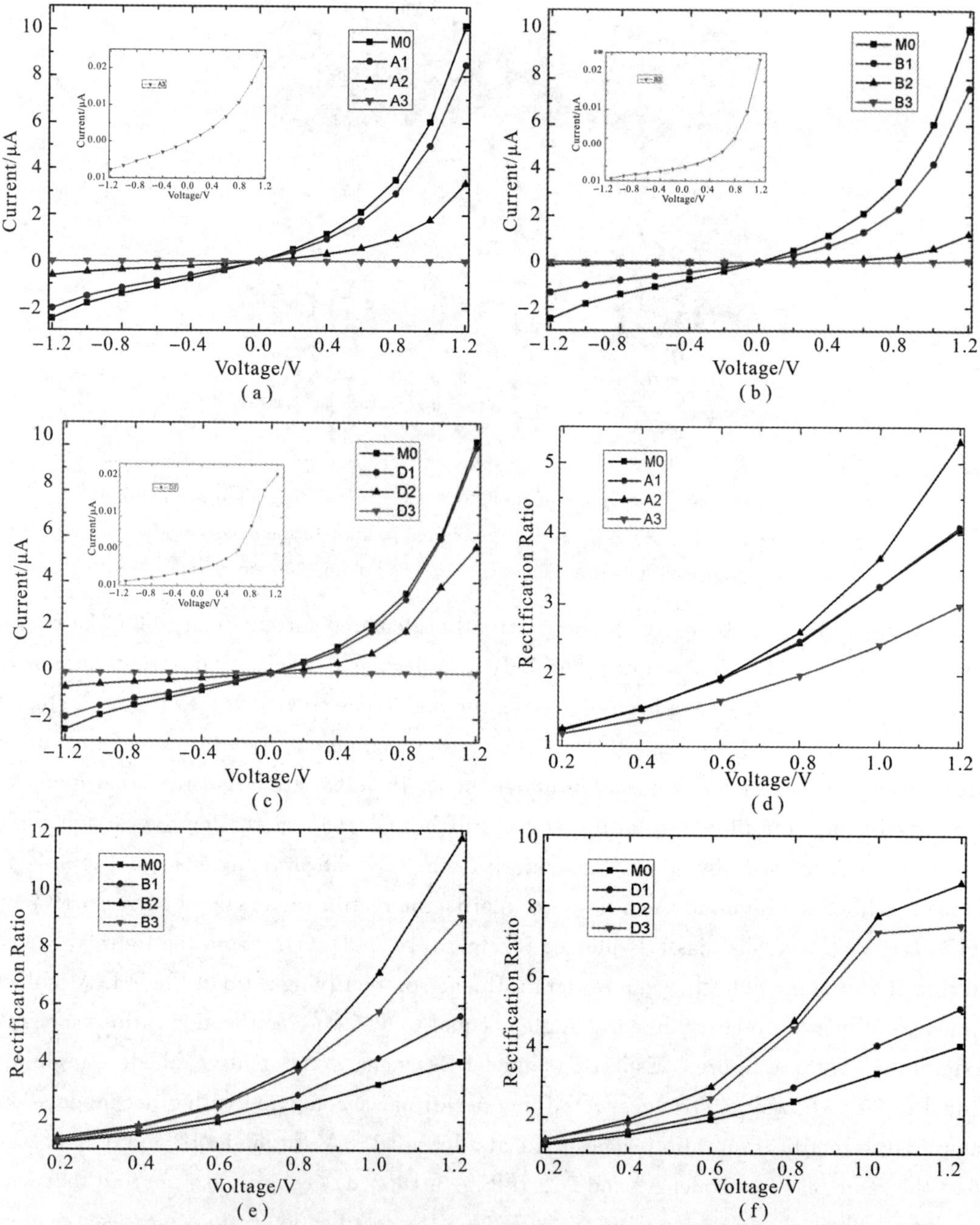

Fig. 3. 18. 2 The self-consistently calculated current and the rectification ratios as a function of bias for the D-B-A molecular device with different torsion angles

(a) (d) the *I-V* characteristics and rectification ratios for M0 , A1 , A2 and A3;

(b) (e) the *I-V* characteristics and rectification ratios for M0 , B1, B2 and B3;

(c) (f) the *I-V* characteristics and rectification ratios for M0 , D1, D2 and D3;

The calculated current for A3, B3 and D3 is plotted in the inset, respectively

The current in the system is calculated by the Landauer-Bütiker formula, which are transmission spectra dependent. Therefore, to understand the dramatic difference in current appearing in the Fig. 3. 18. 2, we calculate the energy dependence of zero-bias transmission spectra. Here we take the models of M0, B1, B2, B3 as the examples and exhibit their transmission spectra in Fig. 3. 18. 3. It can be found that there is no obvious change in the position of two peaks near the Fermi level with the increasing of the torsion angle. But when the torsion angles are small, the value of these two peaks becomes smaller but broader. Especially, when the torsion angle is equal to 90°, the peak near the Fermi energy nearly disappears. Furthermore, in such an open system, the Fermi level E_F of the Au electrode aligns between the highest occupied molecular orbital (HOMO) and the lowest unoccupied molecular orbital (LUMO), and the barrier for the electron transport is intensively relevant to the HOMO-LUMO gap. We can also see from Fig. 3. 18. 3 that the HOMO-LUMO gap is increased with the increasing of the torsion angles. In particular, the HOMO-LUMO gap of B3 is larger than that of others. Moreover, the transmission coefficients also can be related to the modified molecular orbitals, which can be obtained from the molecular projected self-consistent Hamiltonian (MPSH). The MPSH is the self-consistent Hamiltonian of the isolated molecule in the presence of the electrodes, namely the molecular part is extracted from the whole self-consistent Hamiltonian at the contact region. The MPSH of the HOMO and LUMO under zero voltage for four models are also illustrated in Fig. 3. 18. 3. It can be clearly seen from Fig. 3. 18. 3 that both of the HOMO and LUMO states tend to be more and more localized with the increasing of the torsion angles. Especially, when the torsion angle is equal to 90°, the LUMO which is the main transmission channel for the current is almost entirety localized. As a result, the current is decreased by increasing the torsion angle, which is in good agreement with the trends of *I-V* curves shown in Fig. 3. 18. 2.

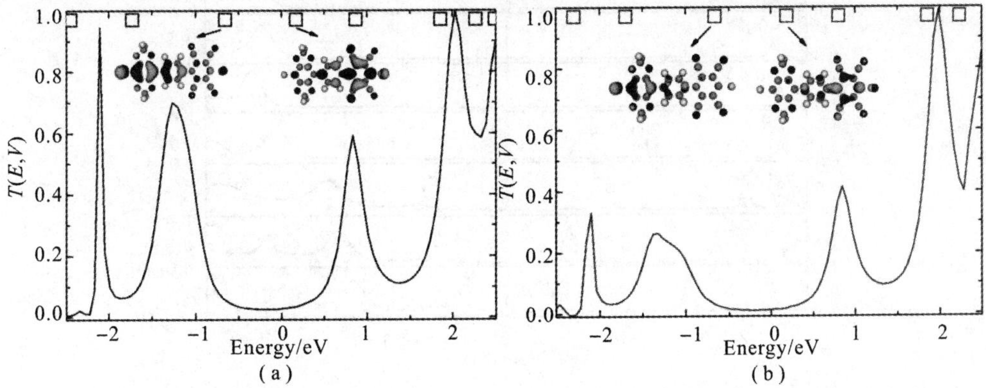

Fig. 3. 18. 3 The transmission coefficients and the MPSH of conduction orbital HOMO and LUMO for (a) M0; (b) B1. The positions of MPSH eigenvalues are marked with squares. In our calculations, we set the energy origin to be the Fermi level E_F

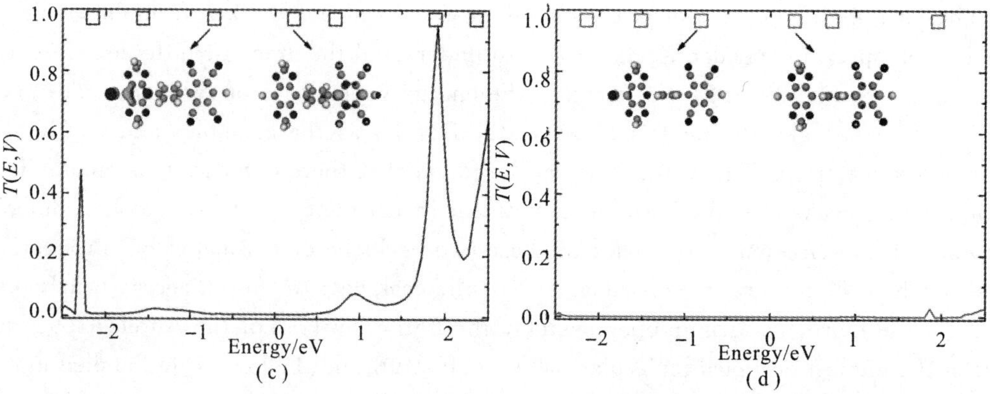

Continued Fig. 3. 18. 3 The transmission coefficients and the MPSH of conduction orbital HOMO and LUMO for (c)B2; (d)B3 under the zero biases. The positions of MPSH eigenvalues are marked with squares. In our calculations, we set the energy origin to be the Fermi level E_F

However, when the bias voltage is applied, the system is driven out of equilibrium. Therefore, in order to explain the origin of the obvious rectifying performance in our models, it is very necessary to investigate the change of the transmission characteristics under the applied bias. Here, we take the model of B2 as an example and the bias dependence of its transmission characteristics at biases of 0. 0 V, ±0. 4 V, ±0. 8 V and ±1. 2 V are presented in Fig. 3. 18. 4. In our calculation, the average Fermi level, which is the average value of the chemical potential of the left and right electrodes, is set as zero. Seen from the Eq. (3. 1. 3), we can expect that only electrons with energies within a range near the Fermi level E_F contribute to the total current. Therefore, a good approximation with the range of the bias window, i. e. ,$[-V/2,+V/2]$ is enough to analyze a finite part of the transmission spectrum.

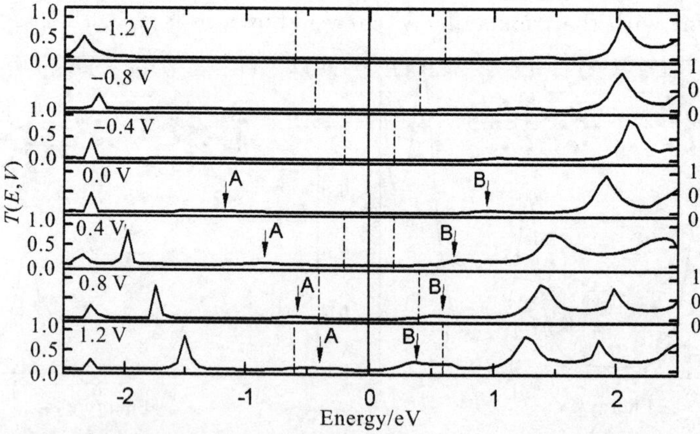

Fig. 3. 18. 4 The transmission spectra of model B2 under various applied voltages of 0. 0 V, ±0. 4 V, ±0. 8 V and ±1. 2 V, respectively. The two main transmission peaks near the Fermi level which denoted by A and B, respectively. The region between the dashed lines is the bias window, and the solid line represents the average Fermi level

It can be seen clearly from Fig. 3. 18. 4 that left peak (A) shifts toward the higher energy with the increasing of applied bias in the positive bias range. However, the right peak (B) is moved to the lower energy. Consequently, when the bias voltages enhance to 1. 2 V, both A and B enter the bias window and produce a large current increase. In contrast, these two peaks move away from the Fermi level E_F upon increasing the applied negative bias. Therefore, the transmission coefficients in the bias window reduce and such a decrease in the transmission coefficients cannot be compensated by the increase in the magnitude of the bias window. As a result, the electronic transfer ability of the device is weakened and further the current is diminished, which leads to the large rectification ratio of model B2 in the Fig. 3. 18. 2(e).

In conclusion, the effect of torsion angle on the rectifying performance in the donor-bridge-acceptor single molecular device has been investigated by using the DFT+NEGF first-principles method. The electronic transmission coefficients and *I-V* characteristics corresponding to different models are calculated and analyzed. The calculated results show that the intermolecular interaction deeply depends on the torsion angles, which plays an important role in determining the rectifying performance of the molecular devices. The large rectification ratio has been explained by the shift in transmission resonance peak across the bias window with varying bias voltage and the spatial distributions of molecular orbitals on the quantum transport through the molecule. The results will be helpful to understand further the possible situation in experiments and design molecular electronic device.

3. 19 The rectifying performance in diblock molecular junctions

Electronic devices based on single molecules have been considered as one of the most potentially promising alternatives to today's semiconductor-based electronics. Use molecular devices to realize the elementary functions in electronic circuits have become an attractive field of research. With the progress of techniques for characterizing and manipulating individual molecules, many interesting physical properties like negative differential resistance (NDR), memory effects, molecular rectification, and electronic switching in molecular devices have been reported. Among these devices, rectifications have attracted great attention because they are basic elements of any modern design of logic and memory circuits. Since the first molecular rectifier was proposed by Aviram and Ratner in 1974, numerous works have been performed on molecular rectifier both experimentally and theoretically. Recently, Morales et al. have developed a new class of molecular rectifier based on a conjugated diblock cooligomer with a directly coupled donor-acceptor structure. Experiments have shown that there is a significant rectifying effect can be observed in this molecular junction. However, in the experiment environment, the small differences in molecular structure, including the conjugation length, dipole moments, and anchoring groups, etc. , can lead to great variations in conductance and current-voltage characteristics. Therefore, a theoretical study is very necessary to predict structures that may have optimized properties by varying chemical structures for a particular application. In this work, we extend the previous work of Morales's and use nonequilibrium Green's function formalism combined first-principles calculations to investigate the effect of different anchoring groups on the electronic transport properties in diblock molecular junction. The calculated results show that the electronic transport properties deeply depends on the anchoring groups, which plays an important role in determining the rectifying performance of the molecular devices.

The molecular architectures we study are illustrated schematically in Fig. 3. 19. 1. The dipyrimidinyl-diphenyl (PMPH) diblock molecule with different anchoring groups, namely the thiol anchoring groups(-S), isocyanide anchoring groups(-CN) and dithiocarboxylate anchoring groups(-CS_2), is bonded to two parallel Au(111) surfaces. M1-M6 correspond to the six different anchoring group geometries.

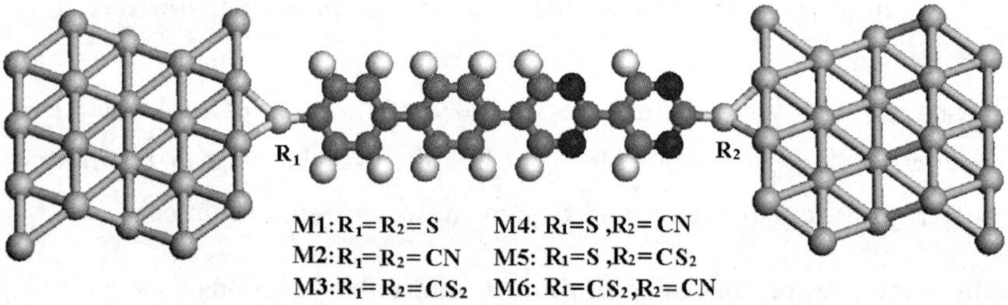

Fig. 3. 19. 1 Schematic illustration of the PMPH diblock molecular device in our simulation. M1-M6 describe the molecule with different anchoring groups connected to two Au electrode

As shown in Fig. 3. 19. 2, the self-consistently calculated current-voltage (*I-V*) characteristics and the rectification ratio for M1 – M6 have been plotted in the bias range from −1. 2 to 1. 2 V in steps of 0. 2 V. From the figures, it can be found that the transport properties and rectifying ratio of the molecular devices are both strongly dependent on the anchoring groups. In the positive bias range, the strongest current occurs in models M2, M5, and weakest in model M1. Furthermore, for models M1, M4, M5 and M6, the current is asymmetric in the whole bias range, and the current under the high positive bias is larger than that under the same negative bias. The degree of asymmetry for *I-V* curve in model M5 is highest. In contrast, the currents through M2 and M3 are almost symmetric under the whole bias range, especially under small bias voltages. As a figure of merit, the rectification ratio of current, $R(V) = |I(V)/I(-V)|$, versus bias V is plotted in Fig. 3. 19. 2(b). Obviously, model M5 shows the strongest rectifying performance compared with other models. For example, the rectification ratio reaches 5.2 at a bias of 1.2 V for model M5 and only 3.4, 1.1 and 2.1 at the same bias for model M1, M4 and M6. However, for model M2 and M3, the currents under both positive and negative bias are almost same, indicating very weak rectification. Furthermore, it is also quite clear from Fig. 3.19.2(b) that initially the rectification ratio of M4 and M6 increases with the increases in external voltage. The increase in rectification ratio due to the variation in voltage is observed up to 0.6 V and 0.8 V for M4 and M6, respectively. Beyond 0.6 V and 0.8 V, there is a rapid decrease in rectification ratio with the increase in bias voltage in both molecular systems.

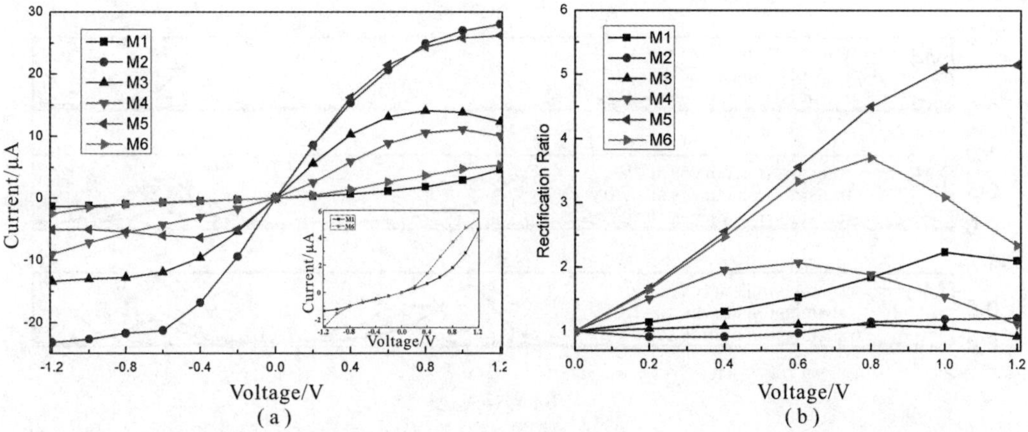

Fig. 3.19.2 *I-V* characteristics and the rectification ratio

(a) the self-consistently calculated current; (b) the rectification ratios as a function of bias for the PMPH diblock molecular device with different anchoring groups. The calculated current for M1 and M6 is plotted in the inset, respectively

The current in the system is calculated by the Landauer-Bütiker formula, which are transmission spectra dependent. When the bias voltage is applied, the system is driven out of equilibrium and the electrode potential change. Therefore, to understand the dramatic difference in current and rectification behavior appearing in the Fig. 3.19.2, we calculate the transmission spectrum at a bias voltage of 0 V and ±1.0 V, respectively. In Fig. 3.19.3, it can be found that the heights and positions of the transmission are more sensitive to the applied bias voltage in M5 and M6 junctions than M1 - M4 junctions. Especially, there is no obviously change in transmission curves at ±1.0 V in M2 and M3 junctions. As a result, the M1 - M4 present much smaller rectification effect, and the rectification behavior of M2 and M3 nearly disappear. In contrast, the main transmission peak of M5 and M6 at positive bias is evidently larger than that at negative bias. Therefore, a larger rectification ratio in M5 and

M6 compared to M1 - M4 junctions is observed.

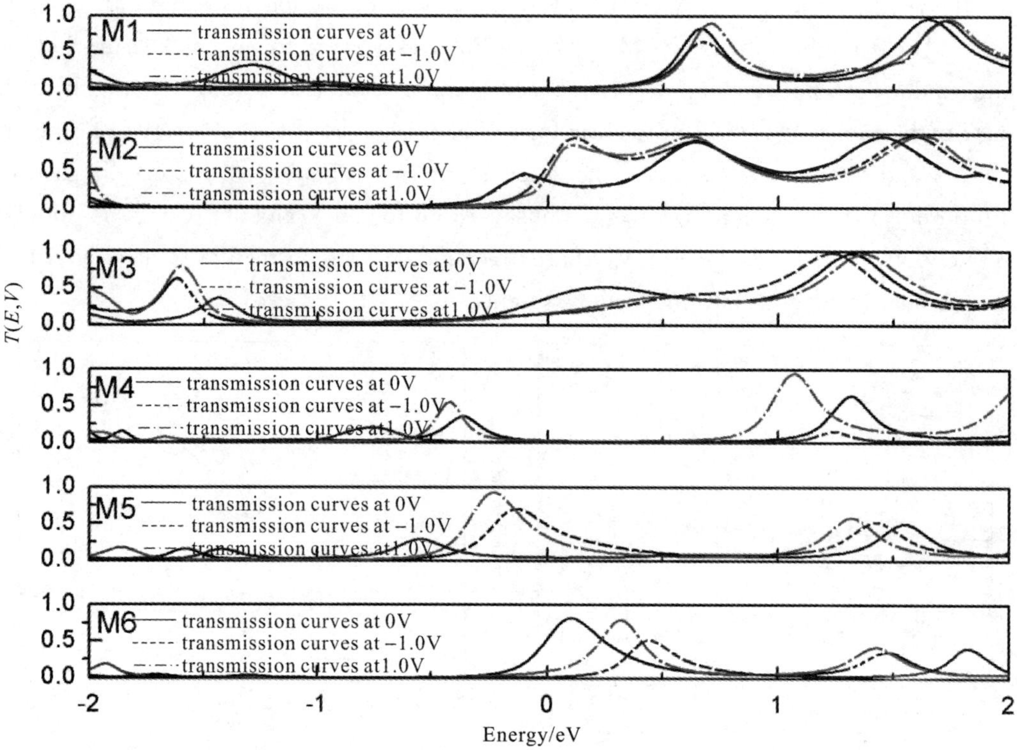

Fig. 3. 19. 3 Transmission curves at zero, −1. 0 V and 1. 0 V bias voltages for M1 - M6 molecular junctions

To get a further insight about the origins of the different rectification behavior of M1 - M6, we analyze the molecular projected self-consistent Hamiltonian (MPSH) which is the self-consistent Hamiltonian of the isolated molecule in the presence of the electrodes, namely the molecular part is extracted from the whole self-consistent Hamiltonian at the contact region. Fig. 3. 19. 4 presents a comparative study of the corresponding MPSH orbital of the main transmission peak which make their contribution to current within the bias voltages of ±1. 0 V. As shown in Fig. 3. 19. 4, for M5 and M6, the energy of the MPSH orbital is remarkable affected by the bias voltage, and the real space distribution is also very sensitive to the bias voltage. As a result, the position of the transmission peak is deeply dependent of the bias voltage, and the peak width and height remarkable change. For example, for M5, the MPSH orbital is complete delocalization at 1. 0 V, while it become completely localized at −1. 0 V. Therefore, the main transmission peak is much stronger at positive bias voltages than that at negative bias. In contrast, the MPSH orbitals of M2 and M3 are complete delocalization both at ±1.0 V. As a result, the currents through M2 and M3 are almost symmetric under the whole bias range, which indicate very weak rectification.

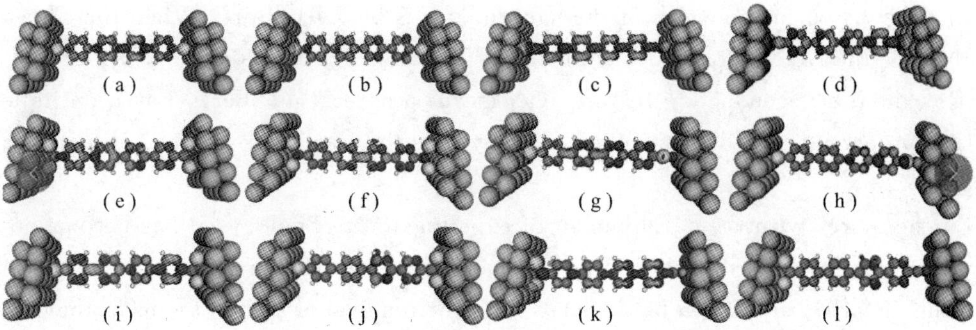

Fig. 3.19.4 MPSH orbitals corresponding to the transmission peak near the Fermi energy at the bias voltages of ±1.0 V

(a)1.0 V M1; (b)−1.0 V M1; (c)1.0 V M2; (d)−1.0 V M2; (e)1.0 V M3;(f)−1.0 V M3; (g)1.0 V M4; (h)−1.0 V M4; (i)1.0 V M5; (j)−1.0 V M5;(k)1.0 V M6;(l)−1.0 V M6

In conclusion, the effect of different anchoring groups on the electronic transport properties in PMPH diblock molecular device has been investigated by using the DFT+NEGF first-principles method. The electronic transmission coefficients and *I-V* characteristics corresponding to different models are calculated and analyzed. The calculated results show that the anchoring group play a significant role on the rectifying performance and can't be neglected in designing the real diblock molecular junctions. The transport properties in molecular devices can be manipulated, enhanced, or suppressed by a careful consideration of the effects of the anchoring group, and such modifications become crucial in optimizing the electron transport properties of chemical structures.

3.20 Effect of chemical doping on the electronic transport properties of tailoring graphene nanoribbons

Following the miniaturization of electronic device, silicon-based devices have gradually reached the limits of their performance. Recently, one-dimensional graphene nanoribbons (GNRs) which cut off from the novel two-dimensional material-graphene have been exhibiting many potentials for next generation of nanoelectronics due to their unique electronic structures and transport characteristics, such as high carrier mobility, quantum hall effect, bipolar electric field effect, tunable bandwidth, and so on. There are two kinds of graphene nanoribbons(GNRs), which can be divided into armchair-type graphene nanoribbons(AGNRs) and zigzag-type graphene nanoribbons(ZGNRs) according to the shapes of edges cut from different directions. For AGNRs, the previous investigations have shown that the width of the armchair-type nanoribbons saturated with H atoms determines the properties of the nanoribbons when the width of the nanoribbons is below 100 nm. When the edge atoms are $3N+2$, the AGNRs exhibit metallicity. When the edge atoms are $3N+1$ or $3N$, the AGNRs exhibit semiconducting features (N means a natural number). Based on its unique electrical properties, AGNRs have been employed as the electrodes for several molecular devices.

Furthermore, with the development of experimental technology, it has become possible to tailor agraphene nanoribbons(GNRs) or even introduce foreign atoms. Jin et al. proved that chain of carbon atoms can be prepared from the removal of carbon atoms of the graphene row by high energy electron irradiation a sputtering process. Li et al. prepared graphene nanoribbons with widths below 10 nm as electrodes and found that GNRs with widths below 10 nm are all semiconductors. Zhang et al. fabricated a new doped graphene material which showed good chemical properties in fuel cells by a thermal annealing method. Zheng et al. introduced nitrogen (N) and boron (B) sequentially into selected locations on the graphene domain. The results show that B, N-graphene greatly improves the electrochemical performance of graphene compared to single-doped graphene and one-step hybrid electrodes. In this work, we designed a molecular junction based on doping tailoring AGNRs. The effects of doping on the electronic transport properties of AGNRs with different widths are investigated by using the non-equilibrium Green's function based on density functional theory. We discovered significantly width-tuned current-voltage (I-V) characteristic. Moreover, the I-V characteristic can be improved by doping.

The model device is illustrated in Fig. 3. 20. 1, a tailoring AGNRs with different widths are sandwiched between two 17-AGNRs electrodes which are shown to be metallic. Here, the widths of tailoring AGNRs with $N=11$ [see Fig. 3. 20. 1(a)] which exhibits metallicity and $N=7$(see Fig. 3. 20. 1(b)) which exhibits semiconducting features are selected, respectively. The device is divided into three region: left electrode, right electrode, and scattering region. Two elements in groups Ⅲ and V, namely B and N, are used as dopants, and they located on the center positions of the tailoring AGNRs. For simplicity, the undoped, B-doped and N-doped AGRNs with width $N=11$ in Fig. 3. 20. 1(a) are named M1, M2 and M3, respectively. The undoped, B-doped and N-doped AGRNs with width $N=7$ in Fig. 3. 20. 1(b) are named M4, M5 and M6, respectively.

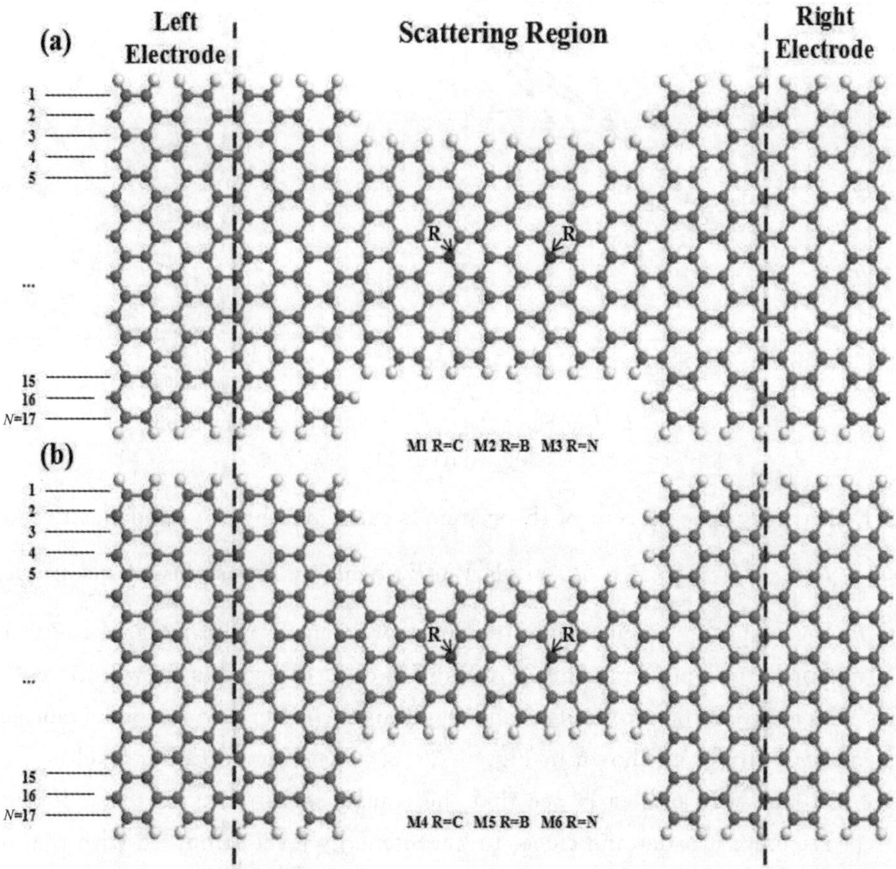

Fig. 3. 20. 1 Schematic of molecular junctions with different width tailoring AGNRS between two 17-AGNRs electrodes

(a) $N=11$; (b) $N=7$

The *I-V* characteristics for different models in bias range from 0 to 2. 0 V are shown in Fig. 3. 20. 2. Most strikingly, the currents of the metallic tailoring AGNRs with $N=11$ are larger than that of semiconducting tailoring AGNRs with $N=7$. The electronic transport properties of the tailoring AGNRs junctions can be improved by doping. From Fig. 3. 20. 2 (a), it can be seen that the currents of M2 and M3 are a little higher than that of M1. However, in Fig. 3. 20. 2 (b), when the width of tailoring AGNRs is $N=7$, the currents of M5 and M6 are remarkably larger than that of M4. For example, the calculated current at 2. 0 V is about 18. 31 μA and 5. 50 μA for M5 and M4, respectively, which is nearly 4 times larger than the current of M4. Furthermore, in view of the *I-V* characteristics, the boron doping in the semiconducting tailoring AGNRs is more useful than the nitrogen doping.

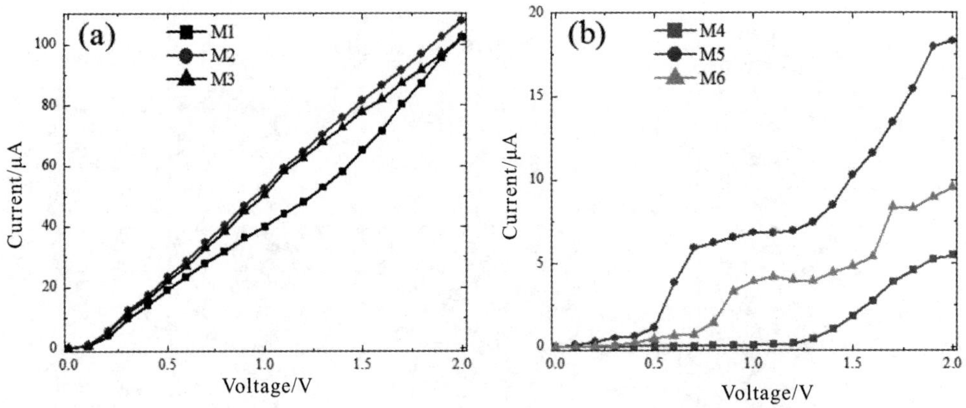

Fig. 3. 20. 2 Current-voltage curves for molecular junctions

(a)M1, M2, M3;(b) M4, M5, M6

In NEGF theory, the current of the system is calculated by the Landauer-Bütiker formula: $I=\frac{2e}{h}\int[f(E-\mu_L)-f(E-\mu_R)]T(E,V)\mathrm{d}E$, which is transmission spectra dependent. Thus, to further explore the electronic transport properties, we calculated the transmission spectra of undoped or doped tailoring AGNRs with different widths between two 17-AGNRs electrodes under zero voltage to understand the dramatic difference in conductance appearing in current-voltage curves as shown in Fig. 3. 20. 3. The average Fermi level is set as zero. From Fig. 3. 20. 3, we can clearly see that the transmission peaks of the tailoring AGNRs with $N=11$ are more broader and closer to Fermi energy level compared with that of the tailoring AGNRs with $N=7$, which leads to higher conductivity in Fig. 3. 20. 2. In addition, in Fig. 3. 20. 3(a), the position and the peaks of transmission for M1, M2 and M3 are almost the same. However, in Fig. 3. 20. 3(b), the transmission coefficient peaks of M5 and M6 are larger than that of M4. Furthermore, the transmission peaks of M5 is more closer to the Fermi energy level compared with M6 and M4. As a result, the large transmission coefficient accounts for the high conductivity in M5.

The varying of the current ofundoped or doped tailoring AGNRs with different widths can be further analyzed by the molecular projected self-consistent Hamiltonian (MPSH) eigenstates. In Table 3. 20. 1 and Table 3. 20. 2, we calculated the MPSH of the highest occupied molecular orbital (HOMO) and the lowest unoccupied molecular orbital (LUMO) to give a visual description of the electronic transport. It can be seen from Table 3. 20. 1 that there is just one delocalized MPSH orbital for all three models: the HOMO-MPSH orbital of M1 localized on the right, the HOMO-MPSH orbital of M2 and the LUMO-MPSH orbital of M3 localized on the central regions, which results in a similar *I-V* characteristic curve shown in Fig. 3. 20. 2(a). However, in Table 3. 20. 2, one can see that there is one delocalized MPSH orbital for M4, M6 and two delocalized MPSH orbital for M5. The localized MPSH-HOMO for M4 and M6 induce the low transmission coefficients in Fig. 3. 20. 3(b).

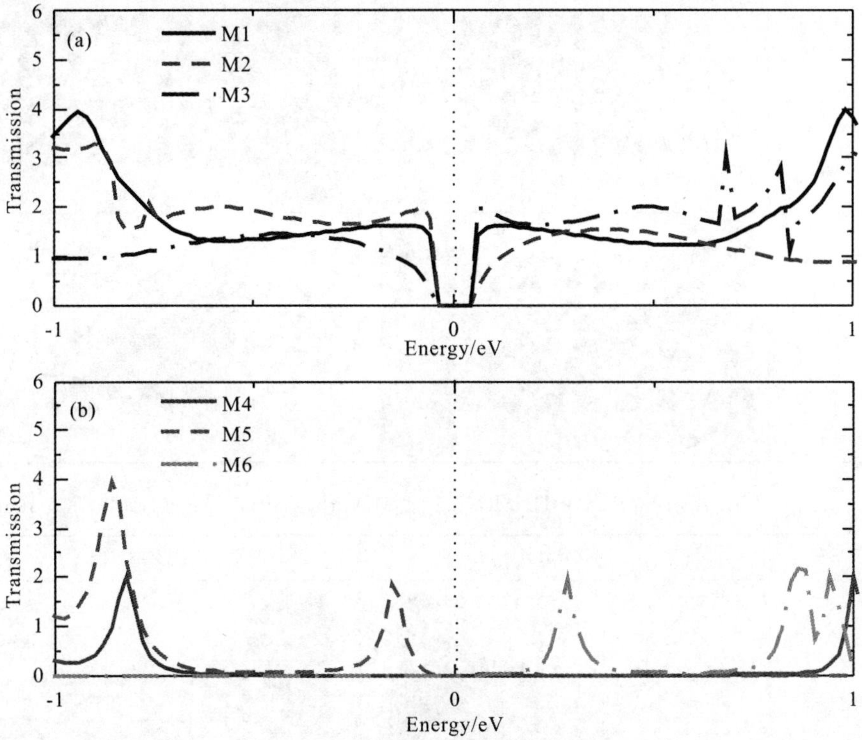

Fig. 3. 20. 3 Transmission spectra of models from M1 to M6 at zero bias. The Fermi level is set to be zero

Table 3. 20. 1 MPSH distributions of HOMO and LUMO orbits of models M1, M2, M3 at zero bias

MPSH	HOMO	LUMO
M1	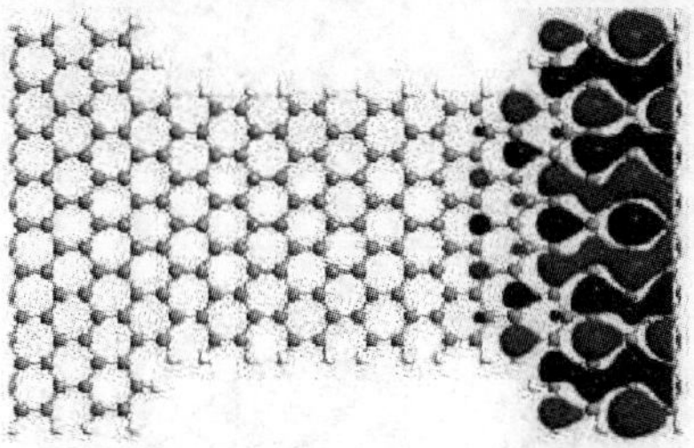	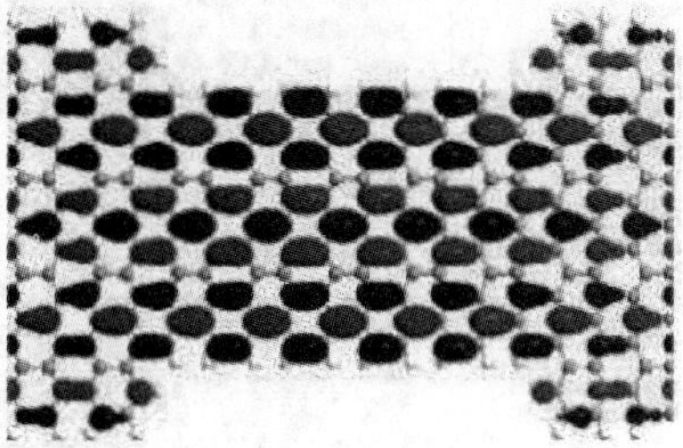

Continued

MPSH	HOMO	LUMO
M2		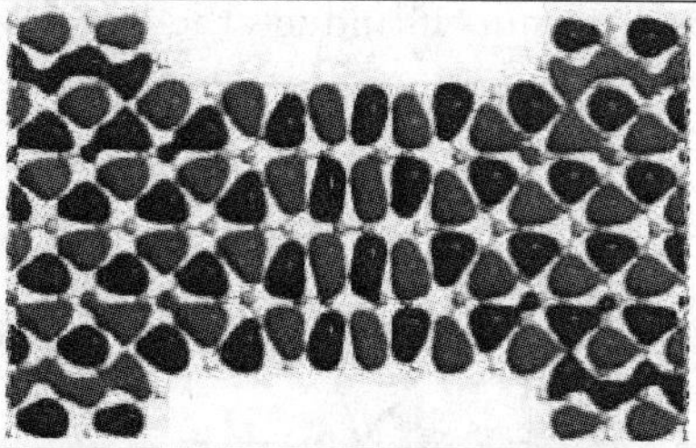
M3	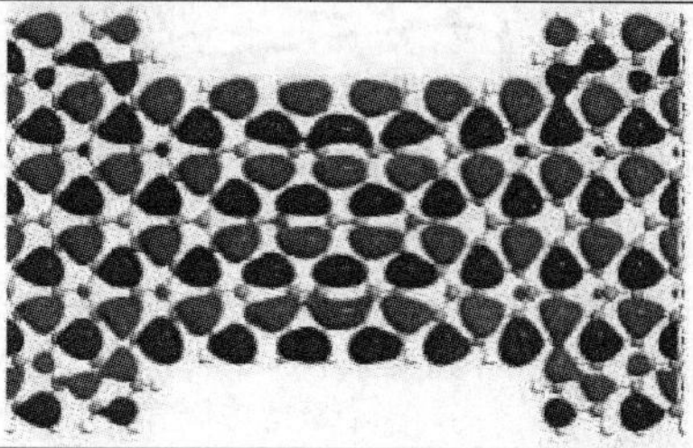	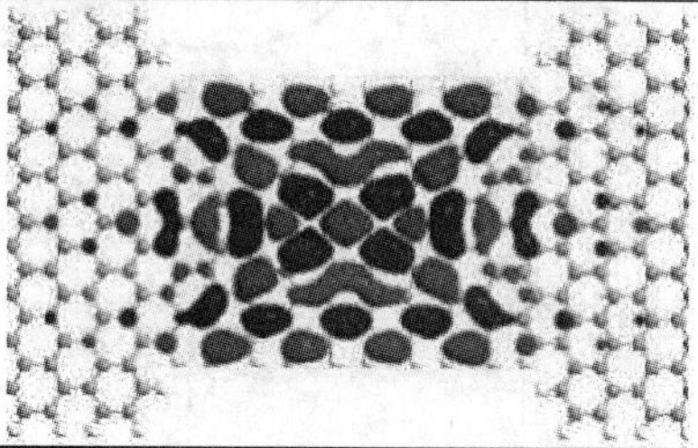

Table 3.20.2 MPSH distributions of HOMO and LUMO orbits of models M4, M5, M6 at zero bias

MPSH	HOMO	LUMO
M4	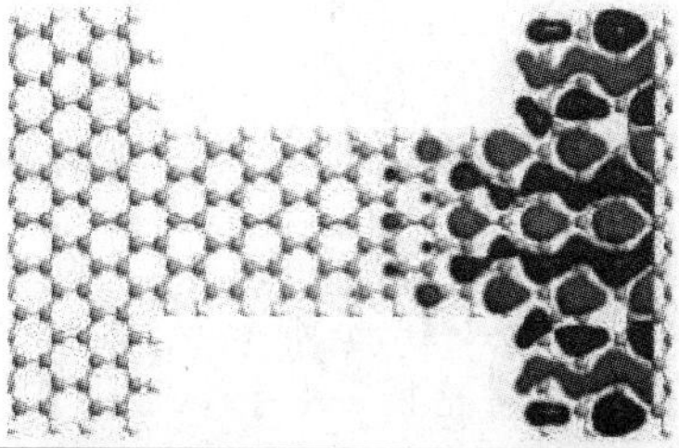	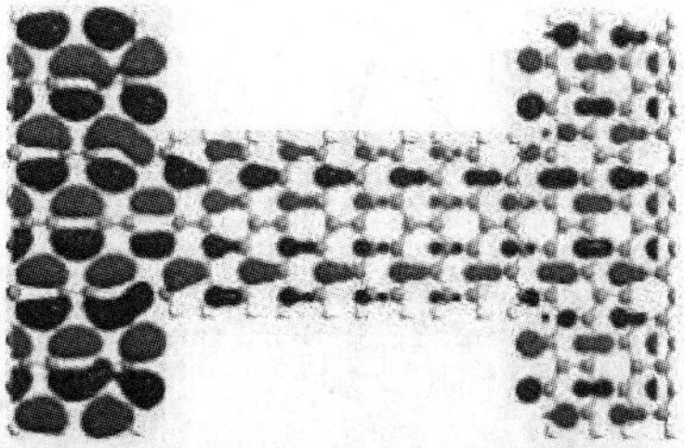
M5	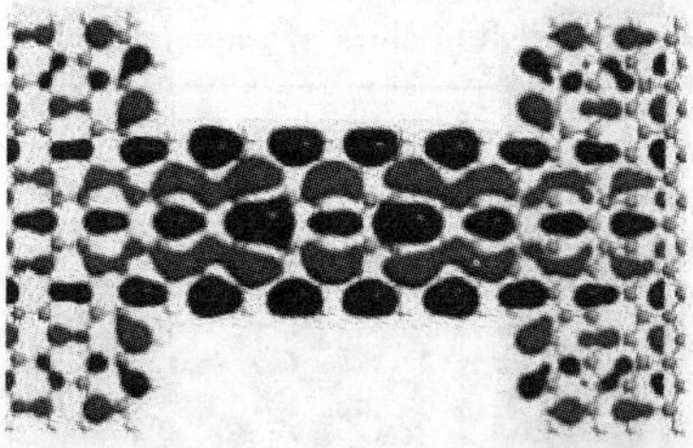	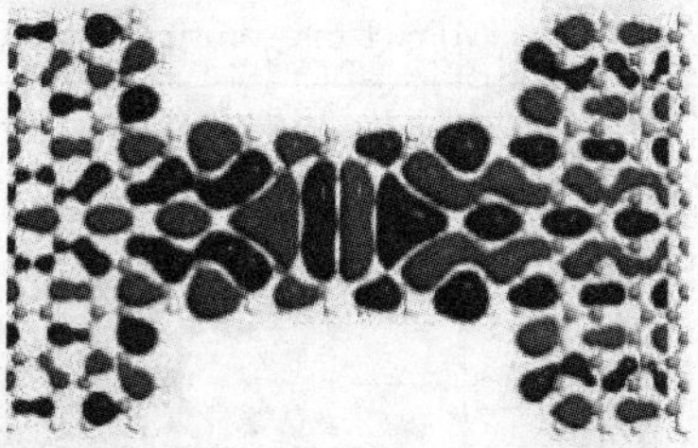
M6	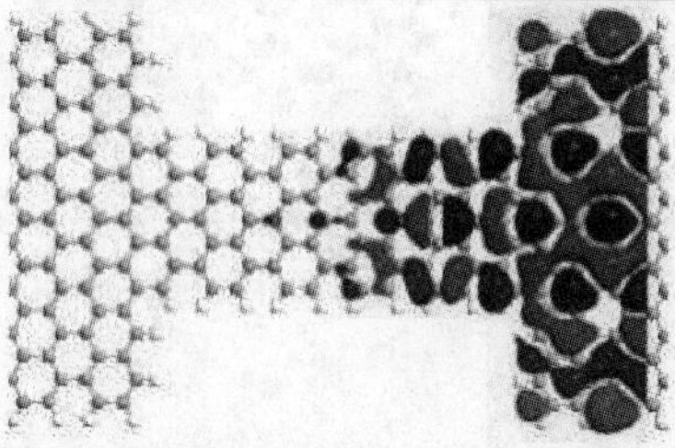	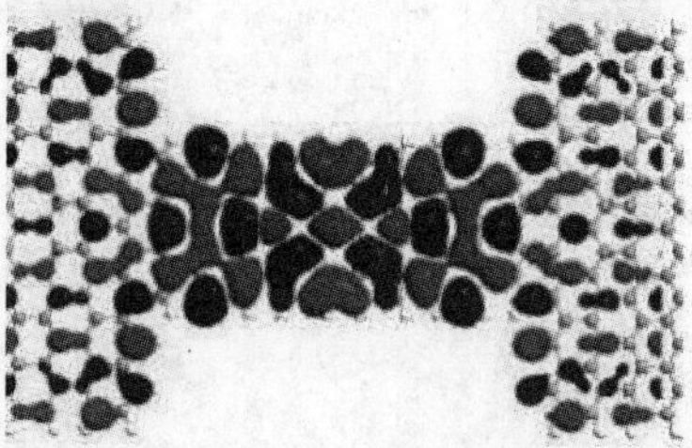

In summary, we explore the electronic transport properties ofundoped or doped tailoring AGNRs with different widths between two 17-AGNRs electrodes by using the non-equilibrium Green's function based on density functional theory. The results indicate that the width and doping of tailoring AGNR plays a significant role in the transport properties of the molecular junction. The currents of the metallic tailoring AGNRs with $N = 11$ are larger than that of semiconducting tailoring AGNRs with $N = 7$. The boron doping in the metallic tailoring AG-NRs can evidently enhance the current more than the nitrogen doping. The current of B-doped tailoring AGNRs with widths $N = 7$ is nearly four times larger than that of undoped one, which can be potentially useful for the design high performance electronic devices.

3.21 The switching behaviors induced by torsion angle in a diblock co-oligomer molecule junction

In recent years, molecular device has been extensively studied since it exhibits a number of novel features such as rectification, negative differential resistance (NDR), spin filtering, switching. In particular, some molecular switch devices have been widely reported. Molecular switch devices have state "0" and state "1" or the open or close state which realized by different current values. In this article, the design of molecular switches using a diblok co-oligomer molecule has the switching properties achieved by changing the angle between the two rings. Experiment found that the diblok co-oligomer molecule which is dipyrimidinyl-diphenyl molecule has stronger rectification. Negative differential resistance and other behaviors of electric transport were reported in some paper. Dipyrimidinyl-diphenyl shows negative differential resistance, switching, and rectifying behaviors induced by contact mode with Au electrodes. However, dipyrimidinyl-diphenyl molecule with tailoring graphene nanoribbon electrodes is investigated in this paper because graphene nanoribbons (GNRs) have unique physical properties, especially, tailoring graphene nanoribbon. GNRs are recently used as electrodes for molecular devices which exhibit negative differential resistance, rectification, spin filtering, molecular switch. Armchair-edgeribbons (AGNRs) and Zigzag-edgeribbons (ZGNRs) show differential physical properties as electrodes for molecular devices, and the molecular device is designed using tailoring AGNRs as electrodes in this work.

In this work, we simulate and calculate the molecular transport of a diblok co-oligomer molecule with tailoring graphene nanoribbon electrodes by using based on density functional theory and NEGF. From the above analysis, we can draw that the current value of the model M1 is much larger than the current value of the model M2, which is the ratio of approximately 5. It reveals that the molecular device shows a remarkable switching characteristics when the angle between the two rings in the molecular rotates from 0° to 90°. It shows that the molecular devices in the future have a broad application prospects.

The diblok co-oligomer molecule is sandwiched between two armchair graphene nanoribbons electrodes as shown in Table 3. 21. 1. According to the right side two benzene rings of the middle molecule in scattering region rotates with respect to the plane formed by the scattering region and the right and left electrodes corresponding to rotation angle 0°and 90°, respectively, we designed two device models, that is, models M1 and M2.

Table 3. 21. 1 Typical molecular junctions with different torsion angles. For simplicity, the 0° and 90° torsion angles between two benzene rings are named M1 and M2, respectively

Left Electrode	Seattering Region	Right Electrode
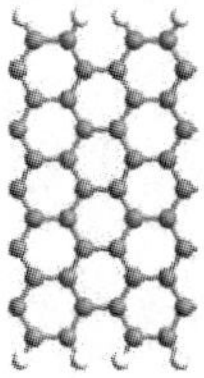	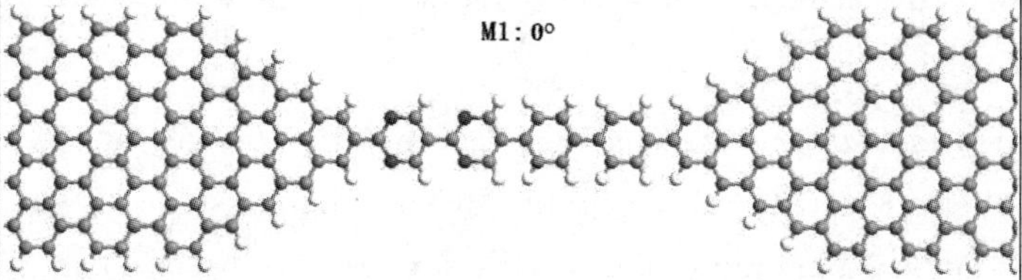	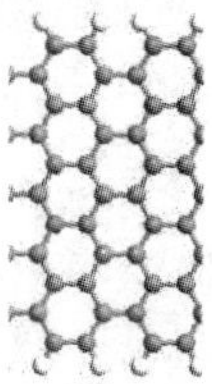
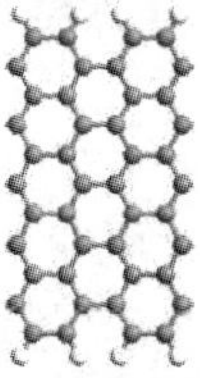	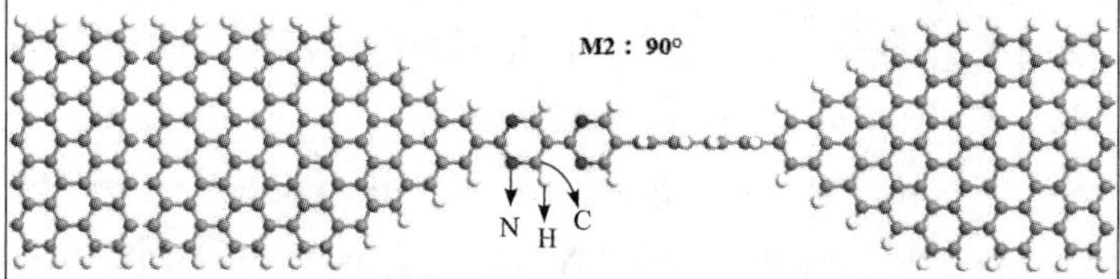	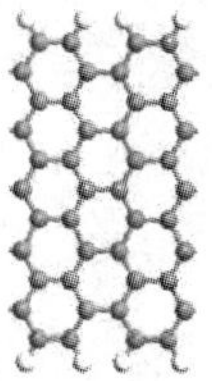

As shown in Fig. 3. 21. 1, *I-V* characteristic curves of molecular junctions with different torsion angles are given under the bias voltage varying from 0 to 0. 8 V. Conductance properties of the molecular devices appear remarkably different when torsion angles between two benzene rings are changed. The current value of M1 is obviously larger than that of M2 under whole bias voltage range in Fig. 3. 21. 1. Thus, the torsion angles between two benzene rings rotates from 0° to 90°, which means that there is a switch from state on to state off. Furthermore, there is the result from high conductance to low conductance, and vice versa. It reveals that the molecular devices with different current-voltage characteristics show the distinct switching behavior. The above inset represents the ratio of the current value of M1 to that of M2, namely the switching ratio of the molecule switch. The switching ratio is approximately 5 given in inset of Fig. 3. 21. 2, which is larger than that Ma et al calculation results for the junctions with metal electrodes .

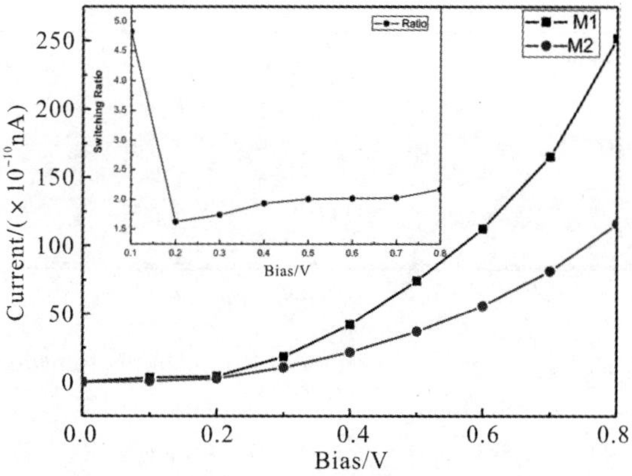

Fig. 3. 21. 1 The *I-V* characteristic curves of molecular junctions with different torsion angles (The inset represents the ratio of the current value of M1 to that of M2, namely the switching ratio of the molecule switch)

In order to explain the different current characteristics of models M1 and M2, it is necessary to analyze the transmission curve of the system for rotation angle 0° and 90°. As shown in Fig. 3. 21. 2, the black line and the red line describe the change transmission of models M1 and M2 with energy under zero voltage, respectively, and the average Fermi level E_F is set as zero. First of all, from the comparison of the transmission curves of model 1 and model 2 in Fig. 3. 21. 2, it is clear that the transmission coefficient of models M1 is much greater than that of models M2. So the current value of models M1 is larger than that of models M2 on the *I-V* curve in Fig. 3. 21. 1. Furthermore, as can be clearly seen from Fig. 3. 21. 2, and the transmission peak of model M1 is closest to the Fermi level corresponding to rotation angle 0°, while that of model M2 is far from the Fermi level for rotation angle 90°. So the current value of rotation angle 0° is larger than that of rotation angle 90°on the *I-V* curve. As a result, the low transmission coefficient resulted from the molecular structure changes accounts for the smaller current value of rotation angle 90°.

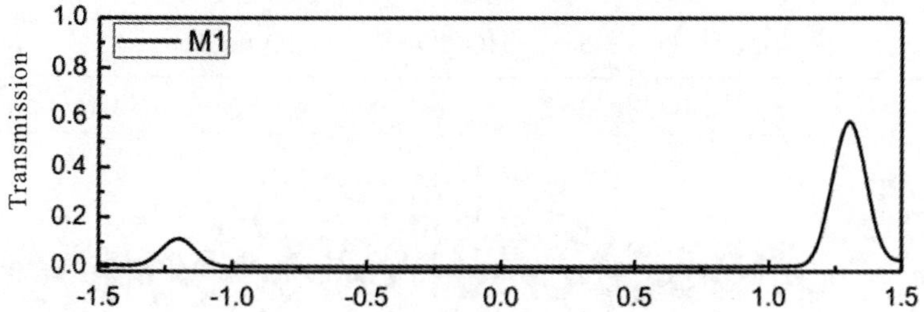

Fig. 3. 21. 2 The transmission spectra of molecular junctions with different torsion angles at zero bias

(The Fermi level is set to be zero in the calculations)

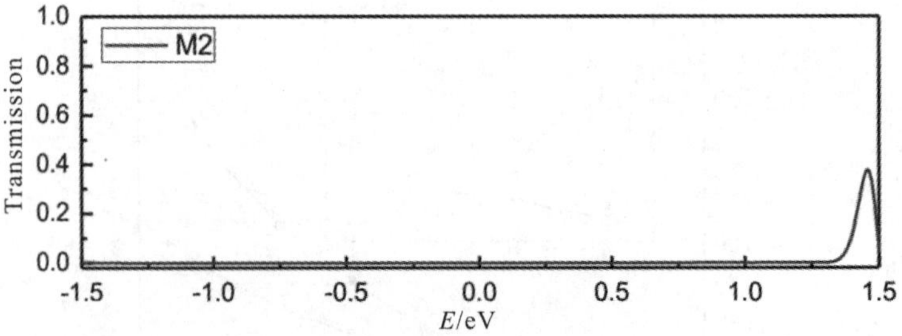

Continued Fig. 3. 21. 2 The transmission spectra of molecular junctions with different torsion angles at zero bias

(The Fermi level is set to be zero in the calculations)

To further illustrate the different current-voltage characteristics of M1 and M2, we calculate the molecular projected self-consistent Hamiltonian (MPSH) of molecular junctions with different torsion angles. The very different current values of M1 and M2 can be understood by a molecular orbital analysis. As shown in Table 3. 21. 2, there are two MPSH orbitals, namely the highest occupied molecular orbital (HOMO) and the lowest unoccupied molecular orbital (LUMO). It is clear that spatial distribution of the HOMO and LUMO for M1 and M2 display remarkably different behavior at zero bias in Table 3. 21. 2. It can be clearly seen that spatial distribution of the HOMO and LUMO is all delocalized for M1 at the molecular backbone, which creates lower molecular barrier and offers good transport channels for electron transport. But on the other hand, spatial distribution of the HOMO and LUMO is all localized for M2, which results in high barrier and the transport of electrons is hindered by the high barrier. And it illustrates the low transmission coefficient of M2. As a result, the current value of the system becomes small when torsion angles between two benzene rings rotates from 0° to 90°, which is in good accord with the *I-V* curve shown in Fig. 3. 21. 1.

Table 3. 21. 2 Spatial distribution of the HOMO and LUMO for M1 and M2 at zero bias

MPSH	HUMO	LUMO
M1		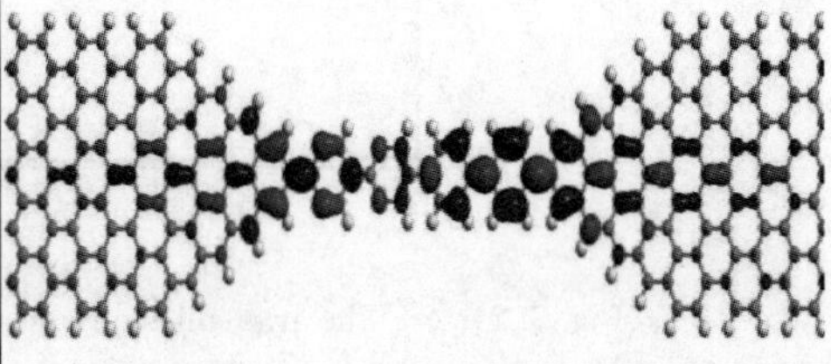

Continued

MPSH	HUMO	LUMO
M2	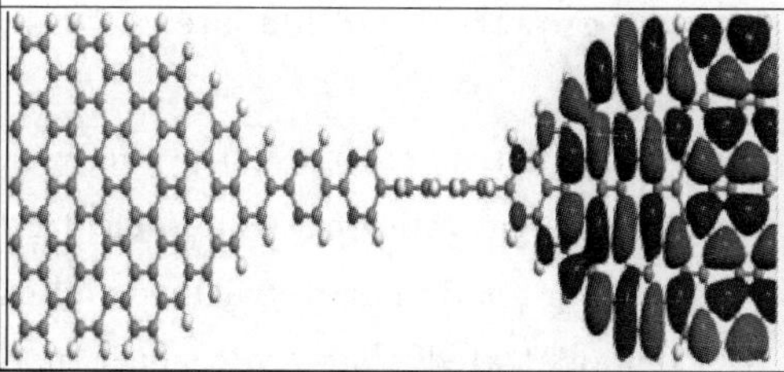	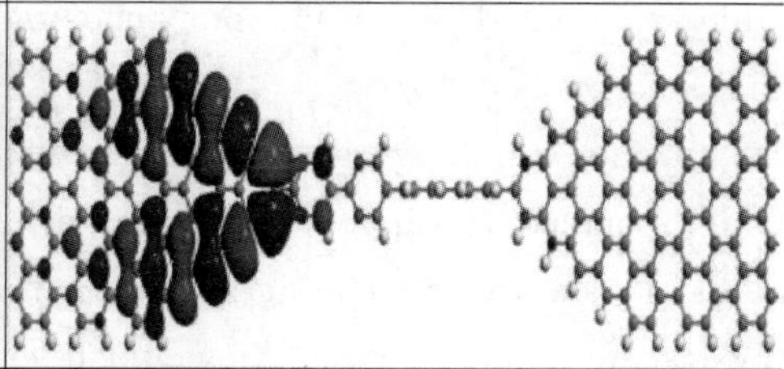

In conclusion, we investigate the effect of torsion angle on the electronic transport properties of a diblock co-oligomer molecule device with tailoring graphene nanoribbon electrodes by applying the DFT+NEGF first-principles method. The results show that the torsion angle plays a significant role on the electronic transport properties of dipyrimidinyl-diphenyl molecular device. When the torsion angle rotates from 0° to 90°, the molecular devices exhibit very different current-voltage characteristics which can realize the on and off states of the molecular switch. The physical origin of the switching behavior is interpreted based on the transmission spectra and molecular projected self-consistent Hamiltonian. The maximum on-off ratio can reach approximately 5, which is larger than the switching ratio of the molecule connected with the Au electrodes, suggesting that this system has attractive potential application in future molecular circuit.

3.22 Effects of different tailoring grapheme electrodes on the rectification and negative differential resistance of molecular devices

As the development of silicon-based microelectronic devices has reached the technological and physical limits, designing molecular devices innano-circuits has become a research hotspot. In recent years, many molecular devices have been discovered with functional properties. Among them, molecular rectifiers and negative differential resistances (NDRs) have achieved attention both in experimental and theoretical investigations due to their wide range of applications in logic circuits. However, some of them show poor rectification ratios because of the weak connection between the molecules and the metal electrodes.

Recently, graphene nanoribbons (GNRs) can be used as electrodes for molecular devices due to their unique electronic structures and transmission characteristics. Furthermore, with the development of experimental technology, tailoring a GNRs has become possible. Jia et al. proved that single diarylethenes covalently sandwiched between tailoring GNRs electrodes demonstrated unprecedented stability (over a year) in experiment. Ci et al. reported a cutting process that cuts with nickel nanoparticles to produce graphene pieces with a particular zigzag or armchair edge. Tapasztó et al. found that GNRs can be tailored by using scanning tunneling microscope (STM) lithography. Song et al. utilized predeposition of via focused ionbeam-assisted chemical vapor deposition for the growth of widthtailored GNRs on insulating substrates.

In this work, we design a molecular junction constructed byoligo p-phenylenevinylene (OPV) molecule capped with symmetric or asymmetric tailoring GNRs electrodes. OPV molecule is a linear π-conjugated molecule which has a good planar structure and can be synthesized in experiments. Ajayaghosh et al. demonstrated that hydrogen-bonded and p-stacked OPV self-assembly play an important role in promoting energy transfer and light harvesting. Crljen et al. investigated the effect of molecule chain length on the electronic transport properties of OPV molecules and discovered that the number of benzene rings in the molecule can change the frontier orbital distribution of the molecule. Tsoi et al. used electrochemical scanning tunnel microscopy to study quinone-modified OPV molecule and found that it could achieve reversible conductance switching function. Here, we study the effects of the tailoring GNRs electrodes on electronic transport properties of OPV molecule by applying the non-equilibrium Green's function (NEGF) formalism combined with density functional theory (DFT).

The models are illustrated in Fig. 3. 22. 1. The OPV molecule anchored with carbon atom are sandwiched between two different tailoring GNRs electrodes. The bonding way of the carbon atom is ···C—C≡C—C···, which is more stable than ···C=C=C=C···. In our calculations, the device is divided into three regions: left electrode, scattering region and right electrode. For simplicity, the models with symmetrical zigzag GNRs electrode, zigzag-armchair-zigzag (zz-ac-zz) GNRs electrode and armchair-zigzag (ac-zz) GNRs electrode are named M1, M2 and M3, respectively. The combination of armchair-zigzag (ac-zz) GNRs and zigzag-armchair-zigzag (zz-ac-zz) GNRs electrodes is proposed for M4. For eliminating dangling bonds, all of the edge carbon atoms of GNRs are saturated by hydrogen atoms.

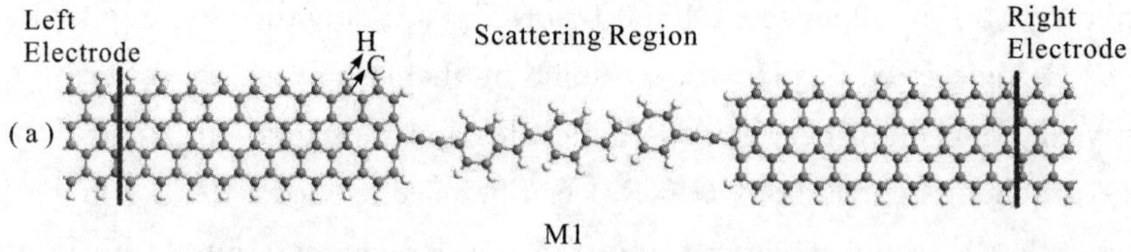

Fig. 3. 22. 1 Schematic diagrams of the devices: the OPV molecule is between two different tailoring GNRs electrodes (The region between the two solid lines is the scattering region)

(a) symmetrical zGNRs electrodes

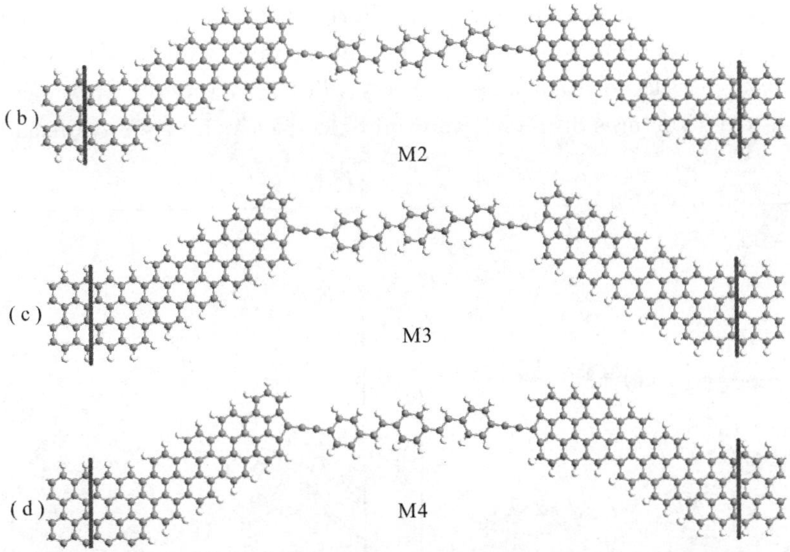

Continued Fig. 3. 22. 1 Schematic diagrams of the devices: the OPV molecule is between two different tailoring GNRs electrodes (The region between the two solid lines is the scattering region)
(b) symmetrical zz-ac-zz GNRs electrodes; (c) symmetrical ac-zzGNRs electrodes; (d) asymmetric ac-zz GNRs and zz-ac-zz GNRs electrodes

The calculated current-voltage (*I-V*) curves of M1, M2, M3 and M4 within bias voltage from −0. 6 to 0. 6 V in steps of 0. 05 are shown in Fig. 3. 22. 2. From Fig. 3. 22. 2(a), it can be seen that the current through the molecular device in the four models is significantly different: for M1, as the bias increases under positive bias, the current gradually enhances, which is similar to negative bias; for M2, the current under negative bias is larger than that under positive bias; for M3, the current increases with bias gradually in a bias range from 0 to 0. 2 V. Then, it slightly decreases to 0. 7 μA at 0. 5 V, which reveals a NDR; for M4, the current under positive bias is obviously larger than that under negative bias when the bias is between −0. 6 V and 0. 6 V. The current increases in the bias range from 0 to 0. 3 V. When the bias is 0. 55 V, the current reaches the minimum value of 1. 89 μA, revealing a NDR.

The asymmetry of *I-V* curves can be expressed by the rectification ratio (RR) which is defined as RR $(V)=\frac{|I(+V)|}{|I(-V)|}$. The Fig. 3. 22. 2(b) shows the rectification ratio curves for four models. From the Fig. 3. 22. 2(b) we can see that RR (V) is close to 1 for M1, so the rectifying behavior is weak; for M2, when the bias voltage is 0. 15 V, the maximum rectification ratio can up to 1. 1; for M3, when the bias voltage is 0. 05 V, the maximum rectification ratio can reach to 1. 2. For M4, when the bias voltage is 0. 4 V, the maximum rectification ratio is 14. 2, so the rectifying behavior is most obvious. In general, the rectification ratio of the molecular junction with asymmetric tailoring GNRs electrodes, namely M4, is larger than that of the molecular junction with symmetric one, namely M1, M2 and M3. The maximum RR (V) of M4 can reach 14.2 at 0.4 V. For M4, the asymmetry of the device is due to the different structures of the left and right electrodes; it may be utilized for rectifying applications.

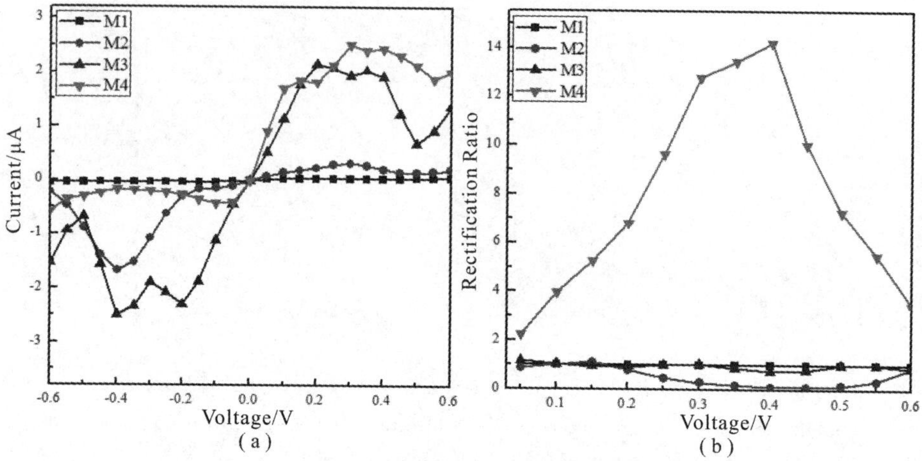

Fig. 3.22.2 The I-V curves and $RR(V)$

(a) the current-voltage curve; (b) the rectification ratio curve for M1, M2, M3 and M4. The bias voltage is from −0.6 to 0.6 V with the steps of 0.05 in the figure

From the Landauer-Büttiker formula, the current through the molecular junctions is calculated by the integral area of the corresponding transmission spectrum over the bias window. The region of the bias window is $[-V/2, +V/2]$, because the Fermi level is set to be zero in our calculations. The bias dependence of the transmission characteristics at biases of 0 V, ±0.2 V, ±0.4 V and ±0.6 V of M1, M2, M3 and M4 is presented in Fig. 3.22.3. In Fig. 3.22.3(a), there is no transmission peak entering the bias window at all biases for M1, so the current is the similar under positive and negative bias, and there is no rectifying behavior. In Fig. 3.22.3(b), when biases are ±0.2 V and ±0.6 V, the transmission peaks entering the bias window are all far from the Fermi level and the peak is low. When the bias is 0.4 V, the transmission peak into the bias window is small and far from the Fermi level for M2. While at −0.4 V bias, the transmission peak into the bias window is high and narrow and close to Fermi energy, so the current under negative bias is bigger than the current under positive bias. According to the definition of rectification ratio, the rectifying behavior of M2 is weak. In Fig. 3.22.3(c), when the biases are ±0.2 V and ±0.6 V, the transmission peak entering the bias window under positive and negative bias is similar. The transmission peak at 0.2 V bias is narrow and close to the Fermi level; the transmission peak at 0.6 V bias is low and far away from the Fermi level. Therefore, the current under positive and negative bias is similar. However, the transmission peak entering the bias window is close to Fermi energy when the bias is ±0. 4 V, but the peak at −0. 4 V bias is slightly higher than the peak of the 0. 4 V bias, so the rectifying behavior of M3 is also weak. One can see that the transmission peak entering the bias window is very low and the value is small at negative bias in Fig. 3. 22. 3(d). Furthermore, the transmission peak of M4 is far away from the Fermi level with the bias voltages increasing at negative bias. In contrast, in the positive bias region, the transmission peak entering the bias window is very narrow and high. Moreover, the position of transmission peak always approaches the Fermi level. As a result, the current under positive bias is significantly larger than that under negative bias, leading to a very obvious rectifying behavior for M4. In addition, we can clearly see that the transmission peak is the highest and closest to the Fermi level at 0. 4 V. Meanwhile, the value of transmission peak is the minimum at −0. 4 V. Therefore, the maximum rectification ratio can be observed at 0. 4 V.

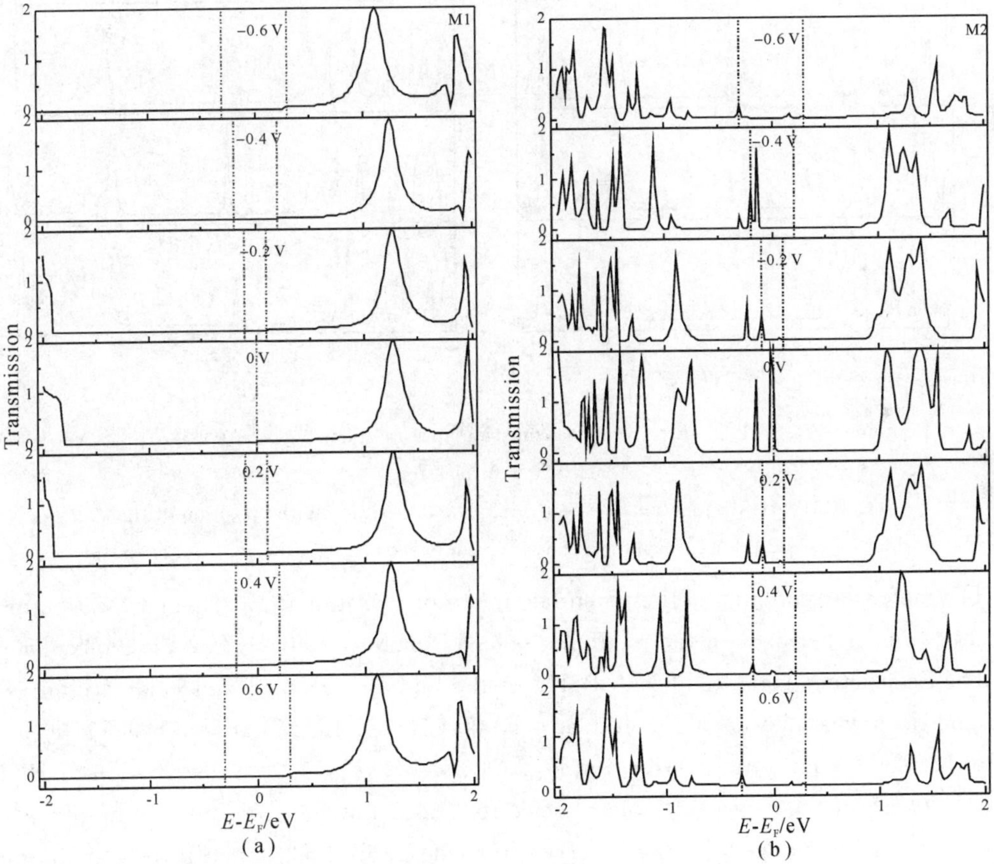

Fig. 3. 22. 3 Electronic transmission spectra under 0 V, ±0. 2 V, ±0. 4 V and ±0. 6 V bias for (a) M1; (b) M2. The dotted line is the position of the bias window [− V/2, + V2], and the energy origin is set to be the Fermi level of the system

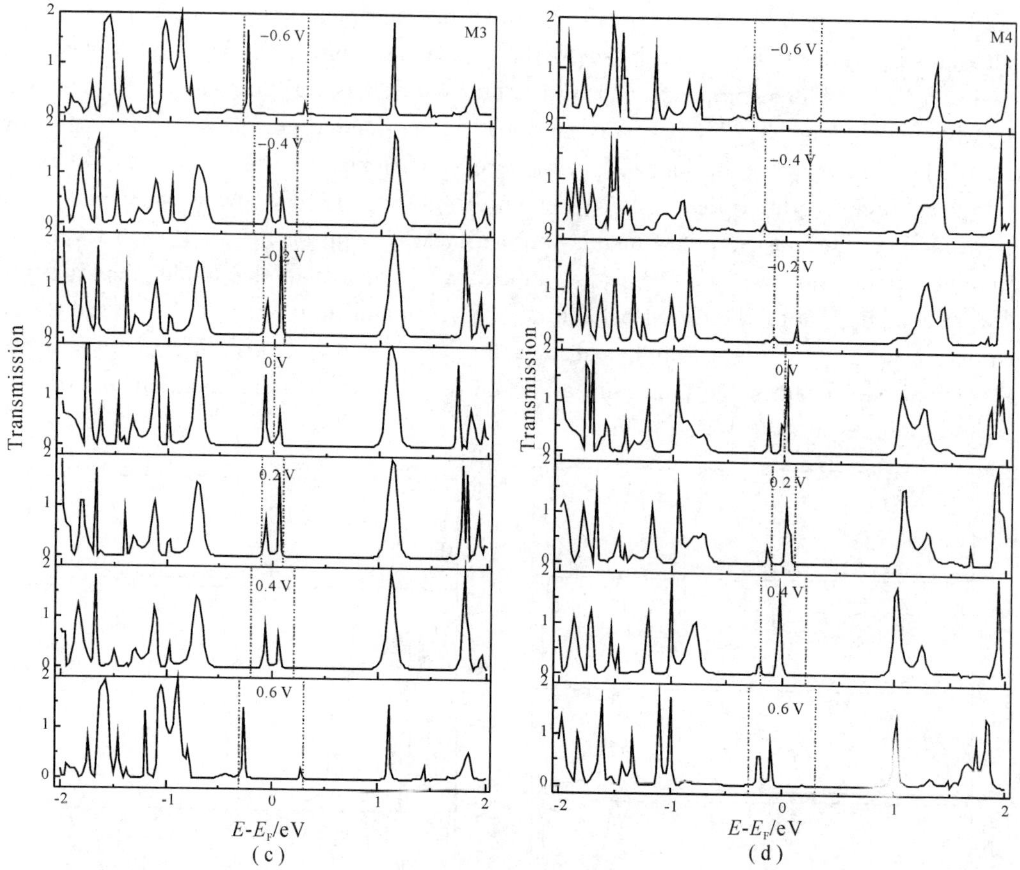

Continued Fig. 3. 22. 3 Electronic transmission spectra under 0 V, ±0. 2 V, ±0. 4 V and ±0. 6 V bias for

(c) M3; (d) M4 junction, respectively. The dotted line is the position of the bias window [$-V/2$, $+V2$], and the energy origin is set to be the Fermi level of the system

The molecular potential is the potential energy of a proton at a particular location near a molecule. To further explain the rectifying behavior of M4, we calculate the molecular projected self-consistent Hamiltonian (MPSH) of the highest occupied molecular orbital (HOMO) and the lowest unoccupied molecular orbital (LUMO) for M4 at zero bias and ± 0. 4 V where the rectification ratio is largest in Fig. 3. 22. 2(b). The spatial distribution of HOMO and LUMO reveal the degree of molecule-electrode coupling and the contribution of each atom to the tunneling peak. It can be observed from the Table 3. 22. 1 that if the distribution of MPSH on the molecular device is concentrated, it appears localization; if the distribution of MPSH on the molecular device is uniform, it appears delocalization. It is very clearly seen that MPSH-LUMO is more delocalized than MPSH-HOMO under zero bias for M4. Meantime, at a bias of 0.4 V, MPSH is almost uniformly distributed on the molecular device, so it appears as delocalized; at a bias of −0.4 V, MPSH-HOMO is distributed on the left electrode, and most of it is distributed in the middle molecule and the right electrode. However, the MPSH-LUMO is the opposite. Thus, it appears localization. As a result, the current under 0.4 V is larger than that under −0.4 V.

Table 3.22.1 MPSH distributions of HOMO and LUMO for M4 junction at 0 and ±0.4 V (the isovalues are all −0.001,976,46). HOMO and LUMO represent the highest occupied molecular orbitals and the lowest occupied molecular orbitals, respectively

	MPSH-HOMO	MPSH-LUMO
−0.4 V	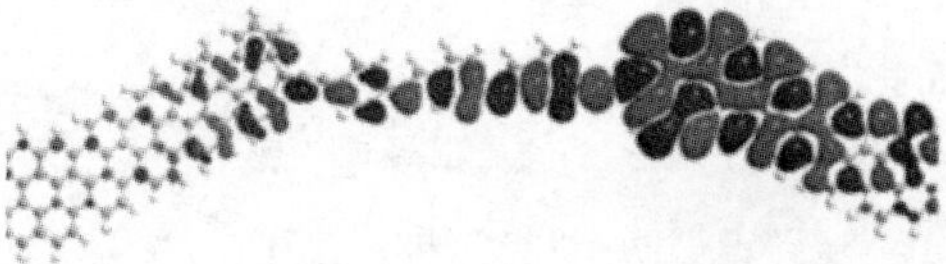	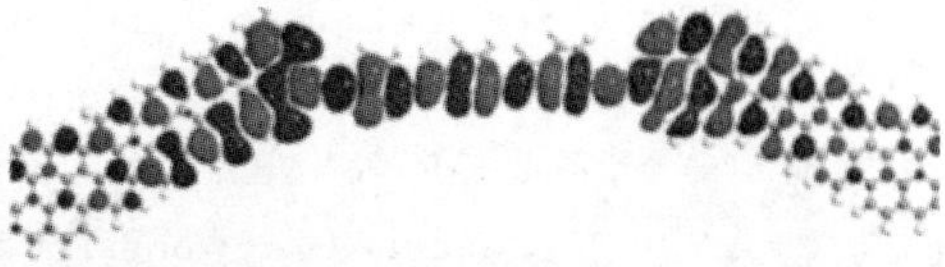
0 V	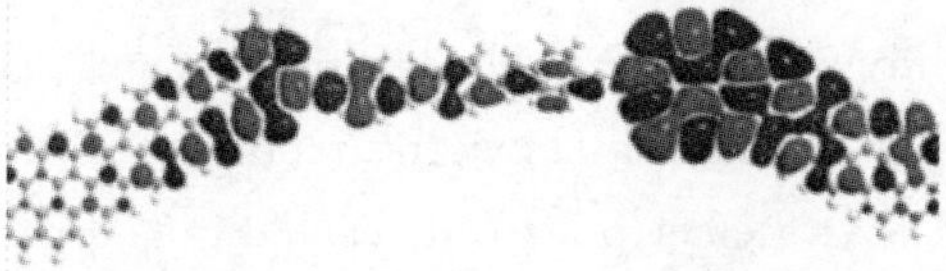	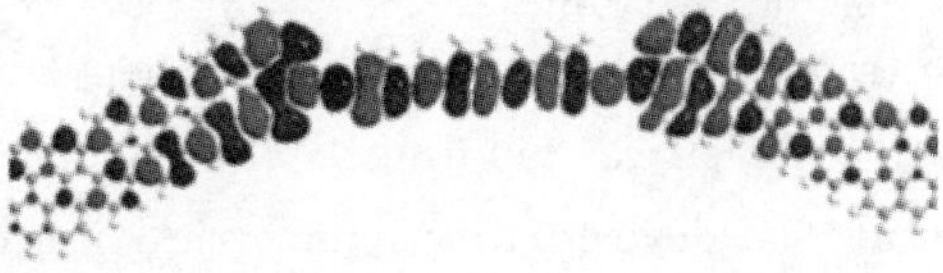
0.4 V	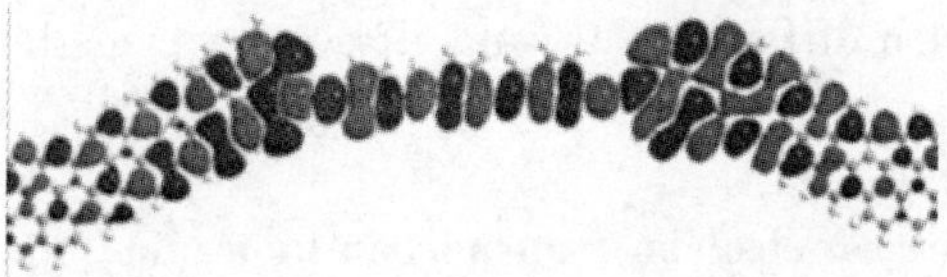	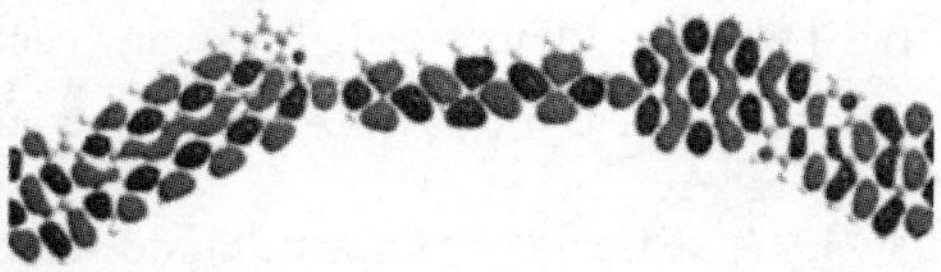

To understand the mechanism responsible behind the NDR effect which can be observed in Fig. 3.22.2(a) for M3 and M4, We calculated the transmission spectra of M3 and M4 under bias of 0.4 to 0.6 V, as shown in Fig. 3.22.4(a) and (b). We can see that HOMO-like and LUMO-like are always in bias window with the applied bias voltage increasing in Fig. 3.22.4(a). However, the peak value of LUMO-like and HOMO-like reduces when the bias increases from 0.4 to 0.5 V and far from the Fermi level, leading to a decrease of the current. Hence, the NDR appears as shown in Fig. 3.22.2(a). When the bias increases from 0.5 to 0.6 V, the peak value of HOMO-like increases and approaches the Fermi level, which results in an increase of the current. Therefore, the NDR disappears. In Fig. 3.22.4(b), we can see that at the bias of 0.4 V and 0.5 V, only the LUMO-like enters the bias window, the LUMO-like at 0.4 V bias is higher than the LUMO-like at 0.5 V bias and close to the Fermi level, so the current is decreased and the NDR occurs. When the bias is 0.6 V, both HOMO-like and LUMO-like enter the bias window and the transmission peak is closer to

Fermi level than the bias of 0.5 V, thus the current increases and the NDR disappears.

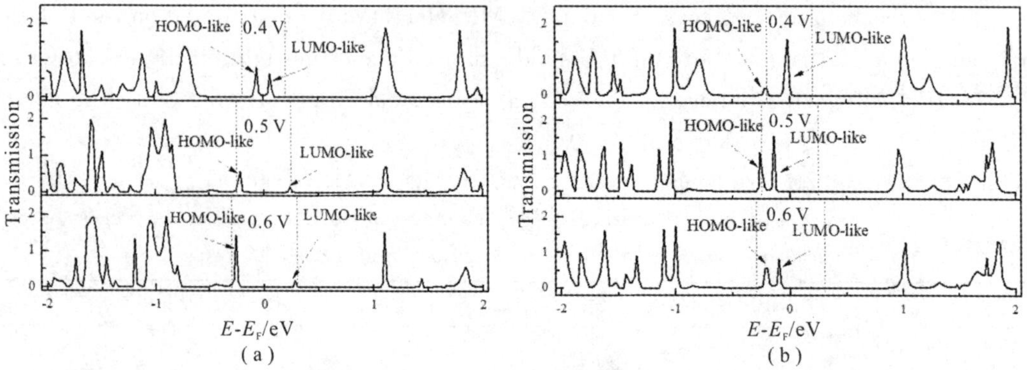

Fig. 3.22.4 Electronic transmission coefficients under the biases $V=0.4$ V, 0.5 V and 0.6 V for (a) M3; (b) M4. The dotted line is the position of the bias window $[V/2, +V/2]$, and the energy origin is set to be the Fermi level of the system

In conclusion, we use the NEGF combined with DFT method to investigate the electronic transport properties ofoligo p-phenylenevinylene(OPV) molecule sandwiching asymmetric and symmetrical tailoring GNRs electrodes. The results show the rectification ratio of the molecular device with asymmetric tailoring GNRs electrodes can reach 14.2. Furthermore, the molecular device with symmetrical armchair-zigzag (ac-zz) GNRs electrode can display NDR behavior. Therefore, these devices with different tailoring GNRs electrode have potential application in the future molecular circuit.

3.23 Effects of contact atomic structure on electron transport in molecular junctions

In recent years, progress in microfabrication and self-assembly technique has made it possible to design a single-molecule device. And the electronic transport properties of single-molecule devices have attracted more and more attention because of their novel physical properties and potential for device application, including single-electron characteristics, negative differential resistance (NDR), electrostatic current switching, memory effects, Kondo effects, etc. Base on these experiments, it become very important and necessary to understand the mechanism of electron transport through single-molecule devices in molecular electronics, especially, to find the contact atomic structure change which can provide more useful stability and high conductance is become one of the critical issues in this part. In many recent experiments, Au was used as leads for electronic current because of its high conductivity, stability and well-defined fabrication technique. A common way to construct a lead-molecule-lead system is by using a break junction, formed either mechanically or electrically. In these break-junction experiments, the atomic structure of the molecule-lead contact is unknown.

In fact, because of the atomic scale roughness of the break surface, different atomic scale structures of the contact may occur in different experiments. Therefore, neither the influence of detailed atomic structure on transport through the molecules nor a path to improved performance is clear. Here theoretical modeling/simulation from empirical parameters may play an important role in understanding, interpreting observed experimental behaviors, or doing predesigns for good contact structures.

In some canning tunneling microscope(STM) and break junctionexperiments, the simplest molecules: benzenedithiol(BDT) have been investigated. These pioneering works have stimulated a large theoretical effort to investigate the conductance of this molecule. Depending on the models and approximations used by different groups, the theoretical conductance values often differ from the experimental data by several orders of magnitude, which call for further study of these relatively simple molecules. It has been pointed out that conductance is sensitive to microscopic details of the molecule-electrode contact. These details include whether the molecule is properly bonded to both electrodes, what the binding site of the sulfur atom locate at the Au electrodes is and how the molecule orients itself with respect to the Au electrode. Therefore based on these cases, the molecular conductance of benzene connected to two Au leads with different contact atomic structure is investigated in this letter. We consider the effects of changing the atomic structure around the contacts, including the presence of an additional Au atom and the different adsorption sites of the sulfur atom, which are simple but important situations in break-junction experiments. We first investigate the transmission coefficient and the molecular conductance within the environment of the electrode under zero bias and then illustrate the current-voltage behaviors, and their dependence on the contact atomic structure. In each case, the first-principles method combine with nonequilibrium Green's function is applied to calculate. And the calculation is performed with electronic transport package, ATK 2. 0, which is developed by Brandbyge and co-workers.

Here we present only the technical details specific to our calculations. The systems we have studied consist of a BDT molecule connected to two Au leads through a sulfur atom located at hollow (H), top (T) and bridge (B) sites of the Au (111) surface respectively. To investigate the role of contact atomic structure, we consider a very simple but possible situation: the presence of an additional Au atom at either one or both contacts, as shown in Fig. 3. 23. 1, and denoted by 1Au and 2Au, use the structural label (111)_1Au, for instance, to denote the system with the Au(111) leads and an additional Au atom at one of its contacts. To show the effect of change in contact atomic structure, we also consider the change of Au-S distance. The purpose of all these considerations is to simulate possible situations in break-junction experiments in which different contact atomic structures may occur because of atomic fluctuations on the break surfaces and the molecule-lead separation can be adjusted by the mechanically controllable break junction (MCB) techniques.

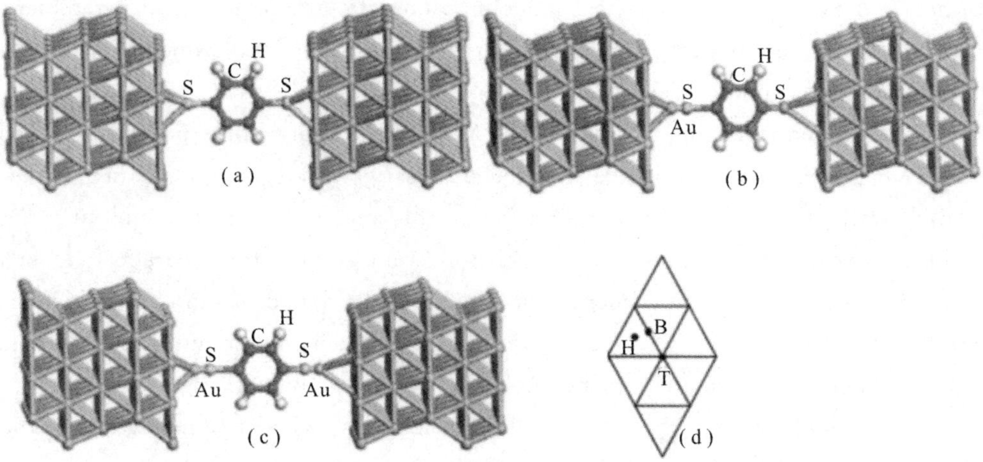

Fig. 3. 23. 1 Configuration of the device in our simulation

(a)(b) (c) the Au leads are in the (111) direction, and there are 0,1 and 2 additional Au atom at the contacts of hollow site, respectively; (d) The hollow, bridge and top sites on Au (111) surfaces

First, the transmission functions under zero bias for three systems including (111)_Au, (111)_1Au and(111)_ 2Au, and the sulfur located at the bridge, hollow and top sites respectively are investigated. When the molecule interacts with the Au(111) electrodes, the molecular levels broaden into a continuum. The eigenstates of the whole lead-molecule-lead system consist of scattering states, which are molecular-orbital-like in the molecule, and Bloch-wave-like in the metal leads. If an orbital is delocalized across the molecule, an electron that enters the molecule at the energy of the orbital has a high probability of reaching the other end, and thus there is a corresponding peak in the transmission probability $T(E, V_b)$, as shown in Fig. 3. 23. 2. It is clearly seen that the additional Au atoms produce a large resonance around the Fermi energy. The $T(E)$ for (111)_Au(see Fig. 3. 23. 2(a)) is quite smooth while that for (111)_1Au [see Fig. 3. 23. 2 (b)] and (111)_2Au [see Fig. 3. 23. 2 (c)] have many sharp features, and there is more transmission peaks are local around the Fermi energy in (111)_2Au than other two cases. The highest-occupied-molecular-orbital (HOMO) specially is more close to Fermi energy following the number increasing of additional Au atoms. Particularly, there is a creation of LUMO-like resonance peak around the Fermi energy when sulfur located at top site in Fig. 3. 23. 2 (c). Another difference during three systems is that the $T(E)$ of top site is most close to Fermi energy , then is the hollow site and bridge site in each system. It means that the top site is more sensitive to additional Au atom, which can be understood by analyzing the molecule-lead coupling and electron transfer for the three sites: in the top site of Au (111) leads, the sulfur atom has six nearest neighboring Au atoms, while there are just three and two for hollow and bridge site respectively, as shown in Fig. 3. 23. 1 (d). This difference in adsorption sites certainly affects the molecule-lead coupling, which may also lead to a difference in the molecule-lead electron transfer. The top site can provide more chance for electron though molecule to another lead, such as the LUMO-like resonance peak around the Fermi energy as shown in Fig. 3. 23. 2 (c).

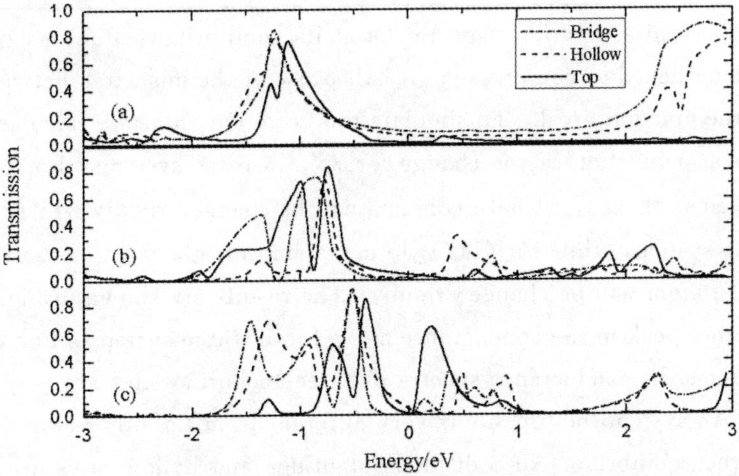

Fig. 3. 23. 2 Comparison among the transmission functions under zero bias of three systems with sulfur located at the bridge, hollow and top sites

(a) (111)_Au; (b) (111)_1Au; (c) (111)_ 2Au

The additional Au atom to the contacts causes a dramatic change in $T(E)$ of three systems, and the equilibrium conductance also evidently change by adding Au atom. The calculated value of equilibrium conductance for different adsorption sites in three systems is shown in Table 3. 23. 1. It can be seen that the top site gives a much larger conductance than the other two sites in each system, which proves the top site is more effective for electron transport again. Another difference shows that an additional Au atom either at one or two contacts increases the conductance distinctly. It means the additional Au atoms change the sign of the electron transfer, i. e. the additional Au atoms lead to more electrons transfer from the leads to the molecule. It also can be found from Fig. 3. 23. 2 (b) and (c) that the additional Au atoms cause the molecular LUMO level to line up.

Table 3. 23. 1 Calculated equilibrium conductance(G)

anchoring atom	lead orientation	additional Au	adsorption site	G
S	(111)	0Au	B	0.01
			H	0.04
			T	0.16
		1Au	B	0.05
			H	0.14
			T	0.21
		2Au	B	0.37
			H	0.39
			T	0.68

The molecule-lead separation may not be at its equilibrium value in experiments, but rather may be lengthened or compressed. It is because of the mismatch between the molecular length and the junction break. To simulate this situation, here we calculate the equilibrium conductance as a function of the change for $d_{\text{Au-s}}$ in three systems. For (111)_0Au and (111)_2Au systems, the $d_{\text{Au-s}}$ of both contacts will be changed rigidly while maintaining the symmetry of the system, while for (111)_1Au system only the $d_{\text{Au-s}}$ of the contact without the additional Au atom will be changed rigidly. The results are shown in Fig. 3. 23. 3. There is a large resonance peak in the conductance curve for all three systems. For (111)_0Au and (111)_1Au systems the conductance spectra of three anchoring sites are very similar, while for the (111)_2Au system the top site is very different from the other two sites. In (111)_2Au system , the equilibrium Au-S distance of bridge and hollow sites are about 1. 4 Å, while that of the top site is about 0. 4 Å. But in (111)_0Au and (111)_1Au systems the equilibrium Au-S distance is more closer compare with that in (111)_2Au, which is 1. 0 Å , 0. 8 Å,0. 6 Å and 0. 8 Å,1. 0 Å,0. 6 Å for bridge, hollow and top site respectively. In three systems, after the equilibrium conductance achieves the resonance peak, it will reduce as the distance increasing.

The large resonance peak in the equilibrium transmission function and LUMO-like resonance peak around the Fermi energy, as shown in Fig. 3. 23. 3 (a) and Fig. 3. 23. 2 (c), suggest the possibility of negative differential conductance under a bias voltage applied in (111)_2Au system, when the molecule locates at the top site. Therefore, as shown in Fig. 3. 23. 4, we calculate the *I-V* spectrum of the top site in (111)_2Au system for biases in the range of 0 to 1. 5 V in steps of 0. 1 V. It shows clearly a large negative differential conductance around $V_b = 0.2 - 0.6$ V as we expected. This especial phenomena can't be observed in hollow and bridge sites, which means the mechanism of the negative differential conductance most relates to LUMO-like resonance peak around the Fermi energy which can't be observed in $T(E)$ of other two sites in (111)_2Au system.

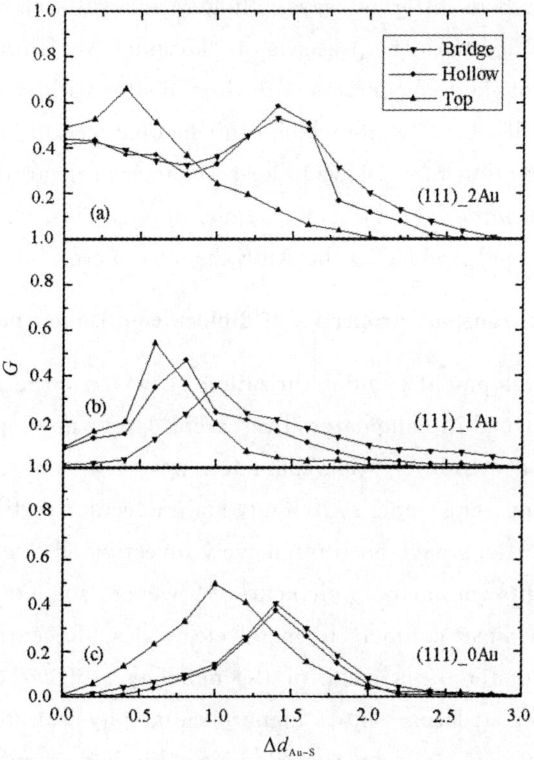

Fig. 3. 23. 3 Equilibrium conductance as functions of the change in Au-S distance (d_{Au-s}), for (a) (111)_2Au system; (b) (111)_1Au system; (c) (111)_0Au system

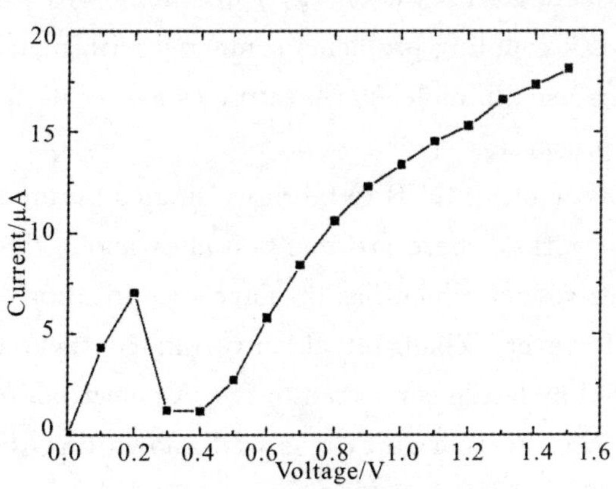

Fig. 3. 23. 4 *I-V* characteristics

In summary, by using density functional theory andnonequilibrium Green's functions method, we calculate from first-principles the molecular conductance of benzene sandwiched between two Au(111) leads in different ways. Numerical results show that the equilibrium conductance will be increased with the presence of additional Au. In particular, when an additional Au atom at each of the two contacts with the top sites will lead to negative differential conductance under applied bias, while which can't be observed in bridge and hollow sites. The mechanism is the creation of a LUMO-like resonance peak around the Fermi energy, which leads to negative differential conductance under applied bias. And the equilibrium conductance of three systems will reduce as the Au-S distance increasing.

3.24 The electronic transport properties of diblock co-oligomer molecule devices

With the further development of miniaturization of devices, the molecular devices have gradually become a new trend in nanomaterial research. Up to now, plenty of molecular devices have been discovered with interesting characteristics such as negative differential resistance (NDR), spin filtering, electronic switching, and molecular rectification. In particular, various molecular rectifications have been intensively investigated due to their wide range of applications in logic circuits and memory elements. However, some of them show poor rectification ratios because of weak contacts to metal electrodes. Recently, graphene as a new two-dimensional crystal materials is made of the monolayer of carbon atoms, due to this unique single atomic layer structure shows many peculiar physical and chemical properties. Graphene nanoribbons (GNRs) can be either semiconducting or metallic depending on its edge geometry, a zigzag edge (zGNR) was shown to be metallic, whereas an armchair-edged GNR (aGNR) is semiconducting with the energy gap scaling with the inverse of the ribbon width, which have less weak coupling problems compared with metallic electrodes materials like Au. When graphene is used in molecular electronics areas, it shows remarkable properties and wide application prospect.

In this work, we propose using GNR to fabricate singledipyrimidinyl-diphenyl molecular device. In some previous works, there are many studies about the dipyrimidinyl-diphenyl molecules attached Au electrodes, which has displayed rectification and negative differential resistance phenomena. However, Zhang et al. investigated the rectification effect of the dipyrimidinyl-diphenyl thiol molecule connected to two Au electrodes, and the maximum rectification ratio value was 3.5. Furthermore, it is well known that the chemical doping in an aGNR plays a pivotal role in determining the conductance behavior. Therefore, we investigate the electronic transport properties of the dipyrimidinyl-diphenyl anchored with carbon atomic chains sandwiched between two asymmetric N-doped aGNR electrodes by applying nonequilibrium Green's function (NEGF) formalism combined with first-principles density functional theory (DFT). The results show that rectifying behavior is strongly dependent on the doping position of electrodes. A higher rectification ratio can be found in dipyrimidinyl-diphenyl molecular device with asymmetric doping of left and right electrodes.

The schematic structures of the molecular devices are illustrated in Fig. 3. 24. 1. The dipyrimidinyl-diphenyl anchored with carbon atomic chains sandwiched between two aGNR electrodes, a linear carbon atomic chain derived from graphene as the narrowest GNR was a good conductor when it was connected between GNRs. As for aGNRs the parallel (perpendicular) case corresponds to an even (odd) number of carbon atoms. So we have taken 7-aGNR and selected the length of the carbon atomic chains with two carbon atoms, which means that the principal plane of the dipyrimidinyl-diphenyl is almost coplanar with the plane of the aGNR electrodes, and carbon atoms choose a bond way (…C—C≡C—C…) which is more stable than (…C=C=C=C…). The entire molecular devices can be divided into three regions: left electrode, central region, and right electrode. There are three typical models: M1, M2 and M3. For M1 both the left electrode and right electrode are not doped with heteroatom, whereas two edge carbon atoms are substituted by N atoms in left electrode with the same way for M2 and M3. In right electrode, one center and one edge carbon atoms is replaced by an N atom for M2 and M3, respectively. All of the edge atoms are saturated with hydrogen atoms to eliminate their dangling bonds.

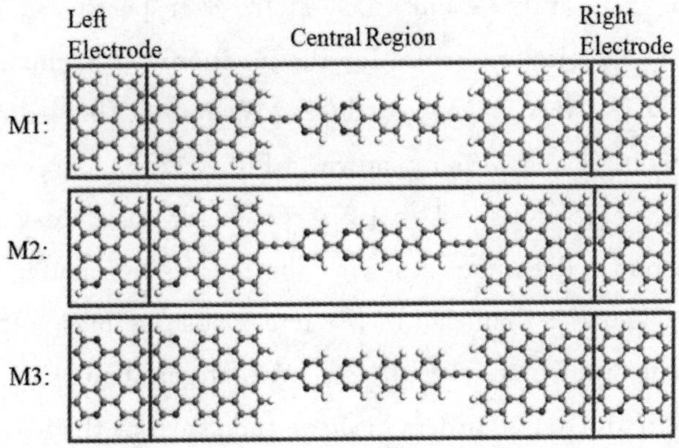

Fig. 3. 24. 1 The models of M1, M2 and M3. For M1 both the left and right electrode are not doped with heteroatom, for M2 and M3 two edge carbon atoms are substituted by N atoms in left electrode with the same way. In right electrode, one center and one edge carbon atoms is replaced by nitrogen atom for M2 and M3, respectively

The current-voltage (I-V) characteristic curves of the M1, M2 and M3 all in the bias range from -1.5 to 1.5 V in steps of 0.1 V are given in Fig. 3.24.2 and Fig. 3.24.3. It should be point out that at each bias, the current is determined self-consistently under the non-equilibrium condition. Obviously, the electronic transport properties of dipyrimidinyl-diphenyl molecules sandwiched between the edge and center N-doped aGNRs are different from each other. The I-V curve of M1 and M2 are plotted in Fig. 3.24.2(a), and we can see that there are little current values for M1 from the picture. However, M2 has large current and strikingly NDR at large bias voltages. For M2 the current I is nearly zero for small bias voltage -0.5 to 0.5 V, and then increase quickly at positive bias until a maximum value, after that decrease with the increase in bias, resulting in the NDR behavior. Eventually, the current value increases gradually and reaches up to 1,084 nA. This indicates that the center-doping in right electrode can give rise to much better NDR performance. Second, it is clear that the I-V curves of M1 and M2 manifest obvious symmetrical and asymmetrical characteristic respectively to the positive and negative bias. The amplitude of I at large positive bias is significantly greater than that at negative bias with same magnitude for M2. To clearly show the asymmetry as shown in Fig. 3.24.2(b), the bias-dependent rectification ratio (RR) is defined as $RR(V) = |I(+V)/I(-V)|$. The maximum RR values in the calculated bias region for M1 and M2 are 2.19 at 0.6 V and 42.90 at 1.5 V respectively. These results are larger than that Song et al calculation results for the junctions with undoped GNRs. It means that the rectification can be strengthened by doping N atom at the center of right electrode. The I-V curve and RR of M1 and M3 are shown in Fig. 3.24.3. From Fig. 3.24.3(a), we can see that at the low bias regions -1 to 1 V, the current is almost zero, but under high voltage value, the current value increases with the increase of voltage. Interestingly, the current of M3 is a little smaller than that of M1 under positive bias voltages. However, under negative bias voltages the current of M3 is a little larger than that of M1. And the current value of M3 is approximately 2 orders of magnitude smaller than that of M2. Moreover, we calculated the RR of M3 in Fig. 3.24.3(b), and the maximum value is 20 at 1 V, it is smaller than that of M2.

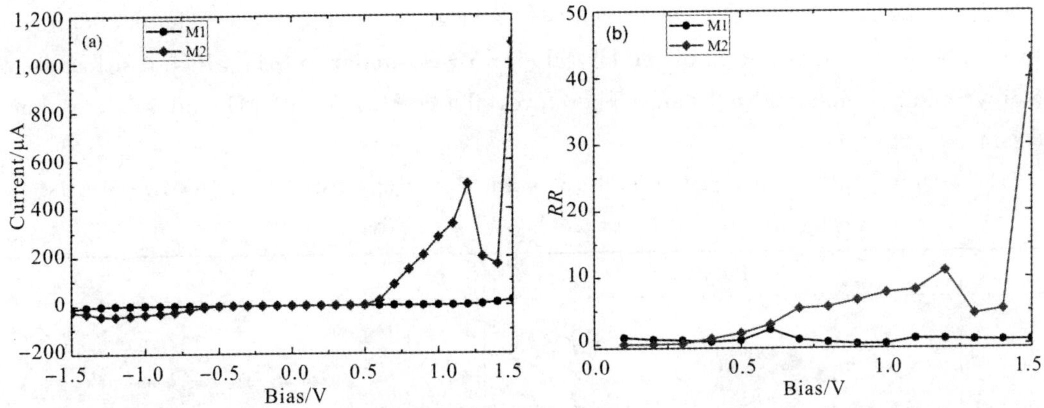

Fig. 3. 24. 2 The current-voltage and rectification ratio curves for molecular junctions M1 and M2

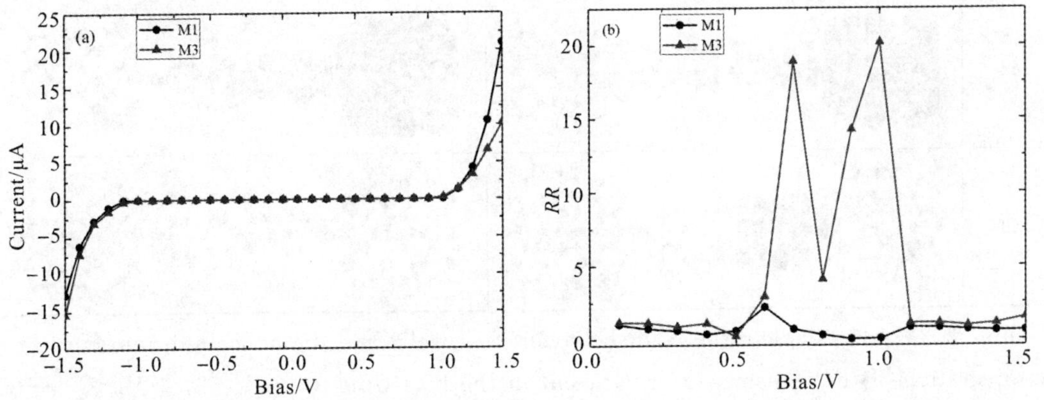

Fig. 3. 24. 3 The current-voltage and rectification ratio curves for molecular junctions M1 and M3

To further illustrate the different current-voltage characteristics of M1, M2 and M3, we calculate the molecular projected self-consistent Hamiltonian (MPSH) of the highest occupied molecular orbital (HOMO) and the lowest unoccupied molecular orbital (LUMO) for M1, M2 and M3 at zero bias in Table 3. 24. 1. From the picture, one can see the spatial distributions of the HOMO for M1 just localize on the dipyrimidinyl-diphenyl molecule. There is no spatial distribution on the left and the right aGNRs. And the LUMO for M1 delocalize on the dipyrimidinyl-diphenyl molecule, but there is slightly spatial distribution on the aGNR. This significant modification of the HOMO and LUMO weakens the electron delocalization and blocks tunneling across the junction. So, the corresponding coefficients on both orbitals are nearly zero, as a result the current value of M1 is very small at all bias region. For M2 the spatial distributions just delocalize on the left aGNR and the dipyrimidinyl-diphenyl molecule. There is no spatial distribution on the right aGNR. That means the transmission coefficients of this orbital are very small. However, the spatial distribution of the LUMO for

M2 is delocalized at all devices which results in a large transmission coefficients and current values. The spatial distribution of the HOMO for M3 is similar to M1, which just localize on the dipyrimidinyl-diphenyl molecule. Therefore, the coefficients of M1 and M2 are almost the same.

Table 3. 24. 1 The molecular projected self-consistent Hamiltonian (MPSH) of HOMO and LUMO for models M1, M2 and M3 at zero bias

	HOMO	LUMO
M1	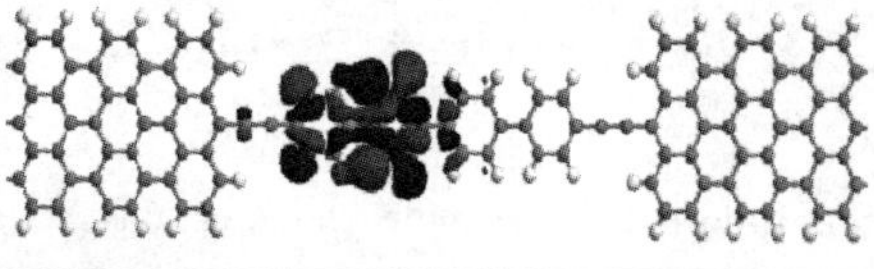	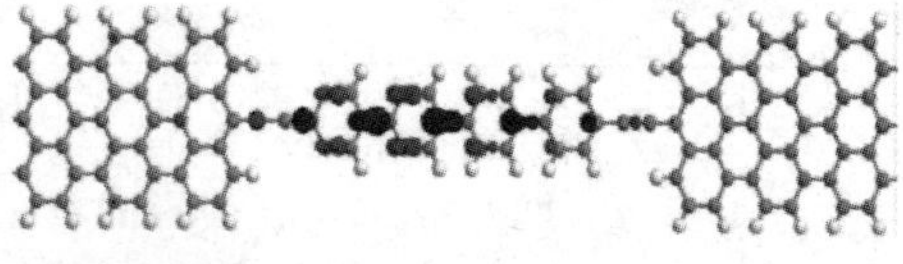
M2	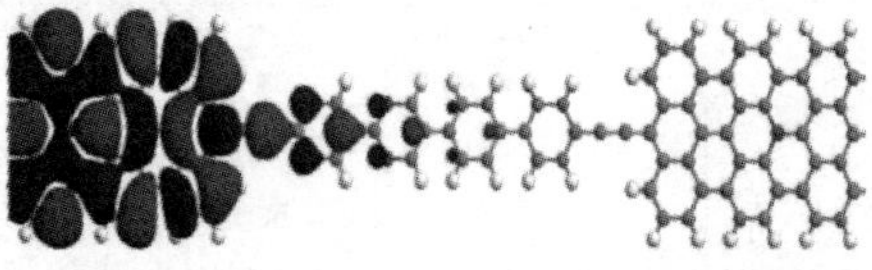	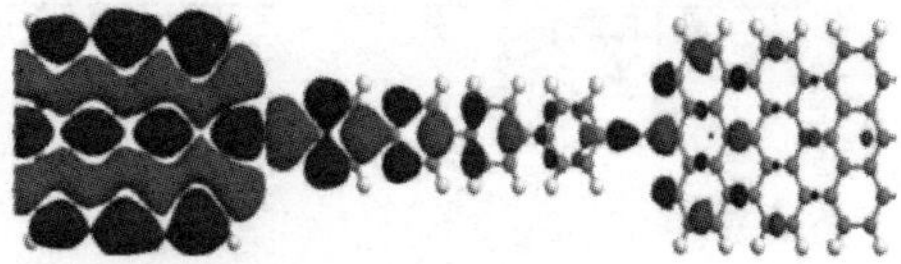
M3	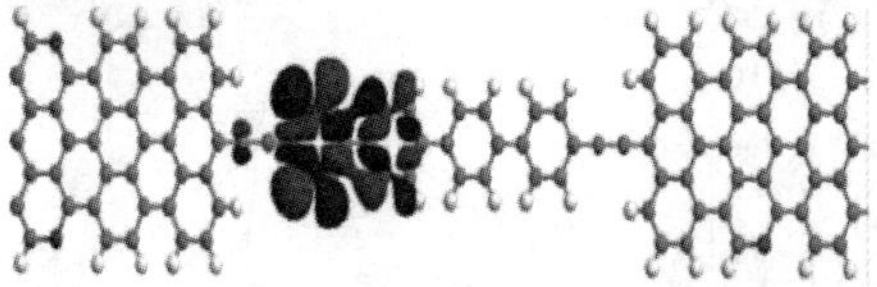	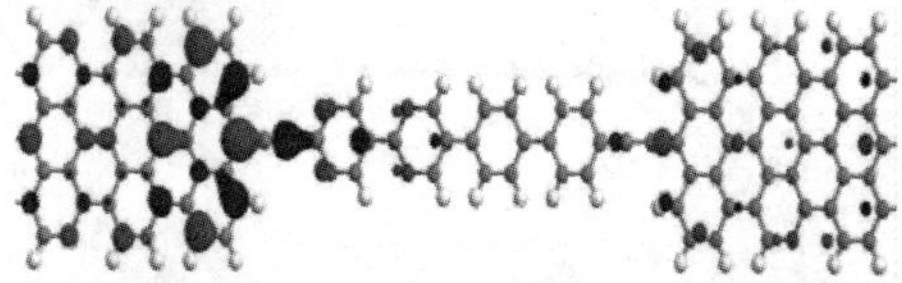

Fig. 3. 24. 4 can explain the NDR behavior very well, and the nonequilibrium current is the integration of the transmission coefficient in the bias window [$-V_b/2$, $+V_b/2$]. Fig. 3. 24. 4 shows the transmission spectra of M1 and M2 at the bias region [0. 3 V, 1. 5 V], we can see that there is probably no transmission peak in the 0. 3 to 0. 6 V, which means the current value is nearly zero at the low bias. With the increasing of the bias from 0. 6 to 1. 2 V, the transmission peak of M2 comes into bias window totally leading to the high currents. When the bias voltage is applied $V_b = 1.4$ V, the transmission peak in the bias window decreases. However, at $V_b = 1.5$ V, the transmission peak increases again. By this time, the transmission peak of M1 is still outside the bias widow, only a slightly peak can be seen in the transmission window until 1. 4 to 1. 5 V.

In conclusion, we have investigated the electrical transport properties of the dipyrimidinyl-diphenyl anchored with carbon atomic chains sandwiched between two N-doped aGNR electrodes by using nonequilibrium Green's function formalism combined with first-principles density functional theory. The calculated results show that the doping position plays a significant role on the electronic transport properties of dipyrimidinyl-diphenyl molecular junctions. For M2, Nitrogen atoms doping at the center of the right electrode can greatly enhanced the current value, and give rise to much better NDR and rectifying behaviors. For M3, Nitrogen atoms doping at the edge of the right electrode, although to some extent there is rectification phenomenon but the current value of M3 is approximately 2 orders of magnitude smaller than that of M2. And the maximum RR value in the calculated bias region for M2 is 42.90 at 1.5 V it is bigger than that of M3 which is 20 at 1 V.

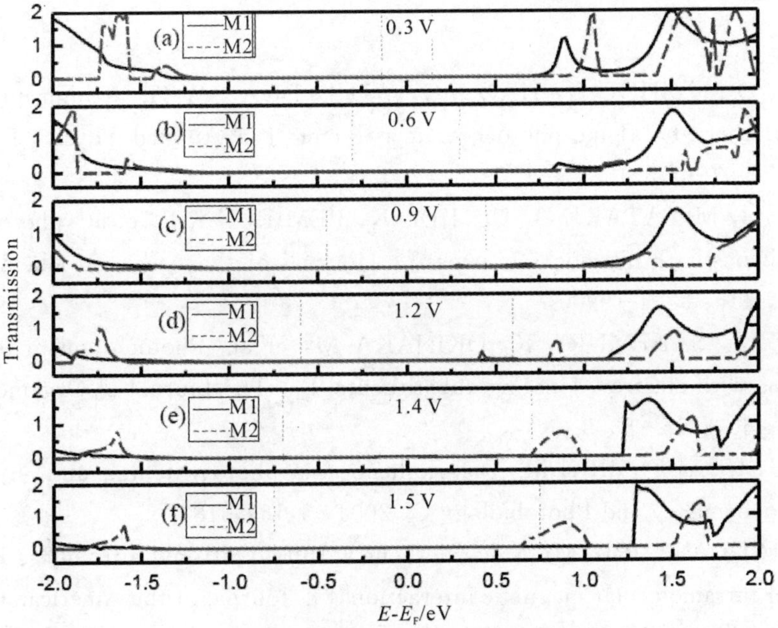

Fig. 3.24.4 Transmission coefficients of M1 and M2 junction under the biases V_b = 0.3 V, 0.6 V, 0.9 V, 1.2 V, 1.4 V and 1.5 V (The dashed-dotted lines indicate the position of the bias window $[-V_b/2, +V_b/2]$, and the energy origin is set to be the Fermi level of the system)

References

[1] YAN Q M, ZHOU G, HAO S G, WU J, DUAN W H. A high quantum yield diarylethene-backbone photochromic polymer[J]. Applied Physics Letter, 2006, 88:173107 - 173115.

[2] IRIE M, MIYATAKE O, UCHIDA K, LEMIEUX R P, et al. photochromic diarylethenes with intralocking arms[J]. Journal of the American Chemical Society, 1994, 116: 9894 - 9900.

[3] IRIE M, SAKEMURA K, OKINAKA M, et al. Photochromism of dithienylethenes with electron-donating substituents[J]. The Journal of Organic Chemistry, 1995, 60:8305 - 8309.

[4] MATSUDA K , IRIE M. Diarylethene as a photoswitching unit[J]. Journal of Photochemistry and Photobiology C, 2004, 5:169 - 182.

[5] MATSUDA K, IRIE M. A diarylethene with two nitronyl nitroxides: Photoswitching of intramolecular magnetic interaction[J]. Journal of the American Chemical Society, 2000, 122:7195 - 7201.

[6] KOSE M. Novel sulfoxide-introducing reaction and photochromic reactions of ethenylsulfinyl derivatives of dithienylethenes[J]. Journal of Photochemistry and Photobiology A, 2004, 165:97 - 102.

[7] NAKATSUJI S. Recent progress toward the expoitation of organic radical compounds with photo-responsive magnetic properties[J]. Chemical Society Reviews, 2004, 33:348 - 353.

[8] TIAN H, YANG S. Recent progress on diarylethene based photochromic switches [J]. Chemical Society Reviews, 2004, 33:85 - 97.

[9] UCHIDA K, MATSUOKA T, KOBATAKE S, et al. Substituent effect on the photochromic reactivity of bis(2-thienyl) perfluorocyclopentenes[J]. Tetrahedron, 2001, 57:4559 - 4565.

[10] STELLACCI F, BERTARELLI C, TOSCANO F. A high quantum yield diarylethene-backbone photochromic polymer[J]. Advanced Materials, 1999, 11: 292 - 295.

[11] KAWAI T, KOSHIDO T, YOSHINO K. Optical and dielectric properties of photochromic dye in amorphous state and its application[J]. Applied Physics Letter, 1995, 67:795 - 797.

[12] LEHN J M. Perspectives in supramolecular chemistry-from molecular recognition towards molecular information processing and self-organization[J]. Angew Chem Int Edit, 1990, 29:1304 - 1319.

[13] KANIS D R, RATNER M A, MARKS T J. Design and construction of molecular Assemblies with large second-order optical nonlinearities. Quantum chemical aspects[J]. Chemical Society Reviews, 1994, 94:195 - 242.

[14] GREEN M L H, MARDER S R, THOMPSON M E, et al. Synthesis and structure of (cis)-[1-ferrocenyl-2-(4-nitrophenyl)ethylene], an organotransition metal compound with a large second-order optical nonlinearity[J]. Nature, 1987, 330: 360 - 362.

[15] CHEN B, WANG M, WU Y, et al. Reversible near-infrared fluorescence switch by novel photochromic unsymmetrical-phthalocyanine hybrids based on bisthienylethene[J]. Chemical Communications, 2002, 123:1060 - 1061.

[16] GIORDANO L, JOVIN T M, IRIE M, et al. Diheteroarylethenes as thermally stable photoswitchable acceptors in photochromic fluorescence resonance energy transfer[J]. Journal of the American Chemical Society, 2002, 124:7481 - 7489.

[17] MALY K E, WAND M D, LEMIEUX R P. Bistable ferroelectric liquid crystal photoswitch triggered by a dithienylethene dopant[J]. Journal of the American Chemical Society, 2002, 124:7898 - 7899.

[18] TSUJIOKA T, KUME M, IRIE M. Optical density dependence of write/read characteristics in photon-mode photochromic memory[J]. Japanese Journal of Applied Physics, 1996, 35:4353 - 4360.

[19] MYLES A J, BRANDA N R. 1, 2-dithienylethene photochromes and nondestructibe erasable memory[J]. Advanced Functional Materials, 2002, 12:167 - 173.

[20] CAMPBELL I H, DAVIDS P S, SMTTH D L, et al. The schottky energy barrier dependence of charge injection in organic light-emitting diodes[J]. Applied Physics Letter, 1998, 72:1863 - 1865.

[21] SCHREIBER M, BUSS V. Origin of the bathochromic shift in the early photointermediates of the rhodopsin visual cycle: A CASSCF/CASPT2 study[J]. International Journal of Quantum Chemistry, 2003, 95:882 - 889.

[22] LIU Y J, HUANG M B. A theoretical study of low-lying singlet electronic states of SF2[J]. Chemical Physics Letters, 2002, 360:400 - 405.

[23] GMEZ I, OLIVELLA S, REGUERO M, et al. Thermal and photochemical rearrangement of bicycle[3. 1. 0]hex-3-en-2-one to the ketonic tautomer of phenol. Computational evidence for the formation of a diradical rather than a zwitterionic intermediate[J]. Journal of the American Chemical Society,2002,124:15375 - 15384.

[24] APLINCOURT P, HENON E, BOHR F, et al. Theoretical study of photochemical processes involving singlet excited states of formaldehyde carbonyl oxide in the atmosphere[J]. Chemical Physics, 2002, 285:221 - 231.

[25] XENIDES D. On the performance of DFT methods in (hyper)polarizability calculations: N-4 (T-d) as a test case[J]. Journal of Molecular Structure: THEOCHEM, 2007, 804:41 - 46.

[26] AOTO Y A, ORNELLAS F R. Predicting new molecular species of potential interest to atmospheric chemistry: The isomers HSBr and HBrS[J]. The Journal of Physical Chemistry A, 2007, 111:521 - 525.

[27] SERRANO-ANDRÉ L, MERCHÁN M. Quantum chemistry of the excited state [J]. Journal of Molecular Structure: THEOCHEM, 2005, 729:99 - 108.

[28] WATTS J D, BARTLETT R J. Iterative and non-iterative triple excitation corrections in coupled-cluster methods for excited electronic states: the EOM-CCSDT-3 and EOM-CCSD(T) methods[J]. Chemical Physics Letterst, 1996, 258:581 - 588.

[29] SATTELMEYER K W, STANTON J F, OLSEN J, et al. A comparison of excited state properties for iterative approximate triple linear response coupled cluster methods[J]. Chemical Physics Letters, 2001, 347:499 - 504.

[30] IRIE M, UCHIDA K, ERIGUCHI T, et al. photochromism of Single crystalline diarylethenes[J]. Chemical Physics Letters, 1995, 10:899 - 900.

[31] KAWAI T, KIN M S, SASAKI T, et al. Fluorescence switching of photochromic diarylethenes[J]. Optical Material, 2003, 21:275 - 278.

[32] IRIE M, FUKAMINATO T, SASAKI T, et al. Organic chemistry: a digital fluorescent molecular photoswitch[J]. Nature, 2002, 420:759 - 760.

[33] HUNTER S, KIAMILEV F, ESENER S, et al. Potentials of two-photon based 3-D optical memories for high performance computing[J]. Applied Optics, 1990, 29:2058 - 2066.

[34] PARTHENOPOULOS D A, RENTZEPIS P M. Three-dimensional optical storage memory[J]. Science, 1989, 245:843 - 845.

[35] IRIE M. Diarylethenes for memories and switches[J]. Chemical Reviews, 2000, 100:1685 - 1716.

[36] KOSHIDO T, KAWAI T, YOSHINO K. Novel photomemory effects in photo-

chromic dye-doped conducting polymer and amorphous photochromic dye layer[J]. Synthetic Metals, 1995, 73:257－260.

[37] KANEUCHI Y, KAWAI T, HAMAGUCHI M, et al. Optical properties of photochromic dyes in the amorphous state[J]. Japanese Journal of Applied Physics, 1997, 36:3736－3739.

[38] FERNANDEZ A, LEHN J M. Optical switching and fluorescence modulation properties of photochromic metal complexes derived from dithienylethene ligands [J]. Chemistry-A European Journal, 1999, 5:3285－3292.

[39] NORSTEN T B, BRANDA N R. Photoregulation of fluorescence in a porphyrinic dithienylethene photochrome[J]. Journal of the American Chemical Society, 2001, 123:1784－1785.

[40] MYLES A J, BRANDA N R. Novel Photochromic Homopolymers Based on 1,2-Bis(3-thienyl)-cyclopentenes[J]. Macromolecules, 2003, 36:298－303.

[41] MULLER C, GIMZEWSKI J K, AVIRAM A. Electronics using hybid-molecular and mono-molecular devices[J]. Nature, 2000, 408:541－548

[42] MALLOUK T, KAWAI T, YOSHINO K. Novel photomemory effects in photochromic dye-doped conducting polymer and amorphous photochromic dye layer[J]. Synthetic Metals, 1995, 73:257－260.

[43] KIM M S, KAWAI T, IRIE M. Synthesis of fluorescent amorphous diarylethenes [J]. Chemistry Letters, 2001, 7:702－703.

[44] KAWAI T, SASAKI T, IRIE M. A photoresponsive laser dye containing photochromic dithienylethene units[J]. Chemical Communications, 2001, 234:711－712.

[45] FUKAMLNATO T, KOBATAKE S, KAWAI T, et al. Three-dimensional erasable optical memory using a photochromic diarylethene single crystal as the recording medium[J]. Proceedings of the Japan Academy, Series. B, Physical and Biological Sciences, 2001, 77B:30－35.

[46] TOUR M P, SVEC W A, WASIELEWSKI M R. Optical control of photogenerated ion pair lifetimes: an approach to a molecular swith[J]. Science, 1996, 274:584－587.

[47] WOLD T, ISEDA T, IRIE M. Photochromism of triangle terthiophene derivatives as molecular re-router[J]. Chemical Communications, 2004, 32:72－73.

[48] CHEN T, KUNITAKE T, IRIE M. Novel photochromic conducting polymer having diarylethene derivative in the main chain[J]. Chemistry Letters, 1999,59:905－906.

[49] STODDART P, HEATHY I, SIMMERER J. Electroluminescence and electron transpot in a perylene dye[J]. Applied Physics Letter, 2000, 71:1332－1334.

[50] DONHAUSER J, PFEIFFER M, WERNER A, et al. Low-voltage organic electroluminescent devices using pin structures[J]. Applied Physics Letter, 2000, 75: 109－111.
[51] YASSAR M, KALLMANN H P, MAGNANTE P. Electroluminescence in organic crystals[J]. The Journal of Chemical Physics, 2001, 38:2042－2049.
[52] BLUM R H. Electroluminescence from polyvinylcarbazole films Electroluminescent devices[J]. Polymer, 2005, 24:748－754.
[53] DINESCU L, WANG Z Y. Synthesis and photochromic properties of helically locked 1,2-dithienylethene[J]. Chemical Communications, 1999, 35:2497－2498.
[54] HE C W, VANSLYKE S A. Organic electroluminescent diodes[J]. Applied Physics Letter, 1987, 51:913－915
[55] AVIRAM S, RATNER M. Thermally irreveraible photochromic systems. A theoretical study[J]. The Journal of Organic Chemistry, 1974, 53:6136－6138.
[56] METZGE M, MIYATAKE O, UCHIDA K, et al. photochromic diarylethenes with intralocking arms[J]. Journal of the American Chemical Society, 1997, 116: 9894－9900.
[57] REED P D. Redox-responsive macrocyclic receptor molecules containing transition-metal redox centers[J]. Chemical Society Reviews, 1997, 18:409－450.
[58] ZHOU B, WANG M, WU Y, et al. Rcversible near-infrared fluorescence switch by novel photochromic unsymmetrical-phthalocyanine hybrids based on bisthienylethene[J]. Chemical Communications, 2002, 56:1060－1061.
[59] ELBING K E, WAND M D, LEMIEUX R P. Bistable ferroelectric liquid crystal photoswitch triggered by a dithienylethene dopant[J]. Journal of the American Chemical Society, 2002, 124:7898－7899
[60] DEKKER I H, DAVIDS P S, SMITH D L, et al. The schottky energy barrier dependence of charge injection in organic light-emitting diodes[J]. Applied Physics Letter, 2003, 72:1863－1865.
[61] TAO M, GRABOWSKA A. Photochromism of salicylideneaniline (SA). How the photochromic transient is created: A theoretical approach[J]. The Journal of Chemical Physics, 2000, 112:6329－6337.
[62] LIU Y J, HUANG M B. A theoretical study of low-lying singlet electronic states of SF2[J]. Chemical Physics Letters, 2002, 360:400－405.
[63] APLINCOURT P, HENON E, BOHR F, et al. Theoretical study of photochemical processes involving singlet excited states of formaldehyde carbonyl oxide in the atmosphere[J]. Chemical Physics, 2002, 285:221－231.

[64] KLENE M, ROBB M A, BLANCAFORT L, et al. A new efficient approach to the direct restricted active space self-consistent field method[J]. The Journal of Chemical Physics, 2003, 119:713 - 728.

[65] SATTELMEYER K W, STANTON J F, OLSEN J, et al. A comparison of excited state properties for iterative approximate triple linear response coupled cluster methods[J]. Chemical Physics Letters, 2001, 347:499 - 504.

[66] VENTRA A B J, RETTIG W, SUDHOLT W. A comparative theoretical study on DMABN: Significance of excited state optimized geometries and direct comparison of methodologies[J]. The Journal of Physical Chemistry A, 2000, 106:804 - 815.

[67] GUO H, GRABOWSKA A. How the photochromic transient is created: A theoretical approach[J]. The Journal of Chemical Physics, 2001, 112:6329 - 6337.

[68] XU K, WAND M D, LEMIEUX R P. Bistable ferroelectric liquid crystal photoswitch triggered by a dithienylethene dopant[J]. Journal of the American Chemical Society, 2002, 124:7898 - 7899.

[69] EMBERLY K, SHIBATA K, KOBATAKE S, et al. Dithienylethenes with a novel photochromic performance[J]. The Journal of Organic Chemistry, 2002, 67: 4574 - 4578.

[70] WANG C K, WANG M, WU Y, et al. Reversible near-infrared fluorescence switch by novel photochromic unsymmetrical-phthalocyanine hybrids based on bisthienylethene[J]. Chemical Communications, 2002, 56:1060 - 1061.

[71] EVERS T, KUME M, IRIE M. Optical density dependence of write/read characteristics in photon-mode photochromic memory[J]. Japanese Journal of Applied Physics,1996, 35:4353 - 4360.

[72] LUO Y, QIU Y, WANG L D, et al. Pure red electriluminescence from a host material of binuclear gallium complex[J]. Applied Physics Letter, 2002, 81:4913 - 4915.

[73] TIVANSKI M, KALLMANN H P, MAGNANTE P. Electroluminescence in organic crystals[J]. The Journal of Chemical Physics, 2006, 38:2042 - 2049.

[74] SEMINARI C, TOKITO S, TSUTSUI T, et al. Organic eletroluminescent devices with a three-layer structures[J]. Japanese Journal of Applied Physics,2006, 27:713 - 715.

[75] TAYLOR J, WANG Y M, HOU X Y, et al. Interfacial electronic structures in an organic light-emitting diode[J]. Applied Physics Letter, 2005, 74:670 - 672.

[76] STOKBRO K, KALLMANN H P, MAGNANTE P. Electroluminescence in organic crystals[J]. The Journal of Chemical Physics, 2005, 38:2042 - 2049.

[77] CHEN H, VANSLYKE S A. Organic electroluminescent diodes[J]. Applied

Physics Letter, 2000, 51:913 - 915.

[78] LI R H. Electroluminescence from polyvinylcarbazole films electroluminescent devices[J]. Polymer, 2004, 24:748 - 754.

[79] LAKSHMI J, QIU Y, WANG L D, et al. Pure red electriluminescence from a host material of binuclear gallium complex[J]. Applied Physics Letter, 2002, 81: 4913 - 4915.

[80] GALPERIN T, UCHIDA K, IRIE M. Asymmetric photocyclization of diarylethene derivatives[J]. Journal of the American Chemical Society,2004, 119:6066 - 6071.

[81] NESS M, MOHRI M. Thermally irreversible photochromic systems. Reversible photocyclization of diarylethene derivatives[J]. The Journal of Organic Chemistry, 2003, 53:803 - 808.

[82] WEI Y, HAYASHI K, IRIE M. Thermally irreversible photochromic systems: Reversible photocyclization of non-symmetric diarylethene derivatives[J]. Bulletin of the Chemical Society of Japan, 2007, 64:789 - 795.

[83] DATTA S. Elastic quantum transport calculations using auxiliary periodic boundary conditions[J]. Physical Review B, 2005, 72: 045417 - 045419.

[84] CALZOLARI A, MARZARI N, SOUZA I, et al. Ab initio transport properties of nanostructures from maximally localized Wannier functions[J]. Physical Review B, 2004, 69: 035108 - 035112.

[85] THYGESEN K S, BOLLINGER M V, JACOBSEN K W. Conductance calculations with a wavelet basis set[J]. Physical Review B, 2003, 67: 115404 - 115408.

[86] FALEEV S V, LÉONARD F, STEWART D A, et al. Ab initio tight-binding LMTO method for nonequilibrium electron transport in nanosystems[J]. Physical Review B, 2005, 71: 195422 - 195426.

[87] WANG L W. Elastic quantum transport calculations using auxiliary periodic boundary conditions[J]. Physical Review B, 2005, 72: 045417 - 045422.

[88] HAVU P, HAVU V, PUSKA M J, et al. Nonequilibrium electron trans-port in two-dimensional nanostructures modeled using Green's functions and the nite-element method[J]. Physical Review B, 2004, 69:115325 - 115329.

[89] FUJIMOTO Y, HIROSE K. First-principles treatments of electron transport properties for nanoscale junctions[J]. Physical Review B,2003,67:195315 - 195319.

[90] LUO Y, WANG C K, FU Y. Effects of chemical and physical modications on the electronic transport properties of molecular junctions[J]. The Journal of Chemical Physics, 2002, 117: 10283 - 10289.

[91] KOSOV D S. Lagrange multiplier based transport theory for quantum wires[J].

The Journal of Chemical Physics, 2004, 120: 7165 - 7169.
[92] KAWAI T, SASAKI T, IRIE M. A photoresponsive laser dye containing photochromic dithienylethene units[J]. Chemical Communications, 2001, 231:711 - 712.
[93] TAYLOR J,GUO H,WANG J. Ab initio modeling of quantum transport properties of molecular electronic devices[J]. Physical Review B, 2001, 63: 245407 - 245412.
[94] BRANDBYGE M,MOZOS J L,ORDEJON P. et al. Density functional method for non-equilibrium electron transport[J]. Physical Review B, 2002, 65: 165401 - 165409.
[95] SOLER J M,ARTACHO E,GALE J D,et al. The SIESTA method for ab initio order-N materials simulation[J]. Journal of Physics: Condensed Matter,2002, 14: 2745 - 2751.
[96] XUE Y Q,DATTA S,RATNER M A. First-principles based matrix Green's function approach to molecular electronic devices: general formalism[J]. Chemical Phyisics, 2002, 281:151 - 157.
[97] CALZOLARI A,MARZARI N,SOUZA I. Ab initio transport properties of nanostructures from maximally localized Wannier functions[J]. Physical Review B, 2004, 69: 035108 - 035113.
[98] VANDERBITT D. Optimally smooth norm-conserving pseudopotentials[J]. Physical Review B, 1985, 32: 8412 - 8419.
[99] TIVANSKI A V, HE Y, BORGUET E, et al. Electronics using hybid-molecular and mono-molecular devices[J]. The Journal of Physical Chemistry B, 2005, 109: 5398 - 5402.
[100] LI Z, KOSOV D S. A photoresponsive laser dye containing photochromic dithienylethene units[J]. The Journal of Physical Chemistry B, 2006, 110: 19116 - 19122.
[101] CORNIL J, KARZAZI Y, BREDAS J L. Potentials of two-photon based 3-D optical memories for high performance computing[J]. Journal of the American Chemical Society, 2002, 124: 3516 - 3522.
[102] HU Y B, ZHU Y, GAO H J, et al. Photoregulation of fluorescence in a porphyrinic dithienylethene photochrome[J]. Physical Review Letters, 2005, 95: 156803 - 156807.
[103] BRANDBYGE M, MOZOS J L, ORDEJON P, et al. Fluorescence switching of photochromic diarylethenes[J]. Physical Review B,2002, 65: 165401 - 165408.
[104] TAYLOR J, GUO H, WANG J. Mono-bisthienylethene ring-fused versus multi-bisthienylethene ring-fused photochromic hybrids[J]. Physical Review B, 2001, 63: 245407 - 245412.
[105] REICHERT J, WEBER H B, MAYOR M, et al. A single photochromic molecu-

lar switch with four optical outputs probing four inputs[J]. Applied Physics Letter, 2003, 82: 4137 - 4142.

[106] GENG H, YIN S W, SHUAI Z G. Photoswitching of intramolecular magnetic interaction: a diarylethene photochromic spin coupler[J]. The Journal of Physical Chemistry B, 2005, 109:12304 - 12308.

[107] MEHREZ H, WLASENKO A, LARADE B, et al. Single-crystalline photochromism of diarylethenes[J]. Physical Review B, 2002, 65: 195419 - 195423.

[108] TANIO N, IRIE M. Photooptical switching of polymer film waveguide containing photochromic diarylethenes[J]. Japanese Journal of Applied Physics, 2001, 33: 1550 - 1553.

[109] TANIO N, IRIE M. Refractive index of organic photochromic dye-amorphous polymer composites[J]. Japanese Journal of Applied Physics, 2001, 33:3942 - 3946.

[110] EBISAWA F, HOSHINO M, SUKEGAWA K. Self-holding photochromic polymer Mach-Zehnder optical switch[J]. Applied Physics Letter, 2003, 65:2919 - 2921.

[111] KAWAI T, FUKUDA N, GROSCHL D, et al. Refractive index change of dithienylethene in bulk amorphous solid phase[J]. Japanese Journal of Applied Physics, 1999, 38:1194 - 1196.

[112] KIM M S, KAWAI T, IRIE M. Synthesis and photochromism of amorphous diarylethene having styryl substituents[J]. Molecular Crystals and Liquid Crystals, 2000, 345:251 - 255.

[113] KIM M S, KAWAI T, IRIE M. Synthesis of amorphous diarylethenes having diphenylethenyl substituents[J]. Chemical Letter, 2000, 10:1188 - 1189.

[114] FERNANDEZ A A, LEHN J M. Optical switching and fluorescence modulation in photochromic metal complexes[J]. Advance Material, 1998, 10:1519 - 1522.

[115] KOGEJ T, BELJONNE D, MEYERS F, et al. Mechanisms for enhancement of two-photon absorption in donor-acceptor conjugated chromophores[J]. Chemical Physical Letter, 1998, 298:1 - 6.

[116] STELLACCI F, BERTARELLI C, TOSCANO F, et al. Diarylethene-based photochromic rewritable optical memories: on the possibility of reading in the mid-infrared[J]. Chemical Physical Letter,1999, 302:563 - 570.

[117] DEBRECZENY M P, SVEC W A, WASIELEWSKI M R. Optical control of photogenerated ion pair lifetimes: an approach to a molecular swith[J]. Science, 1996, 274:584 - 587.

[118] TSUJIOKA T, SHIMIZU Y, IRIE M. Crosstalk in photon-mode photochromic multi-wavelength recording[J]. Japanese Journal of Applied Physics, 1994, 33:

1914 - 1919.
[119] MINIEWICZ A, KOMOROWSKA K, VANHANEN J, et al. Surface-assisted optical storage in a nematic liquid crystal cell via photoinduced charge-density modulation[J]. Organce Electronical, 2001, 2:155 - 163.
[120] FERINGA B L, JAGER W F, LANGE B, et al. The resonance Raman spectrum of cyclobutene[J]. Journal of the American Chemical Society, 1991, 113: 5468 - 5470.
[121] DEBRECZENY M P, SVEC W A, WASIELEWSKI M R. Optical control of photogenerated ion pair lifetimes:An approach to a molecular switch[J]. Science, 1992,257:63 - 65.
[122] ZAHN S, CANARY J W. Electron-induced inversion of helical chirality in copper complexes of N,N-dial-kylmethionines[J]. Science,2000,288:1404 - 1407.
[123] LUCAS L N, KELLOGG R M. Reversible optical transcription of supramolecular chirality into molecular chirality[J]. Science,2004,304:278 - 281.
[124] KOUMURA N, ZIJLATRA R W J. Light-driven monodirectional molecular rotor[J]. Nature, 1999, 401: 152 - 155.
[125] MURAOKA T, KINBARA K, AIDA T. Mechanical twisting of a guest by a photoresponsive host[J]. Nature, 2006, 440: 512 - 515.
[126] MOONEN N N P, FLOOD A H, FERNANDEZ J M. Molecule-Independent Electrical Switching in Pt/Organic Monolayer/Ti Devices[J]. Nano Letter, 2005, 5:2365 - 2372.
[127] DIJK E H, MYLES D J T, HUMMELEN J C. High-density data storage based on the atomic force microscope[J]. Organce Letter, 2006, 8:2333 - 2340.
[128] ZGIERSKI M Z, GRABOWSKA A. Photochromism of salicylideneaniline (SA). How the photochromic transient is created: A theoretical approach[J]. The Journal of Physical Chemistry, 2000,112:6329 - 6337.
[129] KOMILOVITCH P E,BRATKOVSKY A M,WILLIAMS R S. Nondestructive readout of photochromic optical memory using photocurrent detection[J]. Physical Review B, 2002, 66:245413 - 245418.
[130] PATI P, KAMA S P. Photoinduced refractive index change of a photochromic diarylethene polymer[J]. Physical Review B, 2004, 69:155419 - 155425.
[131] LILJEROTHh P,REPP J,Meyer G. Optical switching and fluorescence modulation inphotochromic metal complexes[J]. Science, 2007, 317:1203 - 1208.
[132] IRIE M, FUKAMINATO T, SASAKI T, et al. Organic chemistry: a digital fluorescent molecular photoswitch[J]. Nature, 2002, 420:759 - 760.
[133] MAYASAKA L, JOVIN T M, IRIE M, et al. Diheteroarylethenes as thermally

stable photoswitchable acceptors in photochromic fluorescence resonance energy transfer (pcFRET) [J]. Journal of the American Chemical Society, 2002, 124: 7481 - 7489.

[134] GANJI M D, MOHAMMADI A. Limited photochromism in covalently linked double 1,2-dithienylethenes[J]. Physical Letter A, 2008, 372:4839 - 4845.

[135] NORSTEN T B, BRANDA N R. Photoregulation of fluorescence in a porphyrinic dithienylethene photochrome [J]. Journal of the American Chemical Society, 2001, 123:1784 - 1785.

[136] MYLES A J, BRANDA N R. Novel Photochromic Homopolymers Based on 1,2-Bis(3-thienyl)-cyclopentenes[J]. Macromolecules, 2003, 36:298 - 303.